T0133757

Cybersecurity for Information Professionals

Cybersecurity for Information Professionals

Concepts and Applications

Edited by

Hsia-Ching Chang
Suliman Hawamdeh

CRC Press
Taylor & Francis Group
Boca Raton London New York

CRC Press is an imprint of the
Taylor & Francis Group, an **informa** business

AN AUERBACH BOOK

CRC Press
Taylor & Francis Group
6000 Broken Sound Parkway NW, Suite 300
Boca Raton, FL 33487-2742

© 2020 by Taylor & Francis Group, LLC

CRC Press is an imprint of Taylor & Francis Group, an Informa business

No claim to original U.S. Government works

International Standard Book Number-13: 978-0-367-48681-5 (Hardback)

International Standard Book Number-13: 978-1-003-04223-5 (eBook)

Visit the Taylor & Francis Web site at
www.taylorandfrancis.com

and the CRC Press Web site at
www.crcpress.com

Auerbach Publishing

Contents

Preface

Cybersecurity attacks are on the rise and cybersecurity challenges are real. Cybersecurity has direct and indirect impacts on individuals, organizations, and government institutions. Direct impacts involve individuals and organizations (private and public) with direct connection to the Internet using smart devices such as computers, laptops, smart phones, or Internet of Thing (IoT) devices. Indirect impacts involve people and organizations who might have a direct connection to the Internet, but their data might be hosted by institutions that have different direct connections to the Internet and could be vulnerable to cyberattacks. Cybersecurity threats and challenges are dynamic and complex and tackling them requires both interdisciplinary and multi-disciplinary approach. Given the complexity of cybersecurity issues, awareness and education remain as one of the most effective ways of combating cybersecurity threats. To raise the public awareness of cybersecurity, the U.S. Government designated October as the National Cybersecurity Awareness Month (NCSAM) since 2004. The campaign is co-led by Department of Homeland Security and National Cyber Security Alliance (NCSA) working with the industries, schools, and nonprofit organizations to promote data privacy and security (Department of Homeland Security 2018).

As the transformation into digital citizens is taking place, it is inevitable that we leave digital footprints and traces of our activities as well as unintended digital shadows generated by surveillance cameras and online social networks and applications. Protecting one's personal information in cyberspace and the right to privacy is now a fundamental human right recognized in the UN Declaration on Human Rights. However, the relationship between privacy and security in the Internet environment is challenging due to continued evolution of Internet applications and the increased sophistication of digital technologies. Historically, information professionals including librarians, archivists, and record managers took on the role of gatekeepers and locked important information in files, cabinets, and buildings. In the ancient world, librarians assumed the role of gatekeepers in the form of "the keepers of the tablets" or the "masters of the scrolls." Security of patron's data become of paramount importance when information professionals started to advocate for intellectual freedom and privacy. The new article (Article VII) of the Library Bill of Rights emphasizes the importance of protecting and safeguarding users' information by stating that "All people, regardless of origin, age, background, or views, possess a right to privacy and confidentiality in their library use. Libraries should advocate for, educate about, and protect people's privacy, safeguarding all library use data, including personally identifiable information" (American Library Association 2019). This is a clear undertaking and commitment made by information professionals to protect data and information as one of the core values of the information profession. This will also demand cybersecurity education or information sharing to equip information professionals with cybersecurity knowledge as they come across cybersecurity risks or incidents.

Information professionals have been paying more attention to confidentiality and putting a greater emphasis on privacy over cybersecurity. The cybersecurity breach incidents are soaring, and as a result, cybersecurity risks are high. Utilizing cybersecurity awareness training in the workplace has been one of the most effective methods in promoting cybersecurity culture. It also enhances individuals understanding of the risks and makes them more cybersecurity conscious. However, it is still unknown whether employees' security behavior at work can be extended to their security behavior at home and personal life. While library and information professionals assume the role of gatekeepers to safeguard the organization data and information assets, they can also aid in enabling effective information access and dissemination of cybersecurity knowledge to make users more conscious about the cybersecurity and privacy risks that are often hidden in the cyber universe.

This book introduces the fundamental concepts in cybersecurity and addresses some of the challenges faced by information professionals, librarians, archivists, record managers, students, and the broader audience from relevant and related disciplines. We recommend this book for educators preparing courses in information security, cybersecurity, and the integration of privacy and cybersecurity. The chapters contained in this book represent diverse perspectives of the role information professionals play in the field of cybersecurity. The following are the synopses of each chapter.

Human Aspects of Cybersecurity: Bridging the Weakest Link

Alkhaledi and Hawamdeh's chapter "Cybersecurity Challenges and Implications for the Information Profession" provides historical perspective of the information profession, privacy, and security. The chapter discusses cyberattacks, cybersecurity threats, cybersecurity challenges, cybersecurity awareness, and cybersecurity education. It highlights the role information professionals including librarians, archivists, and record managers play in advocating to protecting privacy and securing patrons information. The chapter emphasizes the point that in order to protect against security threats, there is a need for a holistic approach to security that would include creating awareness among users, building a sound and robust infrastructure, and train a new generation of information and cybersecurity professionals who can deal with both the technical and social aspects of cybersecurity.

Ho's chapter "Trustworthiness: Top Qualification for Cyber Information Professionals" provides an illuminating account that addresses how to prepare qualified cyber information professionals to handle information security with best practices, high ethical standards, and trustworthiness. Ho begins the chapter by discussing the future of cyberattacks along with the knowledge, skills, and abilities (KSAs) that information professionals will need in order to protect and defend the confidentiality, integrity, and availability of information artifacts, systems, and networks. To bolster our conventional perception that humans are the weakest link, Ho suggests that trustworthiness is an essential qualification for cyber information professionals, enabling cyber defenders to defend information assets against both the technological-centric attacks and human-centric threats. With trustworthy handlers, information can be collected, accessed, and processed with ethical considerations, and our human right to information privacy can be protected. Matrices for evaluating trustworthiness are discussed and illustrated to inform the future development of a computational inference engine, that is, a social firewall to enhance the weakest links in the chain of commands.

Al-Suqri, AlKindi, and Saleem's chapter "User Privacy and Security Online: The Role of Information Professionals" discusses the increasing threats to patron privacy in the online information environment and identifies what information professionals and librarians can do to maximize the privacy and security of users in the emergent digital environment. Three important roles are notably identified: promoting digital literacy to minimize the risks when users seek information online, working with IT specialists to develop and implement secure online information systems, and acting as political advocates for the privacy and confidentiality of library users when confronting regulations. This chapter provides key recommendations for ways in which librarians and information professionals can help maximize the privacy and security of users online, while reconciling the conflict with freedom of access to information.

Chang, Jim, and Hawamdeh's chapter "Bridging the Cybersecurity Talent Gap: Cybersecurity Education in iSchools" reviews the current efforts of cybersecurity education and workforce development across various institutions in the United States. To support the development of information professionals, cybersecurity training and education programs among the North America iSchools serve as an example to address the socio-technical aspect of cybersecurity. Chang, Jim, and Hawamdeh analyze the alignment of iSchools cybersecurity curriculum with National Initiative for Cybersecurity Education (NICE) Cybersecurity Workforce Framework. The different extended Open Systems Interconnect (OSI) models are discussed to inform the ongoing training and education to address both technical and social/human aspects of cybersecurity challenges. The NIST's Cybersecurity Learning Continuum is presented as a roadmap to guide future development of educational opportunities for iSchools and other stakeholders.

Markman's chapter "MetaMinecraft: Cybersecurity Education through Commercial Video Games" dives into the ecstasy and agony of utilizing a popular "sandbox survival" video game to introduce threat modeling as a cybersecurity concept for teens and tweens in libraries. Markman presents background research, historical notes, and lessons learned during a yearlong research project to test the effectiveness of Minecraft as a tool for community engagement in public libraries. Through an analysis of Minecraft's online communities and software ecosystem, an overarching view of the commercial video game is presented and paired with common newbie cybersecurity misconceptions including overgeneralization, conflated concepts, biases, and incorrect assumptions. Other examples of emergent gaming ranging from more traditional role-playing games to Virtual Reality (VR) puzzle quests are also presented for the future consideration as potential outreach tools.

Data/Information Aspects of Cybersecurity: Protecting the Assets that Matter

Lomas' chapter "Information Governance and Cybersecurity: Framework for Securing and Managing Information Effectively and Ethically" addresses the complexity of global information dynamics. It introduces the field of information governance (IG) as the solution for individuals, organizations, and nations to manage information. In tandem with the growing potential, value and social significance of information, the online management of a wide range of data has opened up information/data to new forms of theft and attack including information subversion, cybercrime, and cyber warfare. Lomas views cybersecurity as a subcomponent of IG because IG provides the holistic solution to dealing with information challenges in terms of both opportunities and dangers that are implicit within information creation. This chapter sets out IG considerations, tools, and frameworks for securing and managing information effectively and ethically.

Alemneh and Helge's chapter "Providing Open Access to Heterogeneous Information Resources without Compromising Privacy and Data Confidentiality" reviews the current status of data security and argues in favor of balancing the open access aspirations of cultural heritage institutions with the need for privacy and data confidentiality. The escalating cyberattacks can come from insider threats for data breaches or other malevolent tactics by cyber criminals, such as ransomware, viruses, spyware, phishing, and other malicious endeavors. Other types of cyberattacks can result from well-meant initiatives like open access or globally shared data, information, and knowledge. Alemneh and Helge's proposed model statute could offer insight into how the United States federal government, and the 51 states and district legislative bodies could provide legal guidance to address modern data and information misuse.

Zamir's chapter "Cybersecurity and Social Media" addresses that information professionals frequently use social media for user engagements, communications, and building networks; however, cybersecurity risks on social media platforms are often overlooked. Recent social media data breaches make the scenarios more intense. Hackers can easily collect and manipulate user data from social media. Personal identifiable information and confidential digital resources can be harnessed effortlessly through techniques such as phishing, social engineering, and impersonation. Attackers can also distribute malicious contents and short URLs in public conversations over social media platforms. Their attacking styles are getting more sophisticated and advanced. Zamir's chapter discusses common cybersecurity threats over social media platforms and its mitigation techniques. In this regard, it highlights implications for information professionals by recommending strategies to protect confidential user data and use social media safely.

Parker's chapter "Healthcare Regulations, Threats, and their Impact on Cybersecurity" aims to educate information professionals on how to build their own comprehensive security program and address risks in their own organization. Healthcare industry has been especially targeted by data breaches and ransomware attacks. Parker discusses the security and privacy components of the Health Information Portability and Accountability Act (HIPAA), Health Information Technology for Economic and Clinical Health Act (HITECH), the EU General Data Protection Regulation (GDPR), Payment Card Industry – Data Security Standards (PCI-DSS), and 21st Century CURES Act and Trusted Exchange Framework. As a practitioner involved in managing the healthcare records, Parker further examines the key processes and programs an organization needs to proactively defend themselves, presents emergent and current threats to the healthcare environment, and discusses how organizations can address these threats.

Technical Aspects of Cybersecurity: Understanding Socio-Technical Cybersecurity Trends

Chang's chapter "Mobile Cybersecurity: A Socio-Technical Perspective" outlines different layers of the multi-stakeholder mobile ecosystem and the potential cyber threats within the ecosystem. From the technical perspective, the security and privacy risks of mobile apps are discussed using the top-ranked mobile vulnerability categories identified by the Open Web Application Security Project (OWASP)'s mobile security project. In terms of the social and human perspective, this chapter also delineates user interactions with mobile apps as well as mobile users' security awareness and behavior. Chang introduces smartphone security best practices to put forward the idea that protecting mobile cybersecurity at every layer is critical. This chapter concludes with the proverb "prevention is better than cure" and reiterates the importance of layered security to defend mobile cybersecurity.

Hashem's chapter "Psychophysiological and Behavioral Measures Used to Detect Malicious Activities" proposes a multi-modal framework based on the user's psychophysiological measures and computer-based behaviors to distinguish between a user's behavior during regular activities versus malicious activities. He utilized several psychophysiological measures such as electroencephalogram (EEG), electrocardiogram (ECG), and eye movement and pupil behaviors along with the computer-based behaviors such as the mouse movement, mouse clicks, and keystroke dynamics to build a framework for detecting malicious insiders. He conducted human subject experiments to capture the psychophysiological measures and the computer-based behaviors for a group of participants while performing several computer-based activities in different scenarios. He analyzed the behavioral measures, extracted useful features, and evaluated their capability in detecting insider threats. He investigated each measure separately, and then used data fusion techniques to build two modules and a multi-modal framework.

Kinyua's chapter "Cybersecurity in the Software Development Life Cycle" emphasizes that secure software development in an organization is a proactive process of designing, building, and testing software that incorporates security into each phase. Kinyua reviews several commonly used software development processes and explores how security is integrated into every step of the software development process from requirements to testing. The SQUARE process is considered as a methodology for eliciting and prioritizing secure requirements. Kinyua then discusses different approaches to security analysis and design, secure coding practices, security in testing, code reviews, vulnerability testing and penetration testing, and fuzzy testing. In essence, security needs to be built into the software development process from the start in order to reduce security risks introduced at each stage of the secure software development life cycle.

Ding and Awojobi's chapter "Data Security and Privacy" discusses emerging technologies like Internet of Things (IoT), smartphones, mobile applications, and social media. The authors believe that data security and privacy are best addressed using a combination of technological controls, regulatory, and legal policies. They explained the benefits of enforcing controls like password complexity and data retention policies. They explained some of the strategies that social media platforms use to collect data and safeguard private information.

Last but not least, this book addresses the ongoing challenges of cybersecurity. Since very few iSchools have offered cybersecurity programs, it is time for information professionals to step up and contribute to long-term workforce development. We hope this book will encourage more information professionals to design and lead cybersecurity awareness campaigns and cybersecurity hygiene programs to change people's security behavior. It is our hope that this book will aid in promoting the engagement of information professionals becoming the advocates for the cybersecurity field.

References

American Library Association. "New Library Bill of Rights Provision Recognizes and Defends Library Users' Privacy," February 7, 2019. Accessed on March 31, 2019 from www.ala.org/news/press-releases/2019/02/new-library-bill-rights-provision-recognizes-and-defends-library-users.

Department of Homeland Security. "National Cybersecurity Awareness Month," September 26, 2018. Accessed on March 31, 2019 from www.dhs.gov/national-cyber-security-awareness-month.

Contributors

Dr. Daniel Gelaw Alemneh is a faculty member at the University of North Texas (UNT), coordinator of Digital Curation activities, and also teaching at the UNT College of Information. He recently (in 2020) served as a Fulbright scholar at Addis Ababa University in Ethiopia. His research interest includes open access, scholarly communication, digital curation, and ensuring long-term access to cultural heritage resources. He publishes extensively and is actively involved in various professional societies and has served in various capacities, including as a member of the Board-of-Directors of the Association for Information Science and Technology (ASIS&T) and its Executive Committee. Daniel is currently a Fulbright Scholar at Addis Ababa University, Ethiopia and can be reached at daniel.alemneh@unt.edu.

Reem Alkhaledi is a PhD student in the College of Information at the University of North Texas. Reem's area of interest is in electronic health records and cyber security. She graduated with a bachelor's in Advertising and Sociology from the University of Miami in 2006 and obtained a master's degree in Information Science from Kuwait University in 2012, where she discovered her interest in electronic health records and cybersecurity. Reem looks forward in pursuing a career in academia.

Dr. Salim Said AlKindi is an assistant professor at Department of Information Studies, College of Arts and Social Sciences, Sultan Qaboos University. He obtained his PhD in Internet Studies from School of Media, Culture and Creative Arts, Curtin University, Perth, Australia; a master's degree in Information Management and System, Monash University, Faculty of Information Technology, Melbourne, Australia; and a bachelor's degree in Library and Information Science from Sultan Qaboos University, Department of Library and Information Science, Sultanate of Oman. Dr Salim AlKindi research interests are in the area of Information and Knowledge Management, Research Methods, Web-based teaching and learning, Internet/web technologies (e.g., social media, social web), Information Organization and Retrieval, Information Literacy, Web Search Strategies and Information Ethics.

His recent research is conducted in the areas of Online Tools and social media for professional uses, research, and teaching.

Dr. Mohammed Nasser Al-Suqri is an associate professor and the Dean of Postgraduate Studies, Sultan Qaboos University. He received his PhD in Library and Information Management from Emporia State University, School of Library and Information Management, Emporia; a master's degree in Library and Information Science from Pratt Institute, School of Information and Library Science, New York; and a bachelor's degree in Library and Information Science from Sultan Qaboos University, Department of Library and Information Science, Sultanate of Oman. Mohammed Nasser Al-Suqri conducts research in the areas of user studies, knowledge management and sharing, technology adoption theories and models, research methodology, information industries, and impact of new technology on academic libraries.

Biodun Awojobi has been in the Information Technology industry for over 15 years. He received his MBA and Master of Science in Cybersecurity from the University of Dallas. He also has a Bachelor of Science degree in Electronic and Computer Engineering from Lagos State University in Lagos, Nigeria. He has worked in multiple IT software, networking, and security roles across several industries over the years. He worked at Microsoft for six years, before transitioning to Google to focus on Infrastructure, Application Modernization, and Security for the Google Cloud Platform. At Microsoft, he worked in multiple roles over the six years at the company. His last position being a Senior Technical Specialist for Security and Compliance, aligned with the Financial Services industry. He is a Microsoft Certified Trainer with multiple Microsoft certifications, including – Messaging, Productivity, and Security. Biodun is working on his PhD in Information Science at the University of North Texas with focus on Cybersecurity. His research interests are in Privacy and Security with the Internet of Things (IoT).

Dr. Hsia-Ching Chang is an assistant professor in the Department of Information Science, College of Information, at University of North Texas. She is affiliated with the Center for Information and Cyber Security (CICS), an interdisciplinary research center at the University of North Texas. Dr Chang is also a Cybersecurity Policy Fellow at New America, a non-partisan think tank based in Washington, D.C. providing advice to policy makers on emerging topics. Her research interests concentrate on cybersecurity (from a socio-technical perspective), data analytics, social media, knowledge mapping, scientometrics, and information behavior. She holds the Cloud Security Alliance's CCSK (Certificate of Cloud Security Knowledge), the first IT certification for secure cloud computing. She has developed and taught the graduate-level course, Information and Cyber Security, since 2015.

Dr. Junhua Ding is a professor of data science at the University of North Texas. He had been a faculty member of computer science with East Carolina University for 11 years and he also has near 8 years of industry experience with lead biomedical companies. His research interests include machine learning, data analytics, software security, biomedical computation, and software engineering. He received his PhD, MS, and BS in computer science from Florida International University, Nanjing University, and China University of Geosciences, respectively.

Dr. Yassir Hashem is an information technology and cybersecurity specialist with over 18 years of experience in both the educational and industrial sectors. He earned his Ph.D. in Computer Science and Engineering from the University of North Texas (2018). He also received his M.S. degree in computer science from the same university in 2016. He served as a research and teaching assistant at the University of North Texas with research focused on various aspects of cybersecurity and privacy including but not limited to: insider threats, risk assessment and management, cloud computing security, privacy and security of online social networks, context-aware computing and usable security. He also had over ten years of industrial experience in project management and information technology and has been working for several organizations inside and outside the United States.

Dr. Suliman Hawamdeh is a professor in the Department of Information Science, College of Information, University of North Texas. He is a leading authority in the field of knowledge management. Dr Hawamdeh founded and directed several academic programs including the first Master of Science in Knowledge Management in Asia in the School of Communication and Information at Nanyang Technological University in Singapore. He served as the Information Science Department chair from 2010 to 2018 at University of North Texas and founded the Bachelor and Master of Science in Data Science. Dr Hawamdeh is the editor in chief of the *Journal of Information and Knowledge Management (JIKM)* and the editor of a book series on *Innovation of Knowledge Management* Published by World Scientific. He has authored and edited several books in the areas of knowledge management, information science, and data analytics. He has extensive industrial experience and worked as the managing director of ITC Information Technology Consultant from 1992 to 1998. Dr Hawamdeh served in the capacity of chair and founding chair of several conferences including the International Conference on Knowledge Management and the Knowledge and Project Management Symposium. He was the founding president of the Information and Knowledge Management Society in Singapore and the founding president of the Knowledge and Information Professional Association (KIPA) in the United States. He served as a CO-PI from 2012 to 2019 on a $1.2 million NSF grant aimed at educating PhD students in the areas of cybersecurity and information assurance.

Dr. Kris Helge serves as Assistant Dean of Academic Engagement at Texas Woman's University. He also serves as part-time faculty at Texas Woman's University and Rutgers University. He holds a PhD in Information Science from the University of North Texas, JD from South Texas College of Law Houston, MLS from the University of North Texas, and a BA from Baylor University.

Dr. Shuyuan Mary Ho is an associate professor in the School of Information at Florida State University. Her research focuses on understanding the sociotechnical mechanisms of trusted human-computer interaction, specifically the subtle differences of language-action cues in computer-mediated deception and mis-/disinformation. Her work expands the social-computational investigation of cloud forensics, charged language of online influencers, and the effects of news media on the polarization of political opinion. Her work —mostly funded by U.S. National Science Foundation and Florida Center for Cybersecurity, appears in over 50 refereed journal articles and conference proceedings, and has been featured in the popular press as well, including the NPR, Wired, and Forensic Magazine.

Cary K. Jim is a PhD student in the Department of Information Science with a minor in Research, Measurement, and Statistics at the University of North Texas. Her previous work in K-12 education inspired her research interest in educational data science and related issues within K-12 and higher education. She was a Ronald E. McNair Scholar during her undergraduate study at Texas Woman's University and holds an MAT in science education from The University of Texas at Dallas.

Dr. Johnson Kinyua received the BS degree in Electrical and Electronic Engineering from University College London, UK, the MS in Digital Communications from the University of Kent, UK and the PhD degree in Computer Science from the University of Cambridge, U K. He has over 32 years of experience in higher education. He served as departmental Chair for six years and as a Dean for three years at different universities. He is widely experienced in teaching courses in different areas of Computer Science and Information Systems and published widely in peer-reviewed journals and international conferences. He has presented research papers at numerous peer-reviewed international conferences around the world. His teaching areas include: Cybersecurity, Networks, Software Engineering, Database Management Systems, Computer Architecture, Operating Systems, Object-Oriented Programming in Java/C++, Distributed Database Systems, Information systems, Data Mining, Analysis, Modeling and Design, Software Testing and Verification, Distributed Systems, and Functional Programming. His current research interests are in the areas of Software Engineering, Cybersecurity, Data Science, and Distributed Systems.

Dr. Elizabeth Lomas, associate professor of Information Governance, University College London, is an experienced academic and practitioner working in information management policy and practice across private and public sector contexts. This work has included developing information rights law including Data Protection and Freedom of Information law guidance in the UK, as well as ISO 27000 (the information security standard) guidance. In 2018, she chaired work on the UK's archival derogations under GDPR. She is Editor of the international *Records Management Journal* and a member of the Standards' Committees on Privacy and Technology and Archives and Records Management.

Chris Markman is a senior librarian at the Palo Alto City Library in Palo Alto, CA. He started reading *2600: The Hacker Quarterly* when he was a teenager. His current research projects include cybersecurity training, the distributed web, and exploring the use of VR and AR in libraries. You can find him on Twitter @akamarkman and guest blogging around the web.

Mitchell Parker, CISSP, is the Executive Director of Information Security and Compliance/CISO at IU Health. Mitch has done a significant amount of work in researching the effects of cloud and distributed computing, network-based threats, compliance, and privacy and security requirements on connected health devices. Mitch works collaboratively with a number of EMR and biomedical equipment vendors to improve their security postures and provide a better quality of service. He currently resides in Carmel, IN, with his wife, two children, and two cats.

Dr. Naifa Eid Saleem is an associate professor at the Department of Information Studies and the Assistant Dean for Training & Community Service, College of Arts and Social Sciences, Sultan Qaboos University. She received her PhD in Library and Information Science from Exeter University, UK; a master's degree in Library and Information Management from Sheffield University, School of Information Science, UK; and a bachelor's degree in Library and Documents from Sultan Qaboos University, Department of Library and Information Science, Sultanate of Oman. Naifa Eid Saleem conducts research in the area of user studies, library science and the Internet, e-learning, research methodology, Internet ethics, as well as librarian and professional developments.

Dr. Hassan Zamir is an assistant professor at the School of Information Studies at Dominican University. He teaches in the areas of information technology, data science, informatics, and big data. His research interests revolve around social media analytics, big data, and cybersecurity. He is the editor of World Libraries, a Dominican University in-house journal. Zamir is active in leadership services in various professional associations including ASIST and ALISE. Apart from teaching and supporting the expansion of informatics programs at Dominican University, he aims to explore data for social good.

Chapter 1

Cybersecurity Challenges and Implications for the Information Profession

Reem Alkhaledi and Suliman Hawamdeh

Department of Information Science, University of North Texas

Introduction

Technology has changed the way we communicate, exchange information, and conduct business. It has impacted our lives in a big way and made it possible for anyone with a handheld device or mobile phone to access information from different sites and diverse information systems located around the world. Real-time access to information enabled people to engage in business and commerce activities online. The convenience created by the technology comes at a price. It has created an environment where we are more vulnerable to cyberattacks and cybersecurity threats than ever before. The increase in the number of cybersecurity breaches in the past decade highlighted the magnitude and the complexity of the cybersecurity problem. Some of the high-profile cases such as Yahoo 3 billion users data breach in 2013, eBay 145 users in 2014, Uber 57 million users in 2016, the US Democratic party email system hack in 2016, and Equifax 143 million consumers data in 2017 are few examples of the security threats and data breaches that individuals and organizations face today (Armerding, 2018). This could be just the tip of the iceberg of what we would expect in terms of challenges posed by advanced 5G technologies and artificial intelligent–enabled devices.

Most consumers and information service users expect their private and confidential information to be kept confidential, protected, and stored in secure systems. As consumers, we are expected, if not required, to provide personal information to complete certain credit card purchases or online transaction. Even if a privacy policy exists, disagreeing with the policy might means denied access to the services provided, leaving users questioning their rights and the wisdom and value of reading such policies (Menand, 2018). Data security breaches in most cases means customers' valuable information are exposed and probably stolen. Such breaches represent a wakeup call to the fact that everything connected to the Internet could be vulnerable. This includes smartphones, computers, home automation, television sets, cars, and virtual assistant such Alexa, Cortana, and Siri. Land phone communications are replaced by emails, and social media applications such WhatsApp and Facebook have become the norm.

Losing connectivity to the Internet can sometimes bring operations in the organization to a halt. The reliance on Internet as the main mode of communication has increased exposure to cyberattacks. Addressing the Internet vulnerability issues requires a coordinated effort by the government and the private sector to address not only the technical aspect of the problem but rather the regulatory and social issues. Innovation and advances in technology are usually preceded regulations and other measures needed to encounter cybersecurity threats. While technical security measures such as encryption, blockchain, and authentication technologies can be developed and deployed in relatively short time, new regulations, human and social issues move much slower and in most cases operates in a catch-up mode.

Cybersecurity does not affect only large size organizations, but it also affects individuals who are equally vulnerable to cyberattacks. People need to be educated on the importance of personal cybersecurity and learn how to protect themselves from cybersecurity threats. The number of unsecure networks that people connect to it on daily basis is alarming. According to a survey by Symantec released in May 2017 showed that 87% of consumers globally have used readily available unsecured public networks (Symantec, 2017). The survey showed more than 60% of those surveyed globally think their information is safe when using public networks. At the same time 53% of the user cannot tell the difference between secured or unsecured networks. This is not surprising given the increased reliance on digital information for business and entertainment. The survey showed that 59% of those surveyed used the network to access emails, 26% logged into work email, 56% accessed social media, 44% share photos and videos, 25% access banking and financial information, and 17% carried out transaction using personally identifiable information.

The use of unsecure public networks such as hotspots leaves customers vulnerable for cyberattacks through intercepting data packets sent over unsecured networks. Hackers can eardrop on open connections to gather useful information about user activities such as login information or personally identifiable information. Lord (2019) described cybersecurity as the body of technologies, processes, and practices used to protect network devices, programs, and data from attack and unauthorized access. This is important as cyberattacks become more sophisticated, there is a need to look at the problem from a holistic and integrated approach by examining the rules, regulations, procedures, and practices.

Historical Perspective

The Information Profession

The information profession can be described as the field of work or the occupation that deals with information through its lifecycle from creation to disposal. According to Bates (2015), the information discipline is unique and does not fit in the spectrum of traditional disciplines. The information discipline is a metadiscipline similar to the education discipline which deals with a body of knowledge from a particular orientation. Every profession has both academic disciplinary aspects and professional practice aspects. Bates added that one common aspect of the information disciplines is that they all deal with collection, organization, retrieval, and presentation of information in various contexts and on various subject matters.

Historically, librarians, archivists, and record managers had dealt with gathering, processing, organizing, and disseminating information. The evolution of the information profession was shaped overtime with the evolution of information disciplines. Melvil Dewey, who became the librarian at Columbia College in 1883, was offered the first class in library economy in 1887. Dewey is considered one of the pioneers in the field of information in which one of his signature achievements was the creation of the Dewey Decimal Classification system. The field of information science disciplines is inherently interdisciplinary drawing on theories and methods from other fields. Dewey library classification system was based on the classification of knowledge by Sir Francis Bacon (Wiegand, 1998). Dewey helped to establish the American Library Association (ALA) in 1876, and in 1926, the Carnegie Corporation funded the first Graduate Library School at the University of Chicago (Richardson, 2010).

The real transformation of the library and information profession started with the creation of the DDC system to organize the library collections – the first systematic and scientific method to organize the collection by different disciplines. The flexibility and scalability of Dewey classification system made it the default standard for organizing the library material and library collections for long time. The evolution of the information profession into scientific disciplines was driven by the need to go beyond gatekeeping role to a more dynamic and service-oriented profession.

The big shift and transformation of the information profession happened with the invention of computers and the idea of using computer to store and retrieve information. In 1945, Vannevar Bush published an article under the title "As We May Think" in which he talked about mechanizing the process of storing and retrieving books with higher degree of speed and flexibility. Soon after (in 1948), Holmstron described the possibility of using UNIVAC computer in bibliographic searching (Sanderson and Croft, 2012). The real library automation efforts started in 1960s, with the first large information retrieval research group formed by Gerard Salton first at Harvard University and then at Cornell University. Work in the area of information retrieval in the 1970s and 1980s still form the basis for most of the current development in online and database searching. Most notable work in that period is Luhn's term frequency (tf), which is based on the statistical analysis of the keyword in the document. The work by Sparck Jones extended Luhn's tf to include the statistical analysis of the word occurrence in the document and across the collection of documents. She introduced the concept of inverse document frequency which measures the significance of the term in a given document corpus (Jones, 2004).

In 1990s, the field of information was further expanded by the development of several major areas in the field of information that included full text retrieval systems, digitization of paper document, document management systems, hypermedia, multimedia, and virtual reality applications. The birth of the web opened the door to globalization and the global access to information from an integrated diverse resource. This was made possible by the development of hypertext transfer protocol (HTTP), the hypertext markup language (HTML), and the web browsers. The web search engines started to appear in the mid-1990s, and at the turn of the century, we started to see more sophisticated technologies developed and deployed on the Internet. Web 2.0 moved the web from static web pages to more dynamic web applications where access to information in real time is made possible. The concept of knowledge portals and enterprise portals started to emerge by integrating diverse range of applications and by providing single sign on access technologies.

The 2000s era marked the turn of the century in an era characterized by globalization and the increased emphasis on knowledge as a factor of growth. It has given rise to intellectual capital, intellectual property, and the wider concept of the knowledge-based economy. It has also given rise to the birth of online mega corporations such Google, Amazon, Netflix, and Facebook. It created interest in new growth areas in the information field such as knowledge management, knowledge discovery, big data, data analytics, and data science. The relationship between technology, people, and information formed the basis for birth of the iSchools movement or the Information Schools.

The iSchools movement started in 2005 when a number of library and information science schools realized that their teaching and research programs had capacity to reach a broader audience of students and to prepare professionals for work beyond librarianship. Since then the number of schools joined the iSchools movement have increased to more than 80 schools. Since then many of the library and information science schools changed their name by dropping the word library from to reflect the broader nature of the information profession. It is important to note that a number of institutions joined the iSchools consortium were not originally library schools. They are part of the Computing Research Association and business schools with strong management information system programs. The iSchools vision as stated on the iSchool.org website is to expand presence internationally, recognized for creating innovative information solutions and systems to benefit individuals, organizations, and society at large.

The iSchools consortium represents a shift in directions and philosophy from the traditional library and information science education. The shift in direction revolves around broadening the concept of the information field by the diversifications of program offerings and the inclusion of new and emerging areas such as data science, cybersecurity, and knowledge management.

Privacy and the Right to Be Forgotten

The relationship between privacy and security is a complex one. Historically, information professionals including librarians, archivists, and record keepers defended intellectual freedom and championed the user's right to privacy. By doing so, librarians in essence took on themselves the task of protecting user's information and keeping it confidential. This undertaking required that information professionals understand and practice safe records keeping. The American Library Association (ALA), the largest library association in the world, has long held that privacy is a core value of the librarianship profession. The ALA Code of Ethics adopted on June 28, 1995 article II states that "We uphold the principles of intellectual freedom and resist all efforts to censor library resources." This is in line with the Freedom of Information Act and the freedom of expression, as recognized by Resolution 59 of the UN General Assembly adopted in 1946 and Article 19 of the Universal Declaration of Human Rights in 1948 (Magi, 2011; Diaz, 2019).

The right to be forgotten is relatively new and it is an interesting concept given the fact that once information is posted on the web, and gets captured by someone somewhere and saved permanently. This could be the case even if the information might be wrongly posted and removed. In light of this and in response to other privacy concerns raised by data collection on the web, the European Union created the right to be forgotten and the right to erasure as part of the General Data Protection Regulation. The regulation came about in 2014 as a result of the case Google Spain SL, Google Inc v Agencia Española de Protección de Datos, Mario Costeja González.

Privacy refers to the right to control personal information, whereas security refers to how your personal information is protected. There is a delicate balance between privacy and security as both are related. Most web-based services and social media applications provide privacy policies and privacy agreements designed in theory to protect users. However, failure to protect user's information in certain cases undermines users' confidence in these systems. Most users pay little attention the privacy statements or terms of use and the long-term implication of the information collected about them (Menand, 2018).

While technology enhances access to information and provides the needed tools to protect user's data such as advance encryption and privacy setting, cybersecurity remains a challenge to both the users and the information professionals. Threats to privacy in the cyberspace are real and varied. Security threats come in different forms and shapes that include government surveillance tools, sophisticated identity theft, cybercrimes using hacking to exploit vulnerabilities in online applications such banking, e-commerce transaction, online retails databases, and election stations (Armerding, 2018; Miao, 2018). To encounter such threats, there must be a holistic approach to the problem that would include creating awareness among users, building a sound and robust infrastructure, and training a new generation of information and cybersecurity professional who can deal with both the technical and social aspects of cybersecurity.

Cyberattacks and Cybersecurity Threats

Cyberattacks are on the rise and cybersecurity threats are real. Cyberattacks do not only pose problems to individuals but also to private and public institutions, governments agencies, and society at large. The Internet has opened the door to all sorts of hacking and intrusion activities by local players and hostile powers. The most sophisticated types of attacks are those orchestrated by hostile powers and hostile governments in a form of cyberwar. Spying in cyberspace has become a common activity through deploying malicious software and cookies. This sometimes is combined with other form of data collection and monitoring using sophisticated artificial intelligence and predictive analytics tools (Collier, 2019; Joshi, 2019). Information and cybersecurity professional's role is to guard against such activities and threats. This means that information and cybersecurity professionals not only need to understand the risk and the different types of cybersecurity threats that might exist but also need to develop the capacity to work with the tools and technologies needed to encounter such threats.

Phishing Threats

Phishing is a common security practice carried out by sending emails that might appear to be coming from reputable organizations or close friends using their personal contacts. As long as we are connected to the Internet, we are potential targets to different form of phishing attacks. Recent studies have shown that more than 90% of cyberattacks happened using phishing emails (BBB, 2017). In such environment, the hackers by sending fraudulent emails they try to collect credentials that can be used to gain access to sensitive information. Phishing attacks attempt to exploit security vulnerability associated with user's lack of knowledge or user's lack of awareness about the risk by responding to email requests. Sometime the emails are designed to create panic by creating a sense of emergency, causing the person to panic and response to the hacker's request.

In recent years, phishing attacks are becoming more sophisticated and can be tailored using real people information. Much of that information can be stolen from contacts information accessed using various social media applications. Many of the social media applications require users to agree for the app to access their contact information. Users most of the time are left with no choice but to agree and consent for the app to access their contact information or their location. The binary option of yes or no does not leave the user with that many options to protect their contacts or location information. Using contact information hackers are able to design phishing emails and send them on behalf of trusted parties such as coworkers, family members, or friends.

Clicking on a link provided by the phishing email could enable the hacker to gain access to the employee email account and use it to send phishing messages to other employees. Taking over of the employee email can also enable the hacker to gain access to sensitive information According to Moore and Clayton (2007), phishing is a process of enticing people into visiting fraudulent websites and convincing them to enter credential information such as usernames, passwords, addresses, social security numbers, and personal identification numbers.

Ransomware

Ransomware is another type of security threat that is becoming serious. A recent article by CNN dated May 10, 2019 reported a number of ransomware attacks on local US government such as cities, police stations, schools, and government offices (Collier, 2019). The number of ransomware attacks is on the rise. In 2016 the number of attacks was 46. In 2017, the number dropped to 38 but in 2018 the number jumped to 53, indicating a rise in the number of ransomware attacks again.

Ransomware is a type of malware that attackers try to install on the targeted computers. The malware then prevents users from accessing their system or personal information and demands ransom payments. The method by which ransomware is spread is through phishing emails or malicious spam. The email normally has attachments that can act as booby traps. Phishing occurs when the user connects to a fake website by simply clicking on an Internet link that is embedded within the email.

There are three different types of ransomware. These are scareware, screen lock, and encryption ransomware. Scareware uses security alerts and other methods to scare the users of viruses or other malicious apps installed on the computer and it can be removed if you sign up and pay to the services to remove it. Another way is to freeze the screen displaying an FBI message claiming that you have violated certain rules and you need to contact them immediately to resolve the issue. They normally warn you from switching your computer off. One of the nasty methods is to deploy a malware that hijack and encrypt the data (Brewer, 2016). The danger here is unlike the previous two where the computer can be turned off and then a cleanup software is used to remove the malware. In this type of encryption malware once it has encrypted the file, it is impossible to decrypt the files again without paying ransom, resulting in losing data.

Cryptojacking

Cryptojacking is another type of security threat similar to ransomware but mainly affect cryptocurrency. Cryptocurrencies refer to a type of money generated using encrypted codes (Nadeau, 2018). For any currency to have a value it should be in limited supply. Gold is used as a currency due to the limited supply and the difficulty and efforts it takes to mine gold, making it of a higher value metal. Cryptocurrencies are similar in many ways to gold, as bitcoins are hard to mine and require unique skills to discover certain mathematical calculations that are not easy to duplicate.

Cryptocurrency in the form of bitcoin is a relatively new phenomenon and came about in 2009. The value of bitcoins skyrocketed from the time it was used to purchase pizza back in 2010 to now using it to buy luxury cars, real estates, and self-declared billionaire. The value of one bitcoin reached more than $20,000 in 2017, which motivated people around the world to start mining bitcoins (Rooney, 2018). Cryptocurrency as piece of information (a complex mathematical calculation) stored in databases and on computer networks requires higher level of security.

Cryptocurrencies in the form of bitcoins are basically database records protected by the strength of the network security and the carrier in which the financial transaction is performed. Unlike traditional currency which can be backed by gold or government banks guarantees, cryptocurrency is managed by multiple duplicate databases across decentralized networks with no control or government oversight. The decentralized nature of the cryptocurrency makes it hard for anybody or agency to control the number of bitcoins that can be released in the market. Cryptojacking is a process of stealing resources that enable the hackers to locate or mine currency at low or no cost. Cryptojacking software can be deployed in a form of malware that can take advantage of the resources deployed to mine bitcoins.

Cyber Physical Attacks

Cyber physical attacks are viewed in relation to cyber physical systems and the weakness of their computation and communication elements. For example, a user with bad intentions can take control of the computing and communication components of a certain service and devices such as computer centers, home automation devices, cars, or water pumps. Loukas (2015) described cyber physical attacks as "cyber-attacks that have physical effect propagations ... a cyber physical attack is a security breach in cyberspace that adversely affects physical space" (p. 11).

Physical security is critical to the overall security of the services provided online. There are many components of physical security planning and implementation that involve activities such as designing the facilities, risk assessment and asset management, and rules and regulations that govern entry and use of the facilities (Hutter, 2019). Cybersecurity professionals have the responsibility to make sure that the physical security of facilities is under their control. This includes restrictive entry to the facility, authenticating employees looking to gain access to the facilities, and monitoring contractors and other visitors who might try to access the physical site area. The secure areas usually house servers, data storage devices, and other computing resources that are needed to run the organization.

State-Sponsored Attacks

Cyber warfare today is a reality given the increased political tension around the world (Dunn Cavelty, 2012). The increased dependency on the Internet as a form of communication tool, a platform for commerce, and automation vehicle such as Internet of Things (IoT) has created high-value targets. Today every organization, private or public, is connected to the Internet. Such loose public network of billions of machines around the world makes it possible for hackers from around the world to engage in aggressive activities, ranging from stealing information to crippling organization or state infrastructure. Almost all sectors of society including education institutions, health information systems, government department, and private and public corporation rely on technology and Internet communication. Most organization will be brought to a standstill and stop functioning as and when they experience Internet interruption and communication problems. This is applied not only to Internet-based operations such as Amazon or Facebook, but rather to brick and mortar companies that rely heavily on information systems and data repositories essential to their daily operations, and without access to these resources, they cannot operate.

It is easy for states to engage in cyber warfare to inflict damages on adversaries given the low cost, speed, and risk compare to traditional warfare. Some countries have invested heavily in such capabilities, creating a race against time in preparedness and readiness to defend. Most countries developed capabilities to defend against cybersecurity attacks. However, such infrastructure and capabilities can be misused if it falls in the hands of wrong people.

Cybersecurity Challenges

Computer network infrastructure faces security threats on a daily basis, and the number of threats are growing and becoming more sophisticated. According to Wang and Yang (2017), almost every minute, there are more than half a million attack attempts that are taking place in cyberspace. Hackers have become more sophisticated, and protecting information systems form hacking activities has become more difficult. Hackers are constantly looking for vulnerabilities and weaknesses in the system with the motivation of stealing information and undermining the IT infrastructure. Hackers only need to have a single tool or technique to access an information system that has a weakness that they can take advantage of. Most businesses, small or big, are facing difficulty in keeping up with the number of threats due to the growing variety of hardware, devices, applications, and end users. Toward the end of 2016, companies realized the acute shortage in cybersecurity professionals and decided to increase their budgets to meet the challenge. CyberSeek, an initiative funded by the National Initiative for Cybersecurity Education (NICE), reported that the United States has shortfall of 314,000 cybersecurity professionals as of January 2019. The global cybersecurity shortage of highly qualified professionals is projected to reach 1.8 million by 2022 (Crumpler and Lewis, 2019; Frost and Sullivan, 2017).

Internet of Things (IoT) Vulnerability

According to Loukas in 2015, the IoT is the vision of a global infrastructure made of a network of objects. The information in this case is not generated by people but rather by devices and physical objects such as appliances, vehicles, home automation devices, and smart buildings. Consumer devices such as TV sets, cars, and personal devices are the main driver of IoT and it accounts for 5.2 billion units in 2017 and is expected to rise up to 12.86 billion in 2020. Businesses specific business devices such as manufacturing field devices, process sensors for electrical generating plants, and real-time location devices for healthcare are expected to rise from 1.64 billion in 2017 to 3.17 billion in 2020 (Tung, 2017). Powered with the latest and most advanced technology, IoT devices have the ability to communicate and interact via the Internet in a way that can be watched and controlled from a distance. Such devices are used to automate homes, factories, offices, and smart buildings (Rouse, 2016).

IoT device connectivity to the networks depends on the type of IoT application deployed. Just as there are many different IoT applications, there are different connectivity and communications options. Communications protocols include CoAP, DTLS, and MQTT, among others. Wireless protocols include IPv6, LPWAN, Zigbee, Bluetooth low energy, Z-Wave, RFID, and NFC. Cellular, satellite, WiFi, and Ethernet can also be used. Due to the variety of such devices and the danger posed by the rapidly growing IoT attacks, the FBI released the public service announcement "FBI Alert Number I-091015-PSA" in September 2015 (Rouse, 2016). The document outlined the risks associated with the use of IoT devices and made recommendations regarding methods and tools that can be used to protect and secure such devices.

Artificial Intelligence

Making progress in areas like image recognition, voice recognition, natural language processing, and deep learning will enhance our ability to design a more sophisticated intelligent systems capable of defending against cyber. As hacking activities become more sophisticated and complex, the need to deploy AI capabilities to examine and predict vulnerabilities will increase. We could also see an increase in the number of organizations deploying AI tools to monitor and protect against sophisticated attacks.

AI robots have become popular in recent years as tools for collecting data due to low cost. However, autonomous AI tools can inflict long-term damage without the need for human interference. It has the ability to adjust itself to the environment and reinvent itself if need be depending the circumstances (Dietterich and Horvitz, 2015; Harel, Gal, and Elovici, 2017). Mining and analyzing large amount of data gives these tools the ability to adapt to the environment and determine the types of attacks. Based on the information gathered, it could assess the willingness of the targets to pay ransom based on profiling the victims over a long period of time. AI could be used to create trust through engaging targets in long-term relationships designed to gather information through fake social media accounts and other impersonated identities.

Guarding against such capabilities requires organizations to to deploy smarter and more sophisticated tools to deal with malicious AI software. AI systems can be trained to detect malware and viruses with the help of big datasets and machine learning. They can be trained to analyze microbehavior of ransomware attacks to recognize ransomware before it encrypts a system (Joshi, 2019).

AI systems can perform predictive analytics based on the data generated on daily basis and provide alerts of potential future threats.

Social and Cultural Issues

Internet users around the world are increasing in numbers and now they are part of the digital culture. Digital culture is a culture shaped by technology and has transformed the way we communicate and behave in cyberspace. People with different social and cultural backgrounds have different views and attitudes toward privacy and security. According to Miao (2018), the concept of privacy has changed. For example, in the past if you were in your room alone, you did not expect people watching you and you do not have smart machines and devices that might be spying on you. If you were shopping, most of your transaction would have been paid in cash and the merchant has very little knowledge of you. But cybersecurity is not only a technical problem, it is a business and a human issue in which people's perception of security is shaped by social and cognitive issue that involves culture, normative beliefs, and state of mind. Understanding the economical, situational, and behavioral issues governing people's perception of risk is critical to combating cybersecurity attacks.

Hackers are always looking for new ways to steal valuable information of users mainly for monetary gain (Overfelt, 2016). However, stolen information could create social and cultural clash and, in some cases, endangering people's lives. The psychological impact of revealing confidential information related to race, gender, sexual orientation, and health issues has not been adequately studied. In a more conservative culture, the motivation for hacking could be social, ethical, or political, making it more vulnerable for internal security threats. Awareness and the level of education play an important role in tackling the cybersecurity problem. Aloul (2012) reported that the number of Internet users in the Middle East grew by 1825% compared with the growth of 445% in the rest of the world during a 10-years period. He pointed out that one of the major problems is dealing with the number of the "uneducated" users, making this population an easy target for hackers and cyberattacks.

Cybersecurity Awareness and Education

Cybersecurity impact on businesses, government agencies, and society can be assessed from long-term effect on the economy and the cost associated with cyberattacks. The increased reliance on technology and digitization of assets makes it harder to protect intangible assets such as intellectual property from cyberattacks. Intellectual property theft in the form of patents, trade secrets, trademarks, and copyright materials in digital format such as movies, software, and games cost the economy billions of dollars. One study showed that copyright violations and intellectual property theft in movies, songs, software, and video games costs 373,375 jobs and $58 billion in total economic activity (Siwek, 2007). Cyberattacks happen when vulnerability exists, making it necessary for organizations to build robust cybersecurity infrastructure. It is also important to hire qualified information and cybersecurity professionals who make sure policies, regulations, procedure, and practices are in place to deal with potential risks. While organizations normally focus on the technical infrastructure, it is normally the social and ethical aspects of security that most of the time are overlooked.

As stated earlier, organizations are losing billions of dollars yearly due to stolen intellectual properties in the form of trademarks, digital goods, branding, and reputations, and also due to stolen vital confidential information related to products, services, best practices, and procedures. On the other hand, organizations do engage in competitive intelligence and collect information about competitors' products and services in a legal and ethical manner. Such activities are not considered spying as long as it is not a theft of property or trade secret ideas. Cyber theft is an illegal hacking activity that normally cause loss of revenue and can create an unfair advantage for affected parties over a long period of time.

To combat cybersecurity threats, organizations need to hire qualified information and cybersecurity professionals who are to deal with all aspects of security. The shortage in the number of qualified cybersecurity professionals is a challenge that has long-term security implications. It is estimated that, in 2019, the number of vacant positions in cybersecurity will rise to six million globally (Cabaj et al., 2018). The shortage in cybersecurity professionals could be alleviated if we can train information professional in the area of cybersecurity. Today, most organizations focus their attention and limited resources on protecting the technical components of their systems. This approach is insufficient due to the fact that cybersecurity threat started using advanced AI tools as well as social and psychological methods (Lord, 2019). This means that organizations need to investment in technology as well as human capital in the form of training and creating awareness about cybersecurity threats.

The shortage in the number of qualified cybersecurity professionals has prompted various government agencies to embark on funding cybersecurity infrastructure, including the expansion of the workforce with the help of educational institutions. This includes providing funding for student scholarships and capacity building through the creation of formal academic programs. To support these efforts, the Senate Commerce Committee approved the Cybersecurity Act (S. 773) in 2009 to improve and develop the states cybersecurity awareness (Cheung et al., 2011). The act calls for the development of a new generation of information technology (IT) and cybersecurity specialists in order to develop and maintain a sound and effective cybersecurity infrastructure. Internet users who gain access through public network, such as unsecure hotspots, are considered to be the most vulnerable population. Besides understanding the risk of using public networks, users have to change their behavior while using social media and other applications designed to share information (Rahim et al., 2015).

In addition to creating awareness about the dangers of cybersecurity attacks, there is a need for more formal education in which theory can inform practice. By realizing the risks posed by the shortage in highly qualified cybersecurity professionals, the Department of Homeland Security partnered with National Science Foundation (NSF) and the Office of Personnel Management in 2000 to provide educational institutions with funding to support two- and four-years degree programs at colleges and universities. The Scholarship for Service Program (SFS) was created under the Federal Cyber Service Training and Education Initiative to provide funding to colleges and universities for scholarships and capacity building in the information assurance and computer security fields.

The SFS program funded multiple educational institutions over the years with the objectives of creating growth and preparing a cadre of federal information assurance professionals. Figure 1.1 shows the distribution of fund by state awarded to academic institutions. The program offers scholarships that fully cover the costs (tuition and education fees) for full-time students who join an institution that is part of the program. The students in the program also receive stipends of $22,500 for undergraduate and $34,000 for graduate (CyberCorps®, 2019).

Cybersecurity is one of the fast-growing areas in the fields in IT where women are underrepresented. This could be attributed to social, institutional, and personal challenges that women face when it comes to job advancement in cybersecurity. Women seem to experience more social and institutional challenges in the workplace than men. According to Bagchi-Sen et al. (2010), female students feel that they do not have much guidance and mentoring since the majority of faculty members in the computer science and IT departments are males. Being mentored by a faculty is very important for the student's future when it comes to job opportunities; the lack of female mentors might possibly delay their professional development within the field.

In the fields of sciences and engineering, women are underrepresented. "NSF data shows groups identified as being underrepresented in STEM (racial and ethnic minorities, women, and persons with disabilities) still lag behind their majority counterparts in STEM degree attainment and representation in the STEM workforce" (James, Singer, and Elgin, 2017, p. 2). This is no different when it comes to women in the cybersecurity profession. A recent study showed that women make up only 20% of the cybersecurity workforce (Morgan, 2019). The rise of cybersecurity awareness is leading more women to join the field.

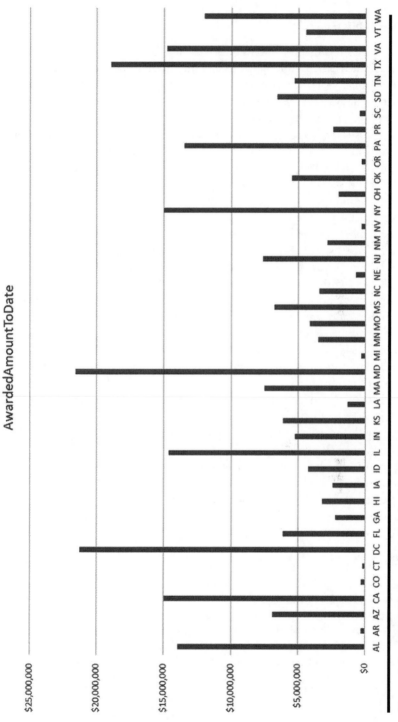

Figure 1.1 NSF-SFS Grant Funding by State.

Conclusion

Historically, information professionals like librarians, archivists, and record managers assumed the role of gatekeepers with the self-mandated tasks of protecting organizational asset, advocating for intellectual freedom and securing patron's information. Advocating for intellectual freedom and privacy required that information professionals protect patron's information from unwanted and unauthorized access. This notion was further supported by ALA Code of Ethics and the Freedom of Information as a basic human recognized by the United Nations Resolution 59 (Magi, 2011; UDHR, 2019). The Library Bill of Rights adopted by the ALA in 1939 clearly states that "*A person's right to use a library should not be denied or abridged because of origin, age, background, or views.*" The bill also recognized the importance of protecting and safeguarding users' information by stating that "*All people, regardless of origin, age, background, or views, possess a right to privacy and confidentiality in their library use. Libraries should advocate for, educate about, and protect people's privacy, safeguarding all library use data, including personally identifiable information*" (www.ala.org; Diaz, 2019).

Today, as much as technology has revolutionized the way people access, manipulate, and use information, it has also increased vulnerability and made it easier for hackers and bad actors to try and gain unauthorized access to information long held sacred by information professionals. Certainly, technology and the Internet did not make it easier for information professionals to do their job. Threats to privacy in cyberspace are real and complex and come in different forms and shapes. To encounter such threats, there must be a holistic approach to dealing with the problem. This would include creating awareness among the users and privide formal education to equip information professional with the skills and competencies needed to combat cybersecurity threats as well as provide sound and robust infrastructure that addresses the technical, legal, and social aspects of cybersecurity. The chapter discussed cybersecurity attacks, threats, and challenges in the context of the information profession and its historical association with the notion of cybersecurity and privacy.

References

Aloul, Fadi A. "The Need for Effective Information Security Awareness." *Journal of Advances in Information Technology* 3, no. 3 (2012): 176–183. doi:10.4304/jait.3.3.

Armerding, Taylor. "The 18 Biggest Data Breaches of the 21st Century." *CSO Online.* December 20, 2018. www.csoonline.com/article/2130877/the-biggest-data-breaches-of-the-21st-century.html (Accessed April 23, 2019).

Bagchi-Sen, Sharmistha, Rao, H.r., Upadhyaya, Shambhu J., and Sangmi, Chai. "Women in Cybersecurity: A Study of Career Advancement." *IT Professional* 12, no. 1 (2010): 24–31. doi:10.1109/mitp.2010.39.

Bates, Marcia J. "The Information Professions: Knowledge, Memory, Heritage." *Information Research* 20, no. 1 (2015): paper 655. http://InformationR.net/ir/20-1/paper655.html.

BBB. "2017 State of Cybersecuriety Among Small Businesses in North America." 2017. https://www.bbb.org/globalassets/shared/media/state-of-cybersecurity/updates/cybersecurity_final-lowres.pdf (Accessed March 16, 2020).

Brewer, Ross. "Ransomware Attacks: Detection, Prevention and Cure." *Network Security* 2016, no. 9 (September 2016): 5–9.

Cabaj, Krzysztof, Domingos, Dulce, Kotulski, Zbigniew, and Respício, Ana. "Cybersecurity Education: Evolution of the Discipline and Analysis of Master Programs." *Computers & Security* 75 (2018): 24–35. doi:10.1016/j.cose.2018.01.015.

Cheung, Ronald, Joseph, Cohen, Lo, Henry Z., and Elia, Fabio. "Challenge Based Learning in Cybersecurity Education." Proceedings of the 2011 International Conference on Security & Management, Las Vegas, NV. http://josephpcohen.com/papers/cbl.pdf (Accessed April 2, 2019).

Collier, Kevin. "Crippling Ransomware Attacks Targeting US Cities on the Rise." *CNN*. May 11, 2019. www.cnn.com/2019/05/10/politics/ransomware-attacks-us-cities/index.html (Accessed May 21, 2019).

Crumpler, William, and Lewis, James Andrew "The Cybersecurity Workforce Gap." 2019. www.csis.org/analysis/cybersecurity-workforce-gap (Accessed May 24, 2019).

CyberCorps®: Scholarship for Service. *National Initiative for Cybersecurity Careers and Studies.* 2019. https://niccs.us-cert.gov/formal-education/cybercorps-scholarship-service-sfs# (Accessed March 21, 2019).

Diaz. E. "New Library Bill of Rights Provision Recognizes and Defends Library Users' Privacy." 2019. Press Relase Tuesday February 12, 2019. http://www.ala.org/news/press-releases/2019/02/new-library-bill-rights-provision-recognizes-and-defends-library-users.

Dietterich, Thomas G., and Horvitz, Eric J. "Rise of Concerns about AI: Reflections and Directions." *Communications of the ACM* 58, no. 10 (2015): 38–40.

Dunn Cavelty, Myriam. "The Militarisation of Cyber Security as a Source of Global Tension (February 1, 2012)." In *Strategic Trends Analysis.* Edited by Daniel Möckli and Andreas Wenger, Zurich: Center for Security Studies, 2012. https://ssrn.com/abstract=2007043.

Frost & Sullivan Briefing. *2017 Global Information Security Workforce Study.* Center for Cyber Safety and Education, 2017. https://iamcybersafe.org/wp-content/uploads/2017/06/europe-gisws-report.pdf (Accessed April 15, 2019).

Harel, Yaniv, Gal, Irad Ben, and Elovici, Yuval. "Cyber Security and the Role of Intelligent Systems in Addressing Its Challenges." *ACM Transactions on Intelligent Systems and Technology* 8, no. 4 (May 2017), Article 49, 12. doi:10.1145/3057729.

Hutter, David. *Physical Security and Why It Is Important.* SANS Institute, 2019. www.sans.org/reading-room/whitepapers/physical/physical-security-important-37120 (Accessed May 10, 2019).

James, Sylvia M., Singer, Susan R., and Elgin, Sarah C. R. "From the NSF: The National Science Foundation's Investments in Broadening Participation in Science, Technology, Engineering, and Mathematics Education through Research and Capacity Building." *CBE-Life Sciences Education* 15, no. 3 (2017). doi:10.1187/cbe.16-01-0059. www.lifescied.org/doi/full/10.1187/cbe.16-01-0059 (Accessed May 21, 2019).

Jones, Karen Spärck. "A Statistical Interpretation of Term Specificity and Its Application in Retrieval." *Journal of Documentation* 60, no. 5 (2004): 493–502. doi:10.1108/00220410410560573.

Joshi, Naveen. "Can AI Become Our New Cybersecurity Sheriff?" *Forbes*, 2019. www.forbes.com/sites/cognitiveworld/2019/02/04/can-ai-become-our-new-cybersecurity-sheriff/#7674288e36a8 (Accesses May 20, 2019).

Lord, Nate. "What Is Cyber Security? Definition, Best Practices & More." *Digital Guardian*, 2019. https://digitalguardian.com/blog/what-cyber-security (Accessed May 21, 2019).

Loukas, George. *Cyber-Physical Attacks a Growing Invisible Threat*. Oxford: Butterworth-Heinemann Is an Imprint of Elsevier, 2015.

Magi, Trina J. "Fourteen Reasons Privacy Matters: A Multidisciplinary Review of Scholarly Literature." *The Library Quarterly* 81, no. 2 (2011): 187–209. doi:10.1086/658870.

Menand, Louis. "Why Do We Care So Much about Privacy? Big Tech Wants to Exploit Our Personal Data, and the Government Wants to Keep Tabs on Us. But 'Privacy' Isn't What's Really at Stake." *New Yorker*, 2018. https://www.newyorker.com/magazine/2018/06/18/why-do-we-care-so-much-about-privacy (Accessed March 15, 2020).

Miao, Christine. "Why You Don't Care about Internet Privacy." 2018. https://medium.com/@christinemiao/you-dont-care-about-internet-privacy-you-should-7b16ef2fcc71 (Accessed May 3, 2019).

Moore, Taylor, and Clayton, Richard. "An Empirical Analysis of the Current State of Phishing Attack and Defense." 2007. www.cl.cam.ac.uk/~rnc1/weis07-phishing.pdf (Accessed March 17, 2019).

Morgan, Steve. "Women Represent 20 Percent of the Global Cybersecurity Workforce in 2019." *Cybercrime Magazine*, April 03, 2019. https://cybersecurityventures.com/women-in-cybersecurity/ (Accessed May 23, 2019).

Nadeau, Michael. "What Is Cryptojacking? How to Prevent, Detect, and Recover from It." *CSO Online*, 2018. www.csoonline.com/article/3253572/internet/what-is-cryptoj (Accessed May 15, 2019).

Overfelt, Maggie. "World's Oldest Hacking Profession Doesn't Rely on Internet." *The Hacking Economy*, 2016. www.cnbc.com/2016/05/13/a-surprising-source-of-hackers-and-costly-data-breaches.html (Accessed March 23, 2019).

Rahim, Noor, Abd, Hayani, Hamid, Suraya, Kiah, Laiha Mat, Shamshirband, Shahaboddin, and Furnell, Steven. "A Systematic Review of Approaches to Assessing Cybersecurity Awareness." *Kybernetes* 44, no. 4 (2015): 606–622. https://libproxy.library.unt.edu/login?url=https://libproxy.library.unt.edu:2165/docview/1686402191?accountid=7113 (Accessed March 2, 2019).

Richardson, John. "History of American Library Science: Its Origins and Early Development." In *Encyclopedia of Library and Information Sciences*. Third Edition. Edited by Marcia J. Bates and Mary Niles Maack, Boca Raton, FL: CRC Press, 2010, 2–6.

Rooney. "Bitcoin Turns 10 – How It Went from an Abstract Idea to a $100 Billion Market in a Decade." *CNBC*, November 03, 2018. www.cnbc.com/2018/10/31/bitcoin-turns-10-years-old.html (Accessed May 21, 2019).

Rouse, Margaret. "What Is IoT Attack Surface? – Definition from WhatIs.com." *IoT Agenda*, 2016. https://internetofthingsagenda.techtarget.com/definition/IoT-attack-surface (Accessed May 21, 2019).

Sanderson, Mark, and Croft, Bruce. The History of Information Retrieval Research, 4 Proceedings of the IEEE, Vol. 100, May 13th, 2012.

Siwek, Stephen. "The True Cost of Copyright Industry Piracy to the U.S. Economy." 2007. www.ipi.org/ipi_issues/detail/the-true-cost-of-copyright-industry-piracy-to-the-us-economy (Accessed May 1, 2019).

Symantec. Norton. WI-FI Risk Report. *Report of Online Survey Results in 15 Global Markets*. 2017. www.symantec.com/content/dam/symantec/docs/reports/2017-norton-wifi-risk-report-global-results-summary-en.pdf (Accessed May 5, 2019).

"The Universal Declaration of Human Rights." OHCHR. www.ohchr.org/EN/UDHR/Pages/UDHRIndex.aspx (Accessed May 2, 2019).

Tung, Liam. "IoT Devices Will Outnumber the World's Population This Year for the First Time." *ZDNet*, February 13, 2017. www.zdnet.com/article/iot-devices-will-outnumber-the-worlds-population-this-year-for-the-first-time/ (Accessed April 21, 2019).

Wang, Yien, and Yang, Jianhua. "Ethical Hacking and Network Defense: Choose Your Best Network Vulnerability Scanning Tool." In 31st International Conference on Advanced Information Networking and Applications Workshops (WAINA). IEEE, Taiwan. 2017. https://ieeexplore.ieee.org/document/7929663/ (Accessed May 21, 2019).

Wiegand, Wayne A. "The 'Amherst Method': The Origins of the Dewey Decimal Classification Scheme." *Libraries & Culture* 33, no. 2 (1998): 175–194.

Chapter 2

Trustworthiness: Top Qualification for Cyber Information Professionals

Shuyuan Mary Ho

Florida State University

Introduction

Our human worth and values are defined by symbols and virtual identities – for example, email address, social media accounts, bank accounts, bitcoins, and so on. In the contemporary interconnected society; these symbols are represented by the "numbers" ($) in bank accounts, by "likes" in social media accounts, grades in the university systems, or scores maintained by the credit unions and background check agencies. These symbols and identities are translated and represented by data and information – whether individually or collectively – and the bits and bytes of data that reside on information systems in cyberspace. Due to connectivity and the user-friendly nature of the networking environment, data and information are left vulnerable and transparent in cyberspace, which leaves our human rights to privacy, worth, and values subject to manipulation and attack.

Cyberspace is that ubiquitous environment where communication occurs and is enabled through interconnected technologies. The term *cyber information professionals* refers to the information handlers in cyberspace that require adequate technical and management knowledge, skills, and abilities (KSAs) in order to identify and understand the dynamic requests and needs of modern information seekers – so as to collect, organize, classify, preserve, archive, retrieve, and disseminate information through up-to-date computer-mediated technology. Unfortunately, threats also manifest in cyberspace. The systems on which information artifacts reside are typically vulnerable and subject to attack. These attacks can occur in the areas of networks, systems, applications, software, and hardware. They emanate from social actors that span from the ill intentioned to naïve users making unintentional errors. These problems create an urgent need for information professionals to acquire up-to-date qualifications and KSAs so as to respond to the risks and threats – external, internal, and coordinated – of information loss, theft, compromise, disruption, and sabotage.

The chapter is outlined to begin with the discussion on a range of the cyberattacks of the future. Based on the varieties of attacks, essential KSAs are raised to provide guidance for cyber information professionals to prepare in order to respond and address the future of cyberattacks. Most importantly, trustworthiness as top qualification should guide information professionals in handling information, data, configuration of systems, and policy reinforcement in cyberspace. The chapter will discuss the approaches to determine trustworthiness, along with a matrix used to evaluate trustworthiness. The inference matrices specifically explain the mechanisms of reviewing observed behaviors for a specific cause based on information types in communication. Evaluation of trustworthiness can be programmable, and thus this discussion opens a new research venue that aims at evaluating social actors' trustworthiness based on their communicative intents and language–action cues in the human–computer interaction of socio-technical systems. The practical adoption of trustworthiness algorithms is discussed to enlighten the workforce on an approach to bridge the weakest (human) link when safeguarding our most valuable information assets. The implications of cyber information professionals' trustworthiness are discussed at the end to conclude the chapter with optimal anticipation of the future.

The Future of Cyberattacks

Cybercriminals use tools and techniques to disrupt services, steal credentials, and gain access to information systems. The cyberattacks of the future will include the following six categories of threats: network-based attacks, Internet-based attacks on websites, advanced malware, social engineering, insider threats, and coordinated attacks.

Network-Based Attacks

Most of the cyberattacks are carried out due to network connectivity. Inevitably, a variety of network-based approaches can be used to exploit and penetrate systems and networks. In the following, we discuss eleven (11) types of cyber-attacks that utilize networking knowledge and techniques.

1. *Eavesdropping*: Attackers can intercept network traffic to obtain sensitive or confidential information such as passwords, credit cards numbers, and so on. The attackers can passively listen on the network for any message transmission on the network. The attackers could be disguised as a normal unit on the network, and actively grab the information by probing or scanning the networks.
2. *Man-in-the-middle Session Hijacking*: Attackers can hijack a session between a client and a server, substituting their source and destination IP addresses to the attackers' machines, and thus hijacking communication between the client and the server.
3. *IP spoofing*: Attackers can send packets with the source IP address of a known, trusted host, convincing a victim system to communicate with the attack machines pretending to be the trusted machines.
4. *Replay*: An attacker can intercept a packet, change its session timestamps or nonce (i.e., a random number), and send the packet at a later time to the victim machine. The victim machines can be confused by the replay technique, not knowing that the packets were delayed, intercepted, and read by the attackers.
5. *Denial-of-service (DoS)*: A DoS or distributed denial-of-service (DDoS) attack rides on the convenience of TCP handshake protocols to launch many forms of TCP SYN flood attacks that aim to disrupt system resources and services. The attackers can take over a device and use it to flood the target systems with connection requests, while not responding with reply packets to those requests. Such request packets would occupy the target systems' bandwidth, and cause the target systems to time-out while waiting for the response packets.
6. *Teardrop attack*: Attackers can also confuse the target systems by changing the length and fragmentation offset field in sequential IP packets so that these packets overlap one another on the target systems. As the target systems attempt to reconstruct the packets, these target systems could thus be confused and crash.
7. *Smurf attack*: Attackers can spoof a target machine's IP address, and send spoofed ICMP echo requests. These spoofed ICMP requests would go to all IP addresses in the network broadcast range, and thus cause huge amounts of network congestion.
8. *Ping-of-death attack*: Attackers can send fragmented, oversized IP packets (i.e., over the maximum of 65,535 bytes) to a victim machine, causing the victim systems to reassemble the packets and experience a buffer overflow crash.

9. *Botnets*: On a large scale, attackers can form a botnet with a wide range of zombie systems to overwhelm the target systems' bandwidth. These bots or zombie systems can be remotely controlled to send out network-based attacks (e.g., based on TCP, IP, and ICMP protocols) against the target systems.

10. *Internet-of-things (IoT) attacks*: With the ever-growing network of physical objects and devices (e.g., thermostats, cameras, cars, electronic appliances, alarm clocks, speaker systems, vending machines, etc.) that utilizes embedded sensor technology (e.g., with an IP address assignment) to interact with external environments, these objects and devices can be taken over as botnets to overflow with distributed DoS attacks against specific targeted victims.

11. *Brute-force password attacks*: Attackers can compose a word bank (e.g., dictionary or birthday) randomly used to crack the victims' password. With modern technology, password files are generally encrypted. Attackers can compose a mammoth rainbow table that contains the hash files of all possible combinations of password and compare the message digest (MD) hashes. Once the hashes match, the attackers can index back to the string and identify the password.

Internet-Based Attacks on Websites

In addition to the networking knowledge and skills, successful attacks also require systems' knowledge.

1. *Drive-by-downloads*: Attackers can infect victims' machines with malware in several ways. First, the attackers can implant malicious scripts into HTTP or PHP codes on the webpages. These scripts could cause victims to either directly download and install malware, or redirect the victims to a falsified website to indirectly collect victims' personal and sensitive information. Second, when a victim uses a browser that contains security flaws and lacks proper patches and updates, and then visits a falsified website (or even an authenticate website with a pop-up window), malware could be downloaded to infect the victims' machines.

2. *SQL injection*: Attackers can attack a database with a malefactor that executes an SQL query via the input data from the client browser to the database server on the Internet. The predefined SQL commands could be inserted into the input fields (e.g., login ID or password) to exploit the database. Using this approach, attackers can view, modify, or delete the database data, or even shutdown the database.

3. *Cross-site scripting (XSS) attack*: Attackers can exploit the vulnerabilities of a website, run scripts via the browser, and inject a payload with malicious

JavaScript into a web server. The impacts of the XSS attacks are two-fold. First, the attackers could surf the web server, gain root access, and traverse to the sensitive data (e.g., password files) on the server. Second, the infected web servers could transmit the payloads and execute malicious scripts on the victims' machines when victims visit the website via a browser. The XSS attacks exist within VBScript, ActiveX, Flash, and JavaScript.

4. *Buffer overflow against application security*: Attackers can write bogus data, malformed inputs, or malicious codes to a program's buffer so as to overrun or exceed the buffer's boundary. The overflowing inputs could overwrite adjacent memory locations, and thus overwrite the program's state and cause unintended behavior of the application (e.g., escalating unlimited access privilege to the information systems).

Malware Attacks

Malware exists in many forms, for example, viruses, worms, and Trojans. It can attach itself to legitimate codes; it can also replicate itself across the network or the Internet. The impacts of malware infection can be devastating, destroying boot-sectors, or permanently deleting systems files. When it is combined with the use of the cryptography, it could encrypt the victims' critical files and ask for a ransom. In the following, we discuss eight (8) different types of malware attacks.

1. *Macro viruses*: These viruses could infect generally Microsoft products such as the Word or Excel applications. Macro viruses attach to the initialization sequences of the Word (.doc or .docx) or Excel (.xls or .xlsx). When these applications are run, the virus executes to replicate itself to attach to other files. Fortunately, these macro viruses can be thwarted with newly updated Microsoft patches.

2. *Boot sector or executable file infection*: Viruses can attach themselves to executable codes (.exe). That is, when the executable files (.exe) are run, the virus code will also execute. When the executable codes attach to the master boot records on the hard disk, it could load the virus into the random access memory and propagate to other disks.

3. *Polymorphic and metamorphic viruses*: The virus could conceal and mutate itself into different codes by encryption algorithms or compression techniques. Polymorphic malware can change part of its own code, while metamorphic malware codes are completely rewritten during each infection. When these types of viruses have a high level of entropy, they will mutate into many modifications or versions of their own source codes to avoid signature-based antivirus detection. An advanced version of mutable virus

can be concealed by compressing the file so that the file size appears to be unchanged, or by altering the date-stamp and timestamp of the file last modification.

4. *Trojan horses*: A Trojan horse generally does not replicate itself like a virus does. But, a Trojan horse can hide itself in a regular program, and establish a backdoor (e.g., opening a port) for attackers' exploitation at a later time. A Trojan dropper is seen as a helper program designed to install a malware or a rootkit to a target system, but Trojan dropper programs themselves do not carry any malicious activities. An example would be a download of a browser program from a third-party website where a browser is installed with a set of backdoors. In this case, the malicious browser is considered a Trojan dropper.

5. *Logic bombs*: Malware could be written to be triggered by a specific logical condition (e.g., when a certain file is executed) or at a specific date and time.

6. *Worms*: Computer worms are different from viruses in that worms do not infect by attaching themselves to a host file. Instead, worms are self-contained programs and worms can propagate via emails or attachments across the networks or the Internet. Computer worms can spread and infect machines at an exponentially accelerated rate – much faster than viruses do.

7. *Ransomware*: Malware that utilizes cryptography to its own advantage. The attackers block victims' access to their own machines and encrypt victims' data files until the ransom is paid to release the decryption keys. A notable example would be a worldwide cyberattack by WannaCry ransomware cryptoworm in May 2017. Files and data on Microsoft Windows operating systems were encrypted for ransom payments in the bitcoin cryptocurrency, and backdoors were installed on infected systems.

8. *Spyware* or *Adware*: Malware that could be unknowingly installed on a victim's browser and collect users' information such as browsing history so as to spy on and profile the victim's Internet access and browsing activities. Or, malware could be automatically downloaded and installed for marketing and advertising purposes.

Social Engineering

Social engineering, another form of deceptive exploitation, can be deployed to seize credentials for unauthorized access.

1. *Phishing*: Attackers can send out phishing emails to a general set of naïve users as victims, hoping a naïve user will click on the links, and offer their personal credentials or information. The victims could also risk downloading malware by clicking on the fake links that pretend to be trusted sites.

2. *Spear phishing*: Attackers can use psychological tricks to compose creative, but falsified phishing emails to more specifically targeted victims. These types of phishing emails appear to be more personable and convincing so that victims can be tricked.

3. *Pharming*: Attackers can redirect Internet users' browsing traffic to a falsified website, and trick users to offer their personal information. Pharming attacks can be accomplished by poisoning a DNS server. Compromised DNS servers can redirect users' requests to a fake or bogus website.

Insider Threat

Insider threat has become one of the most complex organizational problems (Ho, Hancock, Booth, Burmester, et al. 2016, Ho, Kaarst-Brown, and Benbasat 2018). Insiders are mostly composed of employees and contractors with whom extensive access to organizational information assets and systems are granted. As business communication is supported by ubiquitous computer-mediated technological settings, the ubiquitous nature of the computed-mediated technology not only transforms the way people interact with information systems, and with each other in virtual or cloud-based environment, but also grants opportunities for online deception to occur, thus making the shift to intentional betrayal against the organizations difficult to detect. Broadly speaking, there are two categories of insider threats. Negligent insiders can make careless mistakes and errors that allow social engineering to succeed and credential information to leak. More severely, rogue and deceptive insiders with ill intention can compromise and/or steal classified information (e.g., CIA agent Aldrich Ames, FBI agent Robert Hanssen, NSA contractor Edward Snowden, and soldier Chelsea Manning) or launch devastating sabotage against organizations (e.g., publicizing the CIA hacking tools on Wiki-leak results in setting back the national intelligence operations many years).

Coordinated Attacks

In addition to insider threat, another wicked problem that threatens organizational security is coordinated attacks – not only sophisticated in nature involving both social and technological domains, but also harnessing the deception and coordination involving both insiders and outsides – making organizations unable to function and trust one another. One good example is the advanced persistent threat (APT). An APT usually does not focus on sabotaging the systems. Instead, the attackers aim to probe the networks and collect information covertly over a long period of time. An ATP is generally orchestrated by a group of people that conduct Internet-based espionage, and typically aim at large cyber infrastructures, such as financial and banking sector, telecommunication, air transportation systems, utility companies, and so on. Hacker groups will adopt sophisticated techniques and tools (mentioned earlier) to exploit systems and networks, extracting and correlating data for larger attacks that are generally involved with political, business, or monetary motives.

Essential Knowledge, Skills, and Abilities

After having viewed the cyber threats, cyber information professionals should focus next on obtaining the basic KSAs required to handle the attacks of the future. These KSAs enable the critical thinking capability, and equip cyber information professionals to troubleshoot and analyze the cyber threats of the future.

1. *Information access and control*: Basic knowledge of the information systems (e.g., Microsoft, Linux, Android mobile, and proprietary operating systems for IoT), database (e.g., SQL versus noSQL) as well as fundamental knowledge of the TCP/IP networks and Internet security protocols is critical. Generally, user authentication provides the first gate of information access and control. Enterprise-based firewall technology – coupled with intrusion detection/prevention systems (IDS/IPS) technology – can be adopted together with antivirus and malware detection to thwart existing known network-based threats and attacks.

2. *Data encryption*: Basic knowledge of the cryptographic engines, complexity of the key, and algorithmic strengths will help maintain the confidentiality, privacy, and integrity of information. System files and data files can be verified to ensure no unauthorized modification or access, using MD techniques and timestamp validation.

3. *Cyber hygiene practices*: Cyber information professionals should implement good password generation practices so that they are difficult to compromise; likewise, they should have the autonomous power and discernment to perform observable cyber activities and understand the associated risk factors.

4. *Compliance of information policy*: There is no one-size-fits-all policy for organizations. Cyber information professionals should be equipped to quickly recognize a phishing scam so as to prevent data breach. Cyber information professionals should also be proficient and adaptive in manually, verbally, or mentally manipulating data or things (e.g., configuration of systems and networks) so that hardware and software devices function as planned and follow the organizational information security policy.

5. *Collect, analyze, and investigate big data*: To protect and prevent specialized denial and deception operations, cyber information professionals should be able to make sense of, interpret, evaluate, review, correlate, and investigate cyber events. Moreover, access-related information from systems,

networks, and digital evidence may be useful to develop insights and intelligence.

6. *Trustworthy handlers of information, data, and systems*: Most of all, the dependability of the cyber information professionals is a key qualification to secure information assets from external, internal, and coordinated attacks. Third-party vendors should also be subjective to regular audits and investigation by the organization.

Trust versus Trustworthiness

Trust facilitates basic human relationships in our society. The default of trust is two people who can rely on one another. Trust can also be one-sided or group-based largely depending on communication. Trust depicts a type of interdependent relationship between two individuals, two parties, or two groups: a trustor, who may (or may not) trust a trustee. Trust also implies a trustor's willingness to take risks and to be vulnerable. That is, trustors must have a sense of vulnerability toward a trustee for a trust relationship to exist. Trustworthiness, on the other hand, refers to an inferred quality inherent in a trustee. The establishment of trustors' beliefs as to a trustee's trustworthiness is an antecedent to a trust relationship.

Trustworthiness can be defined in several different ways with a spectrum of factors and dimensionality. For example, Lieberman (1981) identifies two factors: *competence* (external or situational cause) and *integrity* (internal or dispositional cause), while Butler (1991) identifies ten factors to measure trustworthiness: availability, competence, consistency, discreetness, fairness, integrity, loyalty, openness, promise, fulfillment, and receptivity. Mayer, Davis, and Schoorman (1995) simplify the concepts to provide a more generic framework with three factors of perceived trustworthiness: ability (competence), benevolence (kindness), and integrity (goodwill/ethics). To be more specific, competence refers to an external (situational) cause, which is the effective application of learned behavior; for example, a person can be competent by acquiring a set of skills. Integrity, on the other hand, refers to an internal (dispositional) cause, which is a person's internal dispositional state; for example, a person is willing to sacrifice his or her own time or energy to make a high integrity contribution regarding an assigned task. Benevolence (kindness) refers to the quality of a trustee that wants to do good, which may benefit the trustor.

Determining Trustworthiness

Trustworthiness may seem easy to understand and define, but it is in reality difficult to quantify and determine due to complicating factors and causes (Ho 2009a). The causes that complicate a determination of trustworthiness can include many reasons, for example, major incidents in life, love, lust, financial crisis, or power. Regardless of these reasons, Ho and Benbasat (2014, 2018) propose that trustworthiness can still be evaluated, attributed, and determined. The attribution of trustworthiness illustrates a basic relationship between two social actors (i.e., Alice and Bob). Supposed Alice and Bob work together in an organization, and Bob can represent a group of coworkers and peers with collective trust (Ho, Ahmed, and Salome 2012). Alice could be someone who holds a critical position and who has authorized access to intelligence or assets in an organization. Bob could refer to a group of peers and subordinates who work closely with Alice. Let us say Bob represents the interests of the organization; in general, Bob would be somewhat dependent on Alice. Let us further assume that the communications between Alice and Bob are meaningful based on their social interactions, and thus trust is built between Alice and Bob when each actor displays the willingness and competence needed to work together over time.

Determining Behavioral Cause Based on Observed Information Types

Here, the inference and the causal attribution of trustworthiness follow the three information types as illustrated by Kelley (1973) and Kelley et al.'s (2003) analysis of variance (ANOVA) model. Table 2.1 illustrates the eight possible behavioral patterns based on three information types. The high–high–high pattern can be attributed to a stimulus (external cause); the low–high–low pattern can be attributed to a person (internal cause); and the low–low–high pattern can be attributed to a circumstance (external cause). For example, a person shouts at a rock concert. If other people who attended the same concert also shouted (high consensus), and this person always shouts whenever attending this concert (high consistency), and s/he only shouts at this particular concert (high distinctiveness), people can make an external attribution that her/his shouting behavior is caused

Table 2.1 Patterns of Information Types in Kelley's ANOVA Model

Information type	Pattern		
Consistency	High	Low	Low
Distinctiveness	High	High	Low
Consensus	High	Low	High
Attribution	External causes -> stimulus	Internal causes -> disposition	External causes -> circumstances

by this particular concert (stimulus). If only this one person shouted at this concert (low consensus), and s/he generally does not shout in other concerts (low consistency), but always shouts whenever attending concert by this group (high distinctiveness), people can make an internal attribution that the shouting behavior is caused by the fact that s/he intrinsically likes this particular performance (disposition). If this person rarely shouts in concert (low consistency), and does not shout in concerts performed by this group (low distinctiveness), but other people shout loudly whenever attending the concert performed by this group (high consensus), then people can make an external attribution that the reason s/he attended this concert was not by will but by persuasion (circumstance).

Taxonomy and Matrices of Attributing Trustworthiness

Ho and Benbasat (2014) further develop a taxonomy of a dyadic attribution model to specifically evaluate trustworthiness (Table 2.2). The hidden cause of an observed behavior attributed to the actor's trustworthiness contains aspects of both internal and external causes. Internal cause would be attributed to a person's disposition or integrity, where a person is held intentionally responsible for the act. External cause would be attributed to a person's external circumstance, uncontrollable to the person, whereby a person would not be held intentionally responsible for the act. The circumstance refers to a type of situation a person is in which leads to a learned behavior. For example, if a person receives a type of training or education, this person would be expected to have competence (i.e.,

Table 2.2   Matrices for Trustworthiness Inference Classifier

Trustworthiness Attribution	Cause	Consistency	Distinctiveness	Group Consensus
Competent (usual)	External	High	Low	High
Integrity (ethical)	Internal	High	Low	Low
Competent (expected)	External	High	High	High
Innovative (unexpected)	Internal	High	High	Low
Incompetent (usual)	External	Low	Low	High
Unreliability	Internal	Low	Low	Low
Mistake (accidental)	External	Low	High	High
Betrayal	Internal	Low	High	Low

ability) as a result of that circumstance. In contrast, if a person takes a pen from a bookstore without paying and is caught, this person would be attributed to have low integrity as a result of his/her disposition. Table 2.2 illustrates the inference matrices of reviewing observed behavior when attributing a cause relating to trustworthiness based on various information types in communication. Any inconsistent and unreliable behaviors could be observed and attributed over time. In other words, social actors' language–action cues can reflect the communicative intent, and can serve as early signals to trustworthiness attribution.

Below illustrates how this inference mechanism works using a series of examples. The iterations of scenario are built on the previous example, and these examples are designed to illustrate the human dynamics underlying perceptions of observed behaviors within larger patterns of profiled behavior as observed by a peer group over time.

For example, a datacenter systems administrator has been tasked with migrating the organizational data records on a MySQL database from Microsoft products to Linux platforms in a cloud environment. Within the performance of this task, a behavior such as staying late in the office and consuming organization's critical resources is observed to be no different from expected behavior (as characterized by low distinctiveness). Suppose that this datacenter systems administrator is able to complete the MySQL database migration to Linux (as characterized by high consistency). His/her peers note that the datacenter systems administrator stays late in the office, consumes critical resources, and has completed this task with satisfactory results (as characterized by high group consensus). In this situation, behavior is attributed to an external cause, whereby this administrator displays usual competent behavior with sufficient ability to complete assigned tasks due to a trained skillset. In other words, this behavior is a learned behavior (competence), interpreted as being influenced by a cause outside of the individual (e.g., training or education).

Suppose that this datacenter systems administrator having committed to convert the MySQL database was able to accomplish such a task with ease due to training (characterized as high consistency). However, a circumstance arises that the cloud server is under severe distributed DoS attacks. An observed behavior is found that this administrator spends extra time in the office and consumes usual critical resources as s/he usually does (characterized as low distinctiveness). However, not many peers know about this challenging situation, and coworkers may have little or no clue about the ethical consideration this administrator has taken on in sacrificing his time and energy, and in securing the MySQL service ports at the gateway – with due diligence (as characterized by low group consensus). The cause of this observed behavior would be attributed to an internal (disposition) cause because each coworker would have a different interpretation about an observed behavior. In this situation, this observed behavior will tend to be attributed to an internal (disposition) cause that this individual is intentionally and full-heartedly held responsible for completing this challenging task. Such an observed behavior may be viewed as an ethical example of trustworthy behavior.

Suppose that this systems migration project is challenging, and this systems administrator needs to make a few phone calls, browse additional website such as the technical blogs for advice, and stay late in the office for a few nights to meet the deadline (as characterized by high distinctiveness). In the end, his co-workers all agree and acknowledge his performance (as characterized by high group consensus). The cause of this administrator's unusual behavior (e.g., staying late in the office, unusual browsing of internal and external websites, etc.) may be attributed to external causation; challenging tasks being overcome by his high competence. If a behavior is characterized as high consistency, high distinctiveness, and with high group consensus, such behavior would tend to be attributed to an external cause that suggests the administrator is highly trained and competent. In other words, this behavior could be expected based on a (profiled) high level of perceived competence.

Suppose that this systems administrator has successfully migrated the MySQL database to the Linux platform in the cloud environment (characterized as high consistency) by identifying some serious process flow problems and voluntarily spending extra time to patch and fix them (as characterized by high distinctiveness). These activities may not have been predefined in his job description and he may not have been adequately trained to perform this migration based on his busy schedules of routine job assignments. Due to the severe challenges of this task, perhaps not every coworker would have the same positive idea about this observed behavior (e.g., works overtime); thus not every coworker would expect that this systems administrator could have been able to complete this difficult task (as characterized by low group consensus). In this situation, this observed behavior would be attributed to internal causality, due to unexpected innovation of this individual. If a behavior were to be characterized as high consistency, high distinctiveness, but low group consensus, this behavior would tend to be attributed to an internal, dispositional cause, suggesting that this individual is unexpectedly innovative.

In another scenario, suppose that the performance of this systems administrator is characterized by drama with a constant streams of errors, and problems that prevent him from successful completion of his appointed tasks (characterized as low consistency). If this administrator may have always had problems in other projects, and typically characterized as not being able to complete assigned tasks as scheduled (characterized as low distinctiveness). If everyone in the group basically observes and agrees that this administrator cannot successfully complete his tasks due to some type of drama (as characterized by high group consensus). The perceptions from coworkers may develop that s/he is unfit for the job. The causation of this administrator's behavior will tend to be attributed to external issues, and perhaps an uncontrollable actor. We may say that if a behavior is characterized by low consistency, low distinctiveness, but with high group consensus, this observed behavior would tend to be attributed to an external causation that suggests the person is not to be held intentionally responsible for the situation. In other words, observers may attribute such a behavior to external nondispositional causation, or possibly incompetence.

Let us say this datacenter systems administrator has committed to the task of database migration, and has difficulty completing other assigned tasks. In this situation, the administrator was unable to perform the database migration task due to personal reasons (characterized as low consistency), and this administrator has never been able to complete the majority of the assigned tasks on schedule (characterized as low distinctiveness). If coworkers are unsure of the outcome of the task performance and may have various interpretations as to why the administrator could not complete the assigned tasks (characterized as low group consensus), the cause of any observed behavior might be attributed to low integrity. To repeat, if a behavior is characterized as low consistency, low distinctiveness, and low group consensus, then such behavior would tend to be attributed to internal causality; s/he would be held intentionally responsible for the act. In other words, observers attribute this behavior to low integrity based on analysis of the actor's disposition over time.

Suppose that this datacenter systems administrator has committed to migrating the MySQL database and has always been able to complete other assigned tasks, but has failed to do so this time (characterized as low consistency). Since the administrator has always done an exceptional job except for this one time (characterized as high distinctiveness), and coworkers agree that the administrator has faced some personal situations (characterized as high group consensus), the coworkers may attribute any failure to external causality, which may reflect an accident or a mistake. If a behavior is characterized as low consistency, but with high distinctiveness and high group consensus, such behavior would tend to be attributed to an external cause regarding competence, and not held responsible for an intentional act. Observers may view this behavior as being outside of the actor's control.

Suppose that this datacenter systems administrator has committed to migrate a MySQL database, but fails to complete this task or other tasks as promised (characterized as low consistency). This administrator has a good reputation in product training, completing task assignments, and meeting deadlines. Lately, s/he has been staying late in the office, consuming critical resources, and still does not complete assigned tasks (characterized as high distinctiveness). In this situation, several coworkers may have different observations and different mental models to evaluate the behavior, and there will be different explanations for why behavior has changed (characterized as low group consensus). The cause of this administrator's different observed behavior may be attributed to internal causality. To summarize this final example, if an observed behavior is characterized by low consistency and high distinctiveness, along with low group consensus, such behavior would tend to be attributed to internal causality; the individual could be held intentionally responsible for the act. In other words, observers may perceive the behavior as driven by betrayal.

To recap, the trustworthiness of cyber information professionals' handling of the information assets are critical to the ability of organizations to defend against the cyber threats of the future. Based on the above examples of behavioral attribution, we may be able to identify unreliable or negligent individuals that could cause loss of credentials (e.g., via social engineering) or loopholes in the networks or systems. Competent individuals could be unaware of simple but significant mistakes (e.g., Trojan horse backdoors) that give hackers a chance to hide in the networks and steal critical information. Even competent individuals could still be unethical, causing the organizations to suffer great loss in information assets, reputation, or financial capital. To summarize, the classification scheme depicted in Table 2.2 provides an illustration on how the behavioral cause can be attributed based on three information types (i.e., consistency, distinctiveness, and group consensus) in terms of determining trustworthiness at different levels and degrees. Ho et al. (2018) discussed the factors – for example, group sensitivity – to further the determination and attribution of trustworthiness.

Practical Adoption and Implications of the Application

Research on analyzing communicative intent and language–action cues has shown great promise in the area of attributing and determining trustworthiness – both subjectively (Ho 2009a, 2009b, 2014, Ho and Lee 2012) and objectively (Ho 2019, Ho, Fu, et al. 2015, Ho and Hancock 2019, Ho, Hancock, and Booth 2017, Ho, Hancock, Booth, Burmester, et al. 2016, Ho, Hancock, Booth, and Liu 2016, Ho, Hancock, et al. 2015, Ho, Hancock, Booth, Liu, et al. 2016, Ho, Liu, Booth, and Hariharan 2016). This chapter identifies the research potential of identifying actors' communicative intent and cues in absolute terms. In the cyber world of the future, information professionals are required to pay attention on the application adoption of evaluating trustworthiness, and understand the implications of an individual's trustworthiness as an absolute qualification for the organization. This will help determine organizational performance and productivity requirements if more business activities move to the cloud environment.

The extension of the trustworthiness determination can contribute to social computational systems that can provide organizational security without intruding on personal privacy. A computational classifier (i.e., a social firewall) could be developed to provide objective analysis of "random samples" of the online communication without collecting and monitoring individual's private information.

Trustworthiness can also have significant impact on organizational hiring or personnel retention practices. When interviewing candidates or evaluating employees, more attention should be focused on the candidates' trustworthiness in terms of integrity and ethical values in addition to competence (external performance indicators) or benevolence (interpersonal) factors. It is important to understand differences in leadership and management style when evaluating key personnel (Ho 2019). Unintentional streams of administrative errors may indicate that an individual may not be suitable or reliable with regard to a certain type of assignment. While organizations must recognize the importance of establishing an ethical culture of trust, a greater challenge for management is evaluating trustworthiness of cyber information professionals with regard to integrity, competence, and benevolence.

Conclusions

Although the cyber information professional's KSAs and competence on the job are important, *integrity* would be more a critical dispositional-based indicator of trustworthiness. An individual may be regarded as incompetent but still be found to be trustworthy. By contrast, a person may be very competent, but not trustworthy. It is quite common that betrayers of organizations appear benevolent and are seen as very nice people. Organizations that wish to put together a cyber defense team should pay more attention to candidates who are trustworthy and can handle information with high integrity, rather than those who are technically savvy but cannot be trusted. Safeguard information assets require collective efforts. Trustworthiness, as a top qualification for cyber information professionals, can be evaluated and measured during interpersonal communication as well as group interaction. Addressing the trustworthiness of human assets will not only solve cybersecurity technical challenges, but also bridges the weakest link of commands. We are in an age where ethical values and moral standards are not clearly defined by society. As such, individuals could be ignorant of ethics, which may negatively impact their decisions in personal life or at work. Poor adoption of standards around ethics may affect and even damage the fabric of trust in society. Today's cyber information professionals should strive to think with social responsibility and act in a trustworthy manner to enable the operations of the organizations, and to sustain the value and structure of our society.

References

Butler, John K. 1991. "Toward understanding and measuring conditions of trust: Evolution of conditions of trust inventory." *Journal of Management*, 17 (3): 643–663.

Ho, Shuyuan Mary. 2009a. "A socio-technical approach to understanding perceptions of trustworthiness in virtual organizations." In *Social computing, behavioral modeling and prediction*, edited by Huan Liu, John J. Salerno and Michael J. Young, 113–122. Tempe, AZ: Springer.

Ho, Shuyuan Mary. 2009b. "Trustworthiness in virtual organizations." In IProceedings of the 15th Americas Conference on Information Systems (AMCIS), at San Francisco. Retrieved from https://aisel.aisnet.org/amcis2019_dc/5.

Ho, Shuyuan Mary. 2014. *Cyber insider threat: Trustworthiness in virtual organizations.* Deutschland, Germany: Lambert Academic Publishing, Saarbrücken.

Ho, Shuyuan Mary. 2019. "Leader member exchange: An interactive framework to uncover a deceptive insider as revealed by human sensors." In Proceedings of the 2019 52nd Hawaii International Conference on System Sciences (HICSS-52), January 8, 2019, at Maui, Hawaii.

Ho, Shuyuan Mary, Issam Ahmed, and Roberto Salome. 2012. "Whodunit? Collective trust in virtual interactions." In *Social computing, behavioral-cultural modeling and prediction, LNCS 7227*, edited by Shanchieh Jay Yang, Ariel M. Greenberg, and Mica Endsley, 348–356. Berlin, Heidelberg: Springer-Verlag. doi: 10.1007/978-3-642-2904-3_42.

Ho, Shuyuan Mary, and Izak Benbasat. 2014. "Dyadic attribution model: A mechanism to assess trustworthiness in virtual organization." *Journal of the Association for Information Science and Technology*, 65 (8): 1555–1576. doi: 10.1002/asi.23074.

Ho, Shuyuan Mary, Hengyi Fu, Shashanka S. Timmarajus, Jung Hoon Baeg, Cheryl Booth, and Muye Liu. 2015. "Insider threat: Language-action cues in group dynamics." In Proceedings of the 2015 ACM SIGMIS Computers and People Research (SIGMIS-CPR'15), 101–104. New Beach, CA: ACM. doi: 10.1145/2751957.2751978.

Ho, Shuyuan Mary, and Jeffrey T. Hancock. 2019. "Context in a bottle: Language-action cues in spontaneous computer-mediated deception." *Computers in Human Behavior*, 91: 33–41. doi: 10.1016/j.chb.2018.09.008.

Ho, Shuyuan Mary, Jeffrey T. Hancock, and Cheryl Booth. 2017. "Ethical dilemma: Deception dynamics in computer-medicated group communication." *Journal of the American Society for Information Science and Technology*, 68 (12): 2729–2742. doi: 10.1002/asi.23849.

Ho, Shuyuan Mary, Jeffrey T. Hancock, Cheryl Booth, Michael Burmester, Xiuwen Liu, and Shashanka S. Timmarajus. 2016. "Demystifying insider threat: Language-action cues in group dynamics." In Proceedings of the 2016 49th Hawaii International Conference on System Sciences (HICSS-49), 2729–2738. Kauai, Hawaii. doi: 10.1109/HICSS.2016.343.

Ho, Shuyuan Mary, Jeffrey T. Hancock, Cheryl Booth, and Xiuwen Liu. 2016. "Computer-mediated deception: Strategies revealed by language-action cues in spontaneous communication." *Journal of Management Information Systems*, 33 (2): 393–420. doi: 10.1080/07421222.2016.1205924.

Ho, Shuyuan Mary, Jeffrey T. Hancock, Cheryl Booth, Xiuwen Liu, Muye Liu, Shashanka S. Timmarajus, and Michael Burmester. 2016. "Real or spiel? A decision tree approach for automated detection of deceptive language-action cues." Paper read at Hawaii International Conference on System Sciences (HICSS-49), 3706–3715. Kauai, Hawaii. doi: 10.1109/HICSS.2016.462.

Ho, Shuyuan Mary, Jeffrey T. Hancock, Cheryl Booth, Xiuwen Liu, Shashanka S. Timmarajus, and Michael Burmester. 2015. "Liar, liar, IM on fire: Deceptive language-action cues in spontaneous online communication." In IEEE Intelligence and Security Informatics, 157–159. Baltimore, MD: IEEE. doi: 10.1007/978-1-4799-9889-0/15.

Ho, Shuyuan Mary, Michelle Kaarst-Brown, and Izak Benbasat. 2018. "Trustworthiness attribution: Inquiry into insider threat detection." *Journal of the Association for Information Science and Technology*, 69 (2): 271–280. doi: 10.1002/asi.23938.

Ho, Shuyuan Mary, and Hwajung Lee. 2012. "A thief among us: The use of finite-state machines to dissect insider threat in cloud communications." *Journal of Wireless Mobile Networks, Ubiquitous Computing, and Dependable Applications (JoWUA), Special Issue of Frontiers in Insider Threats and Data Leakage Prevention*, 3 (1/2): 82–98.

Ho, Shuyuan Mary, Xiuwen Liu, Cheryl Booth, and Aravind Hariharan. 2016. "Saint or Sinner? Language-action cues for modeling deception using support vector machines." In *Social computing, behavioral-cultural modeling & prediction and behavior representation in modeling and simulation (SBP-BRiMS), LNCS 9708*, edited by Kevin S. Xu, David Reitter, Dongwon Lee and Nathaniel Osgood, 325–334. Washington, DC: Springer International Publishing Switzerland. doi: 10.1007/978-3-319-39931-7_31.

Kelley, Harold H. 1973. "The process of causal attribution." *American Psychology*, 28 (2): 107–128.

Kelley, Harold H., John G. Holmes, Norbert L. Kerr, Harry T. Reis, Caryl E. Rusbult, and Paul A. M. Van Lange. 2003. *An atlas of interpersonal situations*. New York, NY: Cambridge University Press.

Lieberman, Jethro Koller 1981. *The litigious society*. New York, NY: Basic Books.

Mayer, Roger C., James H. Davis, and F. David Schoorman. 1995. "An integrative model of organizational trust." *Academy of Management Review*, 20 (3): 709–734.

Chapter 3

User Privacy and Security Online: The Role of Information Professionals

Mohammed Nasser Al-Suqri, Salim Said AlKindi, and
Naifa Eid Saleem
Department of Information Studies, Sultan Qaboos University

Overview

Ensuring the privacy of library users and providing a safe and secure environment
in which to seek and use information resources have always been central values
for information professionals, and encapsulated in guidelines and codes of
practice issued by the International Federation of Library Associations (IFLA)
and those of library associations around the world. The ability to adhere to these
values in day-to-day work, however, is being threatened by the new information
environment, in which users use library computers to search for and use
information resources online via web search engines, or to access the Internet for
other purposes such as online banking, paying bills, or accessing e-government
services. In this context, information professionals face new challenges in their
endeavors to protect the privacy and security of library users.

Growing threats to the online privacy and security of library users include government monitoring and surveillance of online activity in the name of national security legislation: the collection of personal data by third-party organizations for marketing purposes, and the illegal hacking of online information systems for criminal purposes, as well as the online "footprint" left by information system users which can potentially be seen by library staff, IT specialists, or others with access to the system. The chapter discusses these increasing threats to patron privacy in the online information environment, and identify what information professionals and library associations can do and are already doing to maximize the privacy and security of users in the new electronic information environment. Three important roles in particular are identified and discussed: promoting digital literacy, which includes ensuring that users understand threats to privacy and take appropriate steps to minimize these when seeking information online; working with IT specialists to develop and implement secure online information systems; and acting as political advocates for the privacy and confidentiality of library users in the face of increasing regulation which potentially contravenes these values. The chapter concludes with a number of key recommendations for ways in which librarians and other information professionals can help maximize the privacy and security of users online, while reconciling this with other important values such as freedom of access to information.

Growing Threats to User Security and Privacy in the Online Information Environment

A diverse range of factors threatens the security and privacy of users when searching and using online services and resources in libraries. These present information professionals with unprecedented challenges to protect users and uphold the key values of the information profession. A review of relevant literature revealed some of the main factors or trends that are presenting new risks to the privacy or security online library users, as discussed in this section.

First, government surveillance of online activity, ostensibly for purposes of protecting national security or combatting terrorism, has become widespread (Fortier & Burkell 2015; IFLA 2016a). There is often a lack of transparency about this surveillance, with individuals having no idea that their online activities are being monitored. However, this is not a new threat facing libraries; in the United States, surveillance of library users has often been a practice of the FBI for counter-intelligence purposes during times of perceived threats to national security. Matz (2008), for example, refers to the Library Awareness Program of the FBI in the 1980s, which used library records to monitor the reading habits of individuals with Russian or Slavic sounding names. The extent of government surveillance of individuals in the United States and worldwide was also famously revealed by Edward Snowden, the former CIA employee and federal government IT contractor, who in 2013 leaked National Security Agency information about mass online surveillance programs (Clark 2016).

In many countries including the United States since the passing of the Uniting and Strengthening America by Providing Appropriate Tools Required to Intercept and Obstruct Terrorism Act of 2001 (USA PATRIOT Act), libraries are required to provide detailed information on patron activity if requested to do so under a court order, and to unencrypt this if necessary (Kim 2016). In the face of such legislation and surveillance activities, information professionals can no longer offer the guarantees of privacy and anonymity to library users, which were once possible, especially as users are increasingly using online search engines such as Google over which librarians have little control (Pekala 2017) rather than traditional library databases and catalogues.

Second, as libraries themselves increasingly provide digital services and resources and often collaborate with others in establishing extensive shared electronic networks, this also increases the risks to user privacy and security. A large number of actors are typically involved in establishing and maintaining such networks, often including private vendors, and it becomes more difficult to secure systems or identify when user activity is being monitored by third-party organizations. A major breach of the privacy and security of library users was committed by the company Adobe, which collected data on the activity of individuals who used their Digital Editions 4 (DE4) software to read e-books online. It was reportedly sent in unencrypted form to Adobe, enabling anyone monitoring web traffic to see the information (Gallagher 2014).

Researchers have found that third-party vendors who supply digital content to libraries often have unclear or limited privacy policies which are not of the standard typical of libraries in the past (Klinefelter 2016). Furthermore, cloud-based or Library 2.0 services such as BiblioCommons are increasingly being adopted by libraries. These use new business models in which basic services are provided free of charge to users but which are funded through data-driven advertising which requires the collection of personal data (Zimmer 2017). User activity is tracked and analyzed so that personalized advertising or recommendations can be provided to individuals and used in understanding overall user trends, as described in the "IFLA Statement on the Right to be Forgotten" (IFLA 2016b). In this new business environment, multiple parties collect and share data about online activity for marketing purposes, and may also sell this data to brokers who in turn sell it on to other parties who may use this data for legitimate marketing purposes or for criminal activity (Kim 2016; Pekala 2017). Libraries themselves are increasingly collecting and using information about their users in order to provide personalized recommendations or services tailored to the needs of their target population (Hahn 2017). This sometimes involves interacting with users on social media in ways that may also compromise their privacy.

Breaches of personal data can also occur due to gaps in network security on library systems, when a system becomes infected by a virus, or as a result of users becoming victims of phishing scams (Kim 2016). Third parties with malicious intent might intercept information being transmitted across the Internet, especially when using wireless networks or when this is not fully encrypted (Breeding 2015). It was reported in 2017 that one hacker had breached the security systems of libraries in more than 60 leading universities and other institutions in the United States and the United Kingdom, such as Cornell University, Purdue University, the University of Oxford, and the University of Cambridge (Osborne 2017). The risk of hacking is of particular concern when library users take advantage of the availability of library computers and Internet access to use online services such as banking, e-government, and online shopping or to interact on social media sites (Massis 2017). While in many settings this is contributing to a narrowing of the digital divide between those with Internet access at home and those without, the risks to their security are typically greater in the public library setting.

Overall, these security risks are increasing as new technologies evolve and are adopted by libraries. Hahn (2017), for example, discusses the likely impacts on library users as the Internet of Things (IoT) disrupts technologies currently being used by libraries. In this development, miniscule technologies that gather and transmit data are being embedded in nearly all types of items and devices, even including library books, for example, with benefits such as the ability to track borrowing or personalize recommendations to users. However, this is another development which has implications for user privacy, since individuals often have no idea that their activities are being monitored by new interconnected technologies, and for user security, since the developments also often increase the potential for malicious hacking. Privacy and security become more difficult for libraries to protect due to the many third-party tools and service providers typically involved in the interconnected systems comprising the IoT (Hahn 2017).

Privacy and Security as Key Values of Information Professionals

Information professionals have both ethical and legal responsibilities to ensure the privacy and security of their users. These are historically important responsibilities, which are encapsulated in the core values of the profession as set out in professional policies and codes of practice. For example, the American Library Association's (ALA) Code of Ethics states "We protect each library user's right to privacy and confidentiality with respect to information sought or received and resources consulted, borrowed, acquired or transmitted" (2008). Similarly, the IFLA Code of Ethics identifies respect for personal privacy, protection of personal data, and confidentiality as core principles underpinning the relationship between libraries and their users (IFLA 2015).

Traditionally, such provisions have been intended to ensure that library users can browse the library collections and use resources anonymously or at least be assured of confidentiality of their user records. In recent years, these professional codes and guidance have been evolving to reflect the changing information environment and to ensure that information professionals can best protect the users of online library services and resources against threats to their security and privacy. For example, the "IFLA Statement on Privacy in the Library Environment" (IFLA 2016a) recommends that libraries limit or abstain from collecting personal data that would compromise privacy, and also educate staff and users about how to protect their privacy and security online, while the ALA's "Policy Concerning Confidentiality of Personally Identifiable Information about Library Users" (ALA 2004) specifies that personal information relating to the searches and the use of library services and databases must be protected from unauthorized access, including access by government agencies unless a warrant is provided.

Information professionals also have a legal responsibility to ensure that relevant legislation relating to the privacy of personal data is observed. For example, the European Union's General Data Protection Regulation, which came into force in May 2018, requires all organizations to provide full disclosure of data held on EU/EEA citizens and obtain the personal consent to this by the individuals concerned. EU/EEA citizens are covered by this law even when living in or visiting other countries, which means that libraries worldwide, along with other organizations which hold personal data on their customers or users, must abide by this law and seek consent to the storage and use of this data as well as ensuring that the required measures are in place to protect the data from unauthorized access and use (Cox 2018).

At the same time, the new information environment in which library users rely heavily on online services and resources is presenting potential value conflicts for information professionals, especially between the core values of protecting user privacy on the one hand and ensuring freedom of intellectual access and preserving the historical record on the other. Intellectual freedom has always been a cornerstone value of libraries, a point recently reiterated by IFLA in its Global Vision Report Summary which states that "no value was more highly rated than a commitment to equal and free access to information and knowledge" and "We must be champions of intellectual freedom" (IFLA 2018b, p. 1). But measures such as website filtering tools or government surveillance of the online behaviors of library users often hinder intellectual freedom and introduce elements of political control over the activities of library users in ways which clash with the fundamental principles of information professionals.

In the face of these growing and largely unprecedented challenges and value conflicts facing information professionals, a number of important ways in which this group can help protect the security and privacy of library users, while also maximizing intellectual freedom, can be identified and are discussed in the following sections. Specifically, these are defined as promoting digital literacy, implementing technical solutions, and political advocacy.

Promoting Digital Literacy

Since complete privacy and user security cannot be guaranteed in the online information environment, one of the most important roles of information professionals is to ensure that users are provided with adequate information and understanding about threats to their online privacy and security and how to protect themselves from these. These types of knowledge and skills are an important component of what is often referred to as *digital literacy*, or the "the ability to use information and communication technologies to find, evaluate, create, and communicate information, requiring both cognitive and technical skills" (ALA 2013). Although this embraces a broad range of skills required to effectively access and use information online, an increasingly important component of digital literacy involves awareness of cybersecurity issues and risks to personal privacy when using the Internet, as well as the methods and tools that are important in protecting oneself from these, as set out in the "IFLA Statement on Digital Literacy" (2017a).

There are several main components to the role of information professionals in promoting digital literacy skills that enable users to protect themselves online: 1) Developing and disseminating clear privacy policies which are tailored to the online information environment; 2) providing users with information and guidance about the threats to their online privacy and security and how to protect themselves against these, and also about their rights with regard to privacy and intellectual freedom when using online services and resources, and 3) promoting privacy literacy and cybersecurity literacy through participation in conferences and workshops, in which best practices are shared within the profession and with other key stakeholders such as human rights organizations and government representatives.

Researchers have highlighted the importance of reviewing and updating library privacy policies to ensure that these reflect aspects of the new information environment and the new privacy risks associated with this environment (e.g., Hahn 2017; Klinefelter 2016). These may relate, for example to the use of cloud-based services and the ways in which online activity may be monitored by a range of parties. A range of measures for improving an individual's ability to stay safe when using the Internet are set out by Hennig (2018) and might be incorporated into such policies. These include, for example, being able to recognize whether links to websites are real or not, and ways of creating and managing secure online passwords.

Yet empirical studies have provided evidence indicating that very few libraries have updated their privacy policies to take account of the new risks and are not educating users about these or the importance, for example, of opt-in notifications (e.g., Chamberlain & Zimmer 2017; Cotter & Sasso 2016). In a study of the use by libraries of social media to interact with users (Cotter & Sasso 2016) it was found, for example, that while around 75% of libraries surveyed had a social media policy, only 53% of these included reference to privacy. The same study revealed high levels of confusion and uncertainty on the part of librarians about what constitutes a violation of privacy (Cotter & Sasso 2016). It has also been highlighted in the literature that privacy policies must incorporate information about tracking or surveillance tools, and the ways in which user data will be used by the library or other parties, to enable users to make informed decisions about their use of online resources (Fortier & Burkell 2015).

Marden (2017) discusses the development and content of the New York Public Library's new privacy policy, released in November 2016. The process involved conducting a full inventory of the library's systems, databases, and paper-based information gathering; examining the privacy practices of other nonprofit organizations; drawing on the key privacy principles of the ALA; and consulting a wide range of internal stakeholders. The new policy developed as a result of this process provided clear guidance to users on what information the library collects, how it is used, how users can manage their own information, including opt-in and opt-out methods, and how the information is shared with third parties.

Developing the digital literacy skills of library users requires more than updating library privacy policies however; it is also crucial that information professionals are proactive in ensuring that users are aware of and understand the implications of such policies. This is particularly important in the light of UK research which provided evidence of an emerging security and intellectual privacy divide, along socioeconomic lines, between those with and those without the knowledge to protect themselves online (Lloyds Bank 2017). Other UK research revealed very lax practices among Internet users, with many admitting that they do not bother reading privacy statements or that they are happy to divulge personal information in order to achieve the desired outcome from their Internet searches (Ofcom 2015).

On a brighter note, some good practices are emerging among information professionals and related groups for improving digital literacy and increasing awareness of privacy and security issues. In the United States, for example, the Library Freedom Project, a syndicate of librarians, technical specialists, lawyers, and other groups, delivers workshops and conducts other activities intended to ensure that librarians and library users have adequate information and understanding about privacy and intellectual freedom rights and how to use utilize technology to protect these (Kim 2016). In a number of countries including the United States, Canada, Australia, the United Kingdom, France, and Germany, some libraries are using "cryptoparties" in the form of workshops to teach library users the basics of digital privacy and how to protect themselves from online monitoring activities. (IFLA 2015). Good practices also involve simply ensuring that the library policies are effectively disseminated to users. In the case of the

New York Library's privacy policy discussed earlier, this was rolled out on the library website with a clear banner announcing the new policy, and information about it was also emailed directly to more than a million library patrons and others signed up for library information (Marden 2017).

The IFLA as well as many national library associations are actively hosting or participating workshops and conferences for the purpose of promoting digital literacy and online safety. Central to the work of the IFLA in this respect is the Global Vision discussion, a series of workshops and an online platform used to facilitate and promote discussion of these issues among librarians from 190 countries (IFLA 2018a). IFLA participates regularly in international human rights conferences such as RightsCon on topics including cybersecurity and online privacy (IFLA 2018b), as well as other international initiatives. These include annual events such as the European Commission's *Safer Internet Day*, focused on safe use of the Internet by young people (IFLA 2018c), and the *Internet Governance Forums* held by a global multistakeholder group for the purpose of discussing public policy issues relating to safe use of the Internet (IFLA 2017b). It has also produced a *Resource Pack on Digital Literacy*, intended to provide information and guidance to libraries around the world, including how to ensure patrons are protected online, and with case study examples from around the world (IFLA 2018d).

In the United Kingdom, the Government's Internet Safety Strategy, published in 2017, specifically encourages libraries to be directly involved in implementing the strategy, for example, by making online resources and training available and integrating safety messages into existing library services for children and parents (Libraries Taskforce 2017). In one of the library sector's responses to this strategy, the Society of Chief Librarians implemented a series of "family learning and digital roadshows" in which librarians are able to learn about the latest security technologies and how to teach library users to protect themselves online (HM Government 2018).

Implementing Technical and Policy Solutions

The second main role of information professionals in relation to the security and privacy of online users is to work with IT professionals to ensure that best practice technical standards and solutions are in place to secure the online library systems, resources and databases from unauthorized access or use (Massis 2017) and to maximize the privacy of users (Klinefelter 2016).

Information professionals also have a responsibility to their users to ensure that recommended websites or databases, as well as services and resources provided by third-party vendors, meet high standards of privacy and security. An analysis of the privacy policies of leading vendors providing digital content to public libraries in the United States found that while these were often meeting industry standards they were falling short of the privacy standards generally expected of a library setting (Lambert, Parker, & Bashir 2015). It has been flagged up in the literature, however, that there is little guidance available to librarians for use in evaluating whether the resources they might recommend to their users meet high standards of privacy and do not involve monitoring or surveillance of users (Fortier & Burkell 2015). This indicates a need for such guidance to be provided in future.

Despite this current gap in guidance, best practices for securing library systems and protecting the privacy of users is emerging and has been implemented by many libraries. These have not yet been collated as a standard set of international best practices, though the ALA's Intellectual Freedom Committee has issued a number of guidelines to assist libraries and their third-party vendors in developing good practices for online privacy and data management and security, which are available at www.ala.org/advocacy/privacy/guidelines. A wide range of recommended approaches to improving security and privacy online have been documented by Hennig in a recent (2018) Library and Technology Report, in the form of a highly practical guide. For example, Hennig (2018) recommends, based on advice from the Electronic Frontier Foundation, that organizations should initially build a "threat model" or a plan setting out the required level of security for each type of data they collect or use. This involves addressing the questions "what do I want to protect?", "who do I want to protect it from?", "how bad are the consequences if I fail?", "How likely is it that I will need to protect it?", and "How much trouble am I willing to go through to try to prevent potential consequences?" (Hennig 2018, p. 7). Hennig (2018) also discusses measures such as the use of biometric security, for example touch ID and face recognition, to authenticate system users and reduce hacking risks, and the use of ad blockers and private mode browsing for protecting the privacy of individuals online.

Some other specific measures identified from the literature include the use of strict password requirements, and ensuring that patron passwords are not stored as unencrypted text on library systems (Breeding 2016), as well as the use of two factor authentication. This involves asking users not only for a password to log into library systems, but also for a second piece of identifying information usually in the form of a numeric code sent to the user by email or text message (Hennig 2018; Klinefelter 2016). Encryption of user data by the library is also essential, especially when transmitted online or using Wireless technology, in order to protect this against interception by a third party (Hahn 2017).

Other important measures to protect the privacy and security of individual users when using the Internet within libraries include turning off memory functions for passwords and form-filling, the practice of locking accounts after a number of failed log-in attempts, and rebooting of computers after a specified period of inactivity in order to clear personal information (IFLA 2018e; Kim 2016). However, there is still more to be done. For example, Robinson (2017) discusses the move by many governmental and other organizations to move all websites from HTTP to the more secure HTTPS protocol, and recommends that libraries follow this example.

With regard to the collection, storage, and use of user information in general, best practices for libraries include the use of short periods for retention of personal data, and having clear retention policies which are made available to users. Beckstrom (2017) notes that a data retention policy should provide information on how and what information is collected on users by the library itself and by any third-party services used by the library, and on what the user can do to request that their personal information is deleted or not used by the respective organizations. The use of data warehouses where deidentified patron data can be stored is also an emerging best practice. Yoose (2017) provides the example of Seattle Public Libraries, which wanted to find out more about the use of libraries by young people in the millennial population, and were therefore faced with the challenge of how to track online behaviors while also protecting the identities of users. They achieved this by creating a data warehouse to which circulation transactions were exported from the main library database and anonymized and aggregated for the purpose of analysis. Access to the data warehouse is also tightly controlled and limited to those staff who need to maintain or use the data.

Political Advocacy

The third main role of information professionals is as political advocates for the various rights of library users. Libraries and library associations have traditionally been advocates of both privacy and intellectual freedom rights, and this role is becoming increasingly important in the face of increasing surveillance and monitoring of online activity, the growth of privacy legislation, and the inherent risks of the online information environment (Lamden 2015). In the United States, for example, librarians played an important role in opposing demands for library records of individuals to be provided on request under the USA PATRIOT Act, and the Library Freedom project successfully opposed the Department of Homeland Security in its attempt to limit the adoption of encryption technologies by libraries in the United States (Clark 2016). The ALA is also active in advocacy efforts to strengthen individual privacy laws and to promote digital literacy skills among library users (Klinefelter 2016).

Conflicts often arise, however, between measures intended to protect user privacy and the efforts of information professionals to maximize the freedom of library users to access a wide range of resources, or to preserve information for public use. Advocacy efforts are also important, therefore, to ensure that these rights are effectively balanced when developing policies and legislation. An example is provided by the "Right to be Forgotten" legislation set out in the EU's 1995 Data Protection Directive (European Commission 2012), which is intended to increase the personal privacy rights of Internet users. Under this legislation, individual citizens can request that personal or sensitive information about them is removed from the Internet. However, IFLA has argued that librarians have a responsibility to ensure that policymakers understand the potential importance of preserving such information for historical or research purposes or other public interest considerations, particularly when the transparency of this information may be important to the public interest, as in the case of leading politicians for example (Edwards 2017; IFLA 2014, 2016b).

The use of filtering software in libraries is another controversial measure which can restrict intellectual freedom in the interest of national security or the safety and security of library users. As the use of online services and resources by library users increases, many libraries are implementing filtering tools which block access to sites deemed to be illegal or inappropriate (such as pornographic sites or the websites of extremist religious groups), in the interests of ensuring a safe and secure library environment for all users. While addressing some aspects of security, however, such filtering tools can threaten the intellectual freedom rights of users, and have often been shown to be ineffective in the sense of preventing users from making legitimate use of blocked sites for research or educational purposes. Researchers have also reported evidence that library users feel uncomfortable about asking for websites to be unblocked in these circumstances, indicating that such measures not only hinder intellectual freedom but may also have a negative impact on user privacy. Information professionals have an important role to play therefore in the form of advocacy intended to reduce the use of such filtering tools or at least to restrict their use to sites which are illegal or pose a definite national security risk. More generally, it has been noted that information professionals have a professional responsibility to protect the rights of library users to access information free from surveillance (Fortier & Burkell 2015).

The political advocacy role of librarians has come to the fore recently in the case of the United States. Here, the ALA issued releases following the election of President Trump in 2016, making a commitment to support the policies of the new administration, and subsequently retracted these following a backlash from librarians who argued that these conflicted with many of their professional values such as privacy and confidentiality (Zimmer 2017). These developments build on ongoing efforts by librarians in the United States to protect the information freedom rights of users. For example, Clark (2016) cites the example of the Library Freedom Project's successful resistance to the Department of Homeland Security's attempts to limit the use of encryption technologies by libraries, while at individual level, librarians in the United States have won court cases to overturn attempts by the FBI to obtain user data without a warrant or judicial review (Glaser 2015).

Recommendations and Conclusions

This chapter has discussed the increasing threats to the privacy and security of library users when using online services and resources and identified three key roles of information professionals which are important for protecting the privacy and security of online users as well as their intellectual freedom rights. The following key recommendations for information professionals and their professional associations are intended to help ensure that these roles can be effectively fulfilled:

- All information professionals should receive online security and privacy training to ensure that they can effectively contribute to the design of online library services and provide users with appropriate advice and support to enable them to use digital library services and collections securely and with appropriate levels of privacy.
- Information professionals should work together to ensure that their libraries implement best practice standards of security for online systems and resources, and that all online library services and content are compliant with such standards, including those provided by third party vendors.
- The collection of personal information about users and their use of library services should be kept to the minimum required to meet legal or administrative requirements, and such information should be destroyed as soon as it is no longer needed.
- Library privacy policies should be regularly reviewed and updated as necessary to ensure that they are aligned with the online information environment and the threats to user privacy that are inherent in this.
- Information professionals should play an active role in promoting the digital literacy of users by disseminating privacy policies, and providing information on the use of technology to reduce privacy and security risks online.
- Library associations should stay informed of policy and legislative developments which may either threaten the privacy of library users or limit access to information which may be of importance for research or public interest purposes, and should lead or actively support advocacy efforts intended to protect users against such developments.
- Library associations should seek opportunities to be directly involved or consulted in the development of policies or legislation relating to the surveillance of online activity or the protection of personal data, in order to help ensure that these are reasonable and do not contravene intellectual freedom rights or hinder the preservation of important information and historical records.

In conclusion, as the information environment evolves and online searches become established as the main information seeking method of many library users, protecting their privacy and security from the various threats identified in this chapter is becoming one of the most important responsibilities of information professionals. Additionally, the political advocacy role of information professionals and their professional associations is becoming ever more important, particularly in order to ensure that an appropriate balance is maintained between the privacy rights and intellectual freedom rights of users in the new information environment. The discussion and recommendations in this chapter are intended to raise awareness of these roles and the ways in which they can be most effectively fulfilled.

References

American Library Association (ALA). 2004. *Policy Concerning Confidentiality of Personally Identifiable Information about Library Users.* Amended June 30, 2004. www.ala.org/advocacy/intfreedom/statementspols/otherpolicies/policyconcerning.

American Library Association (ALA). 2008. *Code of Ethics.* www.ala.org/united/sites/ala.org.united/files/content/trustees/orgtools/policies/ALA-code-of-ethics.pdf.

American Library Association. Office for Information Technology Policy. Digital Literacy Task Force. 2013. *Digital Literacy, Libraries, and Public Policy: Report of the Office for Information Technology Policy's Digital Literacy Task Force.* https://www.districtdispatch.org/wp-content/uploads/2013/01/2012_OITP_digilitreport_1_22_13.pdf. Accessed March 12, 2020.

Beckstrom, M. 2017. "Use, Security and Ethics of Data Collection." In Newman, B. & Tijerina, B. (Eds.) *Protecting Patron Privacy: A User Guide.* Lanham, MD: Rowman & Littlefield. pp. 35–42.

Breeding, M. 2015. "Smarter Libraries through Technology: Protecting the Privacy of Library Patrons." *Smart Libraries* XXXV (1): 1–7.

Breeding, M. 2016. "Issues and Technologies Related to Privacy and Security." *Library Technology Reports* 52: 4.

Chamberlain, K. & Zimmer, M. 2017. "Privacy Policies and Practices with Cloud-Based Services in Public Libraries: An Exploratory Case of BiblioCommons." *Journal of Intellectual Freedom & Privacy* 2 (1): 23–37.

Clark, I. 2016. *Internet Freedom for All: Public Libraries Have to Get Serious about Tackling the Digital Privacy Divide.* http://eprints.lse.ac.uk/67203/1/Internet%20freedom%20for%20all%20Public%20libraries%20have%20to%20get%20serious%20about%20tackling%20the%20digital%20privacy%20divi.pdf.

Cotter, K. & Sasso, M.D. 2016. "Libraries Protecting Privacy on Social Media: Sharing without 'Oversharing'." *Pennsylvania Libraries: Research & Practice* 4 (2): 73–89.

Cox, K. 2018. *The General Data Protection Regulation: What Does It Mean for Libraries Worldwide?* Association of Research Libraries. https://www.arl.org/wp-content/uploads/2018/05/IssueBrief_GDPR_May2018.pdf. Accessed March 12, 2020.

Edwards, E. 2017. "Libraries and the Right to Be Forgotten: A Conflict in the Making?" *"Privacy" Special Issue of the Journal of Intellectual Freedom & Privacy* 2017 (Spring): 13–14.

European Commission. 2012. *Commission proposes a comprehensive reform of data protection rules to increase users' control of their data and to cut costs for businesses.* file:///C:/Users/Administrator/Desktop/Commission_proposes_a_comprehensive_reform_of_data_protection_ru les_to_increase_users__control_of_their_data_and_to_cut_costs_for_businesses.pdf.

Fortier, A. & Burkell, J. 2015. "Hidden Online Surveillance: What Librarians Should Know to Protect Their Privacy and that of Their Patrons." *Information Technology and Libraries* 32 (3): 59–72.

Gallagher, S. 2014, October 7. *Adobe's E-book Reader Sends Your Reading Logs Back to Adobe – In Plain Text. Ars Technica.* https://arstechnica.com/information-technology/2014/10/adobes-e-book-reader-sends-your-reading-logs-back-to-adobe-in-plain-text/.

Glaser, A. 2015. "Long Before Snowden, Librarians Were Anti-Surveillance Heroes." *Slate.* www.slate.com/blogs/future_tense/2015/06/03/usa_freedom_act_before_snowden_librarians_were_the_anti_surveillance_heroes.html?via=gdpr-consent.

Hahn, J. 2017. "Security and Privacy for Location Services and the Internet of Things." *Library Technology Reports* 53 (1): 23–28.

Hennig, N. 2018. "Privacy and Security Online: Best Practices for Cybersecurity." *Library Technology Reports* 54 (3): 29.

HM Government. 2018. *Government Response to the Internet Safety Strategy Green Paper.* https://assets.publishing.service.gov.uk/government/uploads/system/uploads/attachment_data/file/708873/Government_Response_to_the_Internet_Safety_Strategy_Green_Paper_-_Final.pdf.

International Federation of Library Associations (IFLA). 2016a. *IFLA Statement on Privacy in the Library Environment.* www.ifla.org/files/assets/hq/news/documents/ifla-statement-on-privacy-in-the-library-environment.pdf.

International Federation of Library Associations and Institutions (IFLA). 2014. *IFLA Statement on Access to Personally Identifiable Information in Historical Records.* www.ifla.org/publications/ifla-statement-on-access-to-personally-identifiable-information-in-historical-records.

International Federation of Library Associations and Institutions (IFLA). 2015. *IFLA Statement on Privacy in the Library Environment.* www.ifla.org/files/assets/hq/news/documents/ifla-statement-on-privacy-in-the-library-environment.pdf.

International Federation of Library Associations and Institutions (IFLA). 2016b. *IFLA Statement on the Right to Be Forgotten.* www.ifla.org/publications/node/10320.

International Federation of Library Associations and Institutions (IFLA). 2017a. *IFLA Statement on Digital Literacy.* www.ifla.org/publications/node/11586.

International Federation of Library Associations and Institutions (IFLA). 2017b. *Accessible, Open, Empowering, Lasting: IFLA Engages on Internet Governance.* www.ifla.org/node/20021.

International Federation of Library Associations and Institutions (IFLA). 2018a. *Global Vision Report Summary: Top 10 Highlights and Opportunities.* www.ifla.org/files/assets/GVMultimedia/publications/gv-report-summary.pdf.

International Federation of Library Associations and Institutions (IFLA). 2018b. *IFLA at RightsCon 2018: "Digital Literacy for All: How Can Libraries Help?"* www.ifla.org/files/assets/faife/ochr_privacy_ifla.pdf.

International Federation of Library Associations and Institutions (IFLA). 2018c. *Stay Safe Online? Go to Your Library! IFLA Celebrates Safer Internet Day 2018.* www.ifla.org/node/26419.

International Federation of Library Associations and Institutions (IFLA). 2018d. *IFLA Statement on Digital Literacy: Resource Pack.* www.ifla.org/files/assets/faife/publications/ifla_digital_literacy_resource_pack.pdf.

International Federation of Library Associations and Institutions (IFLA). 2018e. *The Right to Privacy in the Digital Age.* www.ifla.org/files/assets/faife/ochr_privacy_ifla.pdf.

Kim, B. 2016. "Cybersecurity and Digital Surveillance versus Usability and Privacy: Why Libraries Need to Advocate for Online Privacy." *College and Research Libraries News* 77 (9): 442–451.

Klinefelter, A. 2016. "Reader Privacy in Digital Library Collaborations: Signs of Commitment, Opportunities for Improvement." *I/S: A Journal of Law and Policy for the Information Society* 13 (1): 199–244.

Lambert, A., Parker, M. & Bashir, M. 2015. "Library Patron Privacy in Jeopardy: An Analysis of the Privacy Policies of Digital Content Vendors." *Proceedings of the Association for Information Science and Technology*, St. Louis, MO, November 06–10, 2015.

Lamden, S. 2015. "Social Media Privacy: A Rallying Cry to Librarians." *CUNY Academic Works*. http://academicworks.cuny.edu/cl_pubs/52.

Libraries Taskforce. 2017. *Can Libraries Bring Internet Safety to Life?* https://librariestask force.blog.gov.uk/2017/12/15/can-libraries-bring-internet-safety-to-life/.

Lloyds Bank. 2017. *Consumer Digital Index 2017.* www.lloydsbank.com/assets/media/pdfs/lloyds-bank-consumer-digital-index-2017.pdf.

Marden, B. 2017. "The Path to Creating a New Privacy Policy: NYPL's Story." *Special Issue of the Journal of Intellectual Freedom & Privacy* 2017 (Spring): 5–7.

Massis, B. 2017. "Privacy in the Library." *Information and Learning Science* 118 (3): 210–212.

Matz, C. 2008. "Libraries and the USA Patriot Act: Values in Conflict." *Journal of Library Administration* 47 (3/4): 69–87.

Ofcom. 2015. *Annex A: Adults' Media Literacy in the Nations.* www.ofcom.org.uk/__data/assets/pdf_file/0020/63056/nations_adults_media_literacy_annex_2015.pdf.

Osborne, C. 2017. "Lone Hacker Rasputin Breaches 60 Universities, Federal Agencies." *ZDNet.* www.zdnet.com/article/lone-hacker-breaches-60-universities-federal-agencies/.

Pekala, S. 2017. "Privacy and User Experience in 21st Century Library Discovery." *Information Technology and Libraries (Online)* 36 (2): 48–58.

Robinson, M. 2017. "How to Get Free HTTPS Certificates from Let's Encrypt." *Special Issue of the Journal of Intellectual Freedom & Privacy* 2017 (Spring): 11–12.

Yoose, B. 2017. "Balancing Privacy and Strategic Planning Needs: A Case Study in De-Identification of Patron Data." *Special Issue of the Journal of Intellectual Freedom & Privacy* 2017 (Spring): 15–22.

Zimmer, M. 2017. "Introduction: The "Privacy"." *Special Issue of the Journal of Intellectual Freedom & Privacy* 2017 (Spring): 2–3.

Chapter 4

Bridging the Cybersecurity Talent Gap: Cybersecurity Education in iSchools

Hsia-Ching Chang

Department of Information Science, University of North Texas
Cybersecurity Policy Fellow, New America

Cary Jim and Suliman Hawamdeh

Department of Information Science, University of North Texas

Introduction

Cybersecurity has emerged as a global issue in recent years with major challenges to individuals and organizations around the world. The International Systems Audit and Control Association (ISACA) report: *State of Cybersecurity 2019* reveals an expressed concern about the lack of skilled cybersecurity professionals on their teams by approximately 70% of the global respondents. The same report indicates almost 40% of the recent university graduates in cybersecurity are not prepared for the job requirement. The academic goals and objectives of cybersecurity within computer science education may be a frequent misperception because of the interdisciplinary nature of cybersecurity. Other skill sets in relation to human behavior are also important to the design and development of a holistic cybersecurity view which is beyond the traditional effort concentrated on technical training. The Cybersecurity Workforce Framework developed by the National Initiative for Cybersecurity Education (NICE) identifies 33 specialty areas for cybersecurity jobs in which more than half primarily involve nonprogramming tasks (Swire 2018). Information science focuses on information behavior and processes between humans and technologies. As a discipline closely related to computer science and information technology, it can provide a unique advantage to address the human aspects of cybersecurity. The consortium of information schools known as the iSchools are in a good position to address some of the educational issues related to cybersecurity, especially security issues related to information organization, information management, social media, and knowledge management. The iSchools focus on educating students from both technical and nontechnical backgrounds which play an important role to address interdisciplinary training and support for the development of a diversified workforce.

Besides the need of trained cybersecurity professionals, public knowledge of cybersecurity is also important. The innovations of smart devices ranging from smartphones, wearables, smart homes, connected cars, to medical devices are becoming part of our daily life and shaping our behavior slowly toward the foreseeable future. According to the *2018 Study on Global Megatrends in Cybersecurity* conducted by Ponemon Institute with sponsorship from Raytheon, 82% of the IT practitioners predicted a data breach from unsecured Internet of Things (IoT) devices is very likely to occur in the subsequent years. For example, a smartphone could serve as a hub to any connected devices through applications in syncing with wearables to track activities, monitoring doorbell ring, home energy use, and controlling smart devices' functions. It is worth noting that we have been living in the digital age and our next generations have grown up with the use of mobile Internet and smart devices as part of their daily routines. How do we prepare for the cybersecurity risks that could jeopardize our security and safety? The engagement of information professionals with the general public can help raise their awareness of cybersecurity risks. The Pew Research Center designed a cybersecurity knowledge quiz and conducted an online survey to 1,055 adult Internet users who resided in the United States in June 2016. The full report *What the Public Knows about Cybersecurity* concludes that the majority of respondents failed to answer the 13 questions accurately, with an average of 5.5 correct answers. Most Americans may not be certain on how to address cybersecurity or have enough knowledge of cybersecurity protection (e.g., strong passwords or imposed risk in using public WiFi) especially key technical cybersecurity concepts, such as botnet, VPN, and two-factor authentication (Olmstead and Smith 2017). This signifies that most users with smart devices and connected to the Internet are at higher risk of cybersecurity threats.

In addition, a 2017 National Cybersecurity Alliance (NCSA) survey revealed that a cybersecurity/privacy knowledge gap exists between teens and their parents. More than one-third of teens feel that they know more than their parents about cybersecurity/privacy issues. These statistics concur that there is an urgent need to increase the cybersecurity literacy level of the general public in the United States (Furnell and Moore 2014).

Cybersecurity seems to be a far-reaching subject for non-IT professionals due to technical jargon and how it was positioned solely as a technological domain within computer science, networking, and engineering. However, common knowledge of security and privacy is part of the foundational understanding in cybersecurity. It is critical to have both technical and sociobehavioral knowledge of cybersecurity in our increasingly interconnected society. The goal of this chapter is to navigate the different levels of efforts to cybersecurity education and workforce development across the United States. For information professionals, the iSchools in North America are selected examples on how their programs are meeting the interdisciplinary need of cybersecurity training and education. We further analyze the alignment of iSchools cybersecurity curriculum with the NICE Cybersecurity Workforce Framework. At the end, we examine the major opportunities and challenges for information professionals, iSchools, and the future development of cybersecurity education.

Cybersecurity Workforce Development and Training

A recent (ISC)^2 Cybersecurity Workforce Study (2018) shows that there are approximately three million cybersecurity jobs unfilled globally. The study identifies the most needed cybersecurity areas of expertise in eight different areas: security awareness, risk assessment/analysis/management, security administration, network monitoring, incident investigation and response, intrusion detection, cloud computing security, and security engineering. It is noteworthy that the first three areas of expertise are not technical but focus on the management and administration within an organization. This indicates that cybersecurity professionals do need to be aware of social, political, economic, and criminological issues (Leaning and Averweg 2019). The same survey also asks the cybersecurity professionals to self-evaluate the top areas where they need to enhance and improve over the next two years based on their current expertise and future demand. The areas where cybersecurity professionals feel that they need to grow the most are: governance, risk management and compliance (GRC), security analysis, risk assessment/analysis/management, and security engineering. Other top areas where they do not have high expertise and want to improve are cloud computing security, penetration testing, threat intelligence analysis, and forensics. The job title and description of a cybersecurity professional vary by companies and settings. This diverse view on what a cybersecurity professional is could contribute to misalignment of skilled labor to work market. According to the U.S. Bureau of Labor Statistics, *Occupational Outlook Handbook* (2019), information security analysts fall within the computer and information technology occupations. The main role of an information security analyst is to implement computer and network security measures for an organization and their roles are expanding. One aspect of security is to develop emergency or disaster recovery plan of the information system and technology for an organization with a focus to protect their digital assets. This description characterizes some aspects of cybersecurity tasks for a security professional. However, there are other job titles and description of what cybersecurity professionals do on a regular basis. For examples, security analyst, cryptographer, chief information security officer, or security consultant are common job titles used in for recruitment. One example of a cybersecurity position with specific certification and training is ISSO. ISSO stands for information system security officer. This particular position will require knowledge in federal certification and accreditation (C&A) processes and is often the contact person for system security plans and plans of action and milestones. The following sections will further discuss the different levels of training and education for cybersecurity.

Cybersecurity Education Initiatives

National-Level Cybersecurity Centers and Initiatives

There are two major academic centers in the United States that provide guidance on training and education for cybersecurity in the country. The National Centers of Academic Excellence in Cyber Defense Education (CAE-CD) focuses on cyber defense program in higher education and research (CAE-R). Any regionally accredited community colleges, four-years universities and graduate-level institutions can apply for the CAE-CD, CAE-2Y, or CAE-R designations for their academic program by meeting the rigorous criteria and regulations. The National Centers of Academic Excellence in Cyber Operations (CAE-CO) supports the President's NICE to broaden the technical training of skilled workers in supporting a cybersecure nation (NSA, CSS n.d.). Their designation program is applicable to undergraduate and graduate programs with an emphasis in computer science or computer engineering.

K-12 Education and Cybersecurity Initiatives

Traditionally, computer and network training falls under the Career and Technical Education (CTE) program for majority of the American public school. This type of program emphasizes on vocational training for students who do not plan to enter a four-year university after high-school graduation. Due to the STEM education movement and initiatives, many public schools or traditional college preparatory attempt to bring science, technology, and engineering programs into the general curriculum. Other schools may utilize their existing CTE program with modification to increase choices and pathways for students to take computer science, technology, and engineering courses. Computer science, coding, and robotics programs are penetrating many public schools at the elementary and middle school level. They focus on increasing students' interest in STEM and customize pathways to encourage students to consider a career in STEM as early as possible. Cybersecurity is considered a subtopic in relation to the study of computer science and network security in the CTE curriculum. Within K-12 education, cybersecurity education initiatives are supported by different agencies at the national, state, and local level. For example, the United States Department of Homeland Security (DHS) Cybersecurity Education Training Assistant Program (CETAP) provides K-12 teachers cybersecurity curricula and tools for teaching this subject (DHS, NICCS 2019). At the local or state levels, efforts are in place to establish cybersecurity as a core subject with specific curriculum and relevant teacher training to implement it in a regular classroom. The NICE K-12 Cybersecurity Education host their annual conference since 2017 to discuss education, training, and other needs for all stakeholders within K-12 settings (NICE 2019). There is a growing need of cybersecurity curriculum for students in the classroom as well as training for teachers to implement those programs. The school staff and other technology personnel who interact or manage the school information systems also should be trained to handle possible cybersecurity threats.

Community College or Technical School

There are different levels of effort to address the workforce and training needs at the postsecondary level, especially in community colleges and technical schools. In 2002, the National Science Foundation (NSF) and the American Association of Community Colleges (AACC) hosted a three days' workshop to discuss the role of community colleges in cybersecurity education and training of cybersecurity professionals at all levels (AACC 2002). This event covered seven major themes: trustworthy computing, cybercrime, foundations of cybersecurity curricula, cybersecurity literacy, current cybersecurity courses and curricula, hiring cybersecurity professionals, and establishing and maintaining a cybersecurity program. Various representatives presented to discuss the current position of community colleges and the key recommendations to address the training of the cybersecurity workforce as well as responsibilities of the different stakeholders. Another example is the establishment of the National Cyberwatch Center through the support by the NSF (NSF DUE-1204533 and DUE-1601150) to provide mentoring, training, curriculum resource, and services for other community colleges, businesses, and students to strengthen the cybersecurity workforce (National Cyberwatch Center 2019). At the federal level, the US Department of Education, Office of Postsecondary Education, initiated a Pilot Program for Cybersecurity Education Technological Upgrades for Community Colleges (PPCE-TUCC) to support the country's community colleges to expand their cybersecurity education program, especially for lower-income students and schools that lack the resource to maintain such programs (2018).

Professional Certification

There are many types of professional certification for the cybersecurity domain and it is impossible to provide a list of all available certifications in this chapter. There are vendor-specified certification and nonvendor-specified certification depending on the nature of the job requirement and the goal of the company. Many of the postsecondary schools and technical or community colleges are offering information security and/or information technology certifications that the industrial partners would recognize as valid training. Many of these certification programs may require prior computer and networking knowledge and skills before enrollment. Professional certifications issued by private (nonprofit or profit) organizations are often designed for people who have some work experience in a security role. One example is the CISSP (Certified Information System Security Professional), which requires a candidate to have relevant work experience in a specific area. Another type of avenue to get professional certification is through a combined program. There are hybrid programs in which an educational institute includes components of the certification requirement as part of their curriculum. Students who graduated from those hybrid programs will earn a degree (e.g., two or four years) with the qualification and knowledge to participate in a professional certification process. With so many different channels to acquire training and certification, how does the hiring organization determine who qualify for their cybersecurity job openings? Knapp, Maurer, and Plachkinova (2017) reviewed cybersecurity programs and the role of certification in helping candidates to meet qualification for jobs. They found out the industry typically evaluate their candidate's qualification for cybersecurity jobs based on a combination of academic qualification, industry certification and work experience. How can someone learn and become a proficient cybersecurity professional? Assante and Tobey (2011) discussed the development of cybersecurity professionals from amateurs to experts as an ongoing process. The traditional career progression in this area required many years of practical experience and IT knowledge. Eventually those that are knowledgeable and skillful will progress to a mastery level with vast knowledge of digital forensics, operational response, and risk management.

National Initiative for Cybersecurity Education (NICE) Cybersecurity Workforce Framework

Knowledge, skills, and abilities (KSAs) is used to describe a set of competencies that can be demonstrated by a person. In the technology industry, KSA is often used to develop job description, training and development, or evaluation of job performance. A recent study (Jones, Namin, and Armstrong 2018) solicited a list of current practices and views from a group of cyber professionals at two premier hacker conferences, Black Hat 2016 and DEF CON 24. Their findings had shown the top five KSAs expressed by the current professionals are: understanding of network protocols, network security architecture concepts (including how the traffic flows across the network), basic system administration, network, and operating system hardening techniques and overall system and application security threats and vulnerabilities. More than half of the participants indicated that these KSAs were mostly learned from their job, followed by school and self-learning. There is a need to continue to improve skills of the current professionals as well as preparing those who are interested to work in the cybersecurity domain.

NICE Framework 2.0 has been developed and updated with a taxonomy of 7 workforce categories, 33 specialty areas, 52 work roles, and associated KSAs for the cybersecurity domain. Encompassing three components (enhancing awareness, expanding the pipeline, and evolving the field), the NICE framework 2.0 identifies seven work categories: Security Provision, Operate and Maintain, Protect and Defend, Investigate, Collect and Operate, Analyze, Oversee, and Govern (Newhouse et al. 2017). Table 4.1 displays the term and descriptions of each workforce category in an alphabetical order. Three workforce categories appear relevant to what information profession does: Oversee and Govern, Analyze, and Collect and Operate. While Oversee and Govern focus on cybersecurity management, Analyze, and Collect emphasize gathering and evaluating cybersecurity information.

As shown in Figure 4.1, mapping the specialty areas with workforce categories, some of the specialty areas in other categories also seem relevant to information profession. For instance, data administration and knowledge management pertain to Operate and Maintain category; risk management in Securely Provision is related to Oversee and Govern. Analyze competency seems to be the core knowledge areas of security analytics supporting Investigate and Collect and Operate and inform other four workforce categories (Protect and Defend, Oversee and Govern, Operate and Maintain, and Security Provision) to make data-driven decisions. Reciprocally, data administration and knowledge management aid in ensuring the processes align with analytics project management.

Table 4.1   NICE Cybersecurity Workforce Framework Categories and Descriptions

Category	Descriptions
Analyze (AN)	Performs highly specialized review and evaluation of incoming cybersecurity information to determine its usefulness for intelligence.
Collect and Operate (CO)	Provides specialized denial and deception operations and collection of cybersecurity information that may be used to develop intelligence.
Investigate (IN)	Investigates cybersecurity events or crimes related to information technology (IT) systems, networks, and digital evidence.
Operate and Maintain (OM)	Provides the support, administration, and maintenance necessary to ensure effective and efficient information technology (IT) system performance and security.
Oversee and Govern (OV)	Provides leadership, management, direction, or development and advocacy so the organization may effectively conduct cybersecurity work.
Protect and Defend (PR)	Identifies, analyzes, and mitigates threats to internal information technology (IT) systems and/or networks.
Securely Provision (SP)	Conceptualizes, designs, procures, and/or builds secure information technology (IT) systems, with responsibility for aspects of system and/or network development.

Source: Department of Homeland Security, Office of Cybersecurity and Communications, National Initiative for Cybersecurity Careers and Studies (2019a).

Cybersecurity Education in North America iSchools

The turn of the century is characterized by globalization and the increased emphasis on knowledge as a factor of growth. This has given rise to intellectual capital, intellectual property, and wider concept of the knowledge-based economy. It has also given rise to the birth of online mega corporations such as Google, Amazon, Netflix, and Facebook. The shift from brick and mortar institutions to data- and technology-driven institutions created growing interest in new areas within the information field such as knowledge management, knowledge discovery, big data, data analytics, and data science. The increased emphasis on intangible assets and the relationship between technology, people, and information formed the basis for birth of the iSchools movement or the Information Schools. It started in 2005, when a number of library and information science schools realized that their

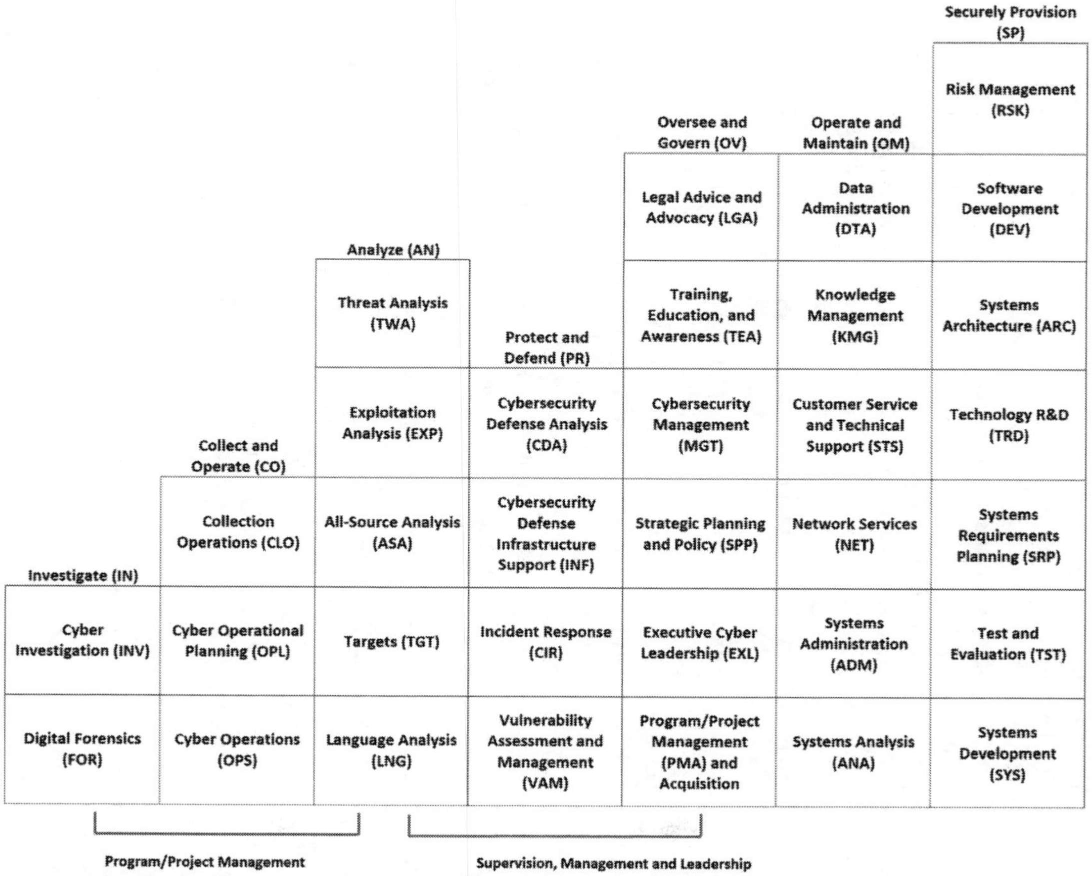

Securely Provision (SP)	Operate and Maintain (OM)	Oversee and Govern (OV)	Protect and Defend (PR)	Analyze (AN)	Collect and Operate (CO)	Investigate (IN)
Risk Management (RSK)	Data Administration (DTA)	Legal Advice and Advocacy (LGA)	Cybersecurity Defense Analysis (CDA)	Threat Analysis (TWA)	Collection Operations (CLO)	Cyber Investigation (INV)
Software Development (DEV)	Knowledge Management (KMG)	Training, Education, and Awareness (TEA)	Cybersecurity Defense Infrastructure Support (INF)	Exploitation Analysis (EXP)	Cyber Operational Planning (OPL)	Digital Forensics (FOR)
Systems Architecture (ARC)	Customer Service and Technical Support (STS)	Cybersecurity Management (MGT)	Incident Response (CIR)	All-Source Analysis (ASA)	Cyber Operations (OPS)	
Technology R&D (TRD)	Network Services (NET)	Strategic Planning and Policy (SPP)	Vulnerability Assessment and Management (VAM)	Targets (TGT)		
Systems Requirements Planning (SRP)	Systems Administration (ADM)	Executive Cyber Leadership (EXL)		Language Analysis (LNG)		
Test and Evaluation (TST)	Systems Analysis (ANA)	Program/Project Management (PMA) and Acquisition				
Systems Development (SYS)						

Supervision, Management and Leadership

Program/Project Management

Figure 4.1   The Cybersecurity Workforce Framework.

Adapted from Partnership for Public Service, and Booz Allen Hamilton (2015, 8).

teaching and research programs had the capacity to reach a broader audience and to prepare students for work beyond librarianship. The iSchools represent a shift in directions and philosophy from the traditional library and information science education.

Since the inception of the consortium of iSchools, the number of universities that joined the iSchools movement had increased to more than 80 schools from around the world. Many iSchools modify their representation by changing their name of library and information science by dropping the word "library" to reflect a broader nature of the information profession. Despite the name change, most of the current iSchools still focus on preparing librarian, archivist, or curators for different organizations and institutions. Most of the graduate-level library science programs of the current iSchools in North America are accredited by the American Library Association.

It is important to note that a number of institutions that joined the iSchools movement were not purely library science schools. Some of the institutions are engineering and computer science schools as part of the Computing Research Association and some are business schools. This shift in direction of broadening the information field is evident from the inclusion of computer and business schools as well as diversifying the iSchools degree program offering to include new and emerging areas such as data science, cybersecurity, and knowledge management. The iSchools organization vision as stated on iSchools.org (2014) is to expand internationally, recognized for creating innovative information solutions and systems to benefit individuals, organizations, and society at large.

iSchools Cybersecurity Education Program Status

As of Spring 2019, there are 47 iSchools in North America from the United States and Canada with four types of memberships: iCaucus, Basic, Supporting, and Associate Member (Table 4.2). To further assess the existing cybersecurity training and education among the iSchools in North America, we evaluated the current cybersecurity program information at the undergraduate and graduate levels.

Undergraduate Level Programs

At the undergraduate level, we could find program concentration or minor that relates to cybersecurity topics. There are two universities out of the North America iSchools that offer a bachelor-level program with cybersecurity as the degree title. One program is a Bachelor of Science degree in emergency, preparedness, homeland security, and cybersecurity. The other program is a Bachelor of Science in cybersecurity analytics and operations (CYAOP). These

Table 4.2   iSchools in North America and their Member Status

Member Status	Number of iSchools in North America
iCaucus	29
Supporting	3
Basic	8
Associate Member	7
Total	47

titles reflected a combination of topics in relation to cybersecurity. Besides the two universities with the indication of cybersecurity in their degree program, there are no other formal cybersecurity degrees at the undergraduate level among the iSchools, except being referenced as a concentration or minor within the degree program. Table 4.3 displays the list of current North America iSchools and their offerings of cybersecurity-related studies.

Graduate-Level Master Programs

At the graduate levels, there are more options for students and professionals to pursue cybersecurity-related training and degree. At the master's level, there are two universities formally granting degrees in "cybersecurity." Another unique program colists information and cybersecurity in the program name. Additionally, two universities provide information security relevant programs. While one program focuses on information security policy and management as well as information security and assurance, the other specifically stresses identity management and security. There are 12 universities that offer master-level degree of security-related topics within information science or information management. Different versions of cybersecurity programs are evident in the variety of names and description presented by the iSchools. For example, there are concentration options in information assurance, cyber intelligence, or management and policy. There are other offerings with technical bases such as information forensics, intelligence analysis, or cybersecurity in computing. Eight of the universities in the iSchools also offer certification at the graduate level. The certification programs vary by admission requirement, cost, and duration. We noticed the same blending of information security or system security with cybersecurity at the certification level. There is no minor represented at this level; details of each program of the iSchools at the master level can be found in Table 4.4.

Table 4.3   Bachelor-Level Cybersecurity-Related Program, Concentration, and Minor Among iSchools

iSchools	Division	Degree	Major	Concentration	Minor
University of South Florida	School of Information	BS	Information Studies	Information Security	Information Security
Dominican University	School of Information Studies	BS	Informatics	Cybersecurity	Informatics with Cyberse-curity Focus
University of Albany	College of Emergency Preparedness, Homeland Security, and Cybersecurity	BS	Emergency Preparedness, Homeland Security, and Cybersecurity	–	–
Drexel University	College of Computing and Informatics	BS	Computer and Security technology	Computing Security	Security Technology
The Pennsyl-vania State University	College of Information Sciences and Technology	BS	Security and Risk Analysis (SRA)	–	–
The Pennsyl-vania State University	College of Information Sciences and Technology	BS	Cybersecurity Analytics and Operations (CYAOP)	–	–

Graduate-Level Doctorate Programs

At the doctoral level, the representation of the cybersecurity program is similar to the undergraduate level, in which there is no degree titled as "cybersecurity" formally. Most of the iSchools that offered a security related PhD program used information system or information science as the main degree title and offer with security or privacy as a concentration or minor area. Table 4.5 displays the related programs available at the doctoral level among the iSchools.

As we evaluated the current program offerings in relation to cybersecurity, we noticed the naming of each program has similarities and differences. It is an interesting representation of the view of cybersecurity education and how the degree title displayed the selected aspects of cybersecurity as part of the teaching and learning process in each degree. Table 4.6 displays the count in parenthesis of the terms used in each degree title at the undergraduate and graduate level.

Table 4.4   Graduate Level Master's Degree Options and Certificate Offerings

iSchools	Division	Degree	Major	Concentration	Certificate
University of South Florida	School of Information	MS	Intelligence Studies	Strategic Intelligence Cyber Intelligence	Strategic Intelligence Cyber Intelligence
Dominican University	School of Information Studies	MS	Information Management	Cybersecurity	–
University of Albany	College of Emergency Preparedness, Homeland Security, and Cybersecurity	MS	Information Science	Intelligence Analysis	Emergency Preparedness, Homeland Security, and Cybersecurity
Carnegie Mellon University – Heinz College	School of Information Systems and Management and School of Public Policy and Management	MS	Information Security Policy and Management (MSISPM) Information Security and Assurance (MSIT Online)	–	Executive Education Certificate: Chief Information Officer (CIO), Chief Risk Officer (CRO), Chief Information Security Officer (CISO) Certificate
Drexel University	College of Computing and Informatics	MS	Cybersecurity	Data Science Health Informatics Information Systems Library and Information Science Software Engineering	–
Georgia Institute of Technology	College of Computing	OMS MS	Cybersecurity	Information Security	Cybersecurity
Indiana University – Bloomington	School of Informatics, Computing	MS	Secure Computing	–	–

(*Continued*)

Table 4.4   (Cont.)

iSchools	Division	Degree	Major	Concentration	Certificate
	and Engineering				
Syracuse University	School of Information Studies	MS	Information Management	Information Security Management	Information Security Management
The Pennsylvania State University	College of Information Sciences and Technology	MPS	Information Sciences (Master of Professional Studies)	Cybersecurity or Information Assurance Information Security and Forensics (Online)	Security Certificate (NSA) or Post-Bac Certificate: Information System Cybersecurity
University of California, Berkeley	School of Information	MICS	Information and Cybersecurity	–	–
University of Pittsburgh	School of Computing and Information	MS	Information Science	Information Security	
The University of Texas at Austin	School of Information	MS	Identity Management and Security (MSIMS)	–	–
University of Washington	The Information School	MS	Information Management	Information Security	–
University of Wisconsin-Milwaukee	School of Information Studies	MS	Information Science and Technology	Information Security	–

Alignment of iSchools Cybersecurity Curriculum with NICE Cybersecurity Workforce Framework

As examined in the previous section, three iSchools have cybersecurity programs at the master level. The Master of Science in Cybersecurity at the University of Drexel is a joint program between College of Computing and Informatics and College of Engineering where the Electrical and Computer Engineering (ECE) Department in Drexel's College of Engineering took the

Table 4.5   Graduate-Level Doctorate Degree and Concentration in Relation to Cybersecurity

iSchools	Division	Degree	Major	Concentration
University of Albany	College of Emergency Pre-paredness, Homeland Security, and Cybersecurity	PhD	Information Science	Information Assurance
University of Wisconsin	School of Library and Information Studies	PhD	Information Studies	Digital Privacy, Safety, and Secur-ity Studies
Carnegie Mello Univer-sity, Heinz College	School of Information Sys-tems and Management and School of Public Policy and Management	PhD	Information Systems and Management	Information Secur-ity and Privacy
The Pennsyl-vania State University	College of Information Sciences and Technology	PhD	Information Sciences	Security and Privacy
University of Pittsburgh	School of Computing and Information	PhD	Information Science	Information Secur-ity, Privacy (as coursework in foundation)
University of North Texas	College of Information	PhD	Information Science	Cybersecurity

lead to develop the curriculum with three different tracks: computer science track, electrical and computer engineering track, and information science track (Drexel University Online 2019). The Master of Science in Cybersecurity at Georgia Institute of Technology is offered by the three units: School of Computer Science (CE), School of Electrical and Computer Engineering (ECE), and School of Public Policy (PUBP). The three schools provide specialized cybersecurity tracks, including information security, energy systems, or policy (Georgia Institute of Technology 2019). The Master of Information and Cybersecurity (MICS) from the University of California (UC), Berkeley, seems to be the only unique program designed by the School of Information. The MICS program consists of three foundation courses, six advanced courses, and one capstone course (UC Berkeley School of Information 2019). Therefore, the UC Berkeley's MICS curriculum sets a benchmark for cybersecurity education in the iSchools. The NICE framework is a national-level initiative and designed for communicating cybersecurity education, training,

Table 4.6 Counts of Cybersecurity-Related Program Title at the Undergraduate and Graduate Levels

Bachelor Level	Master Level	PhD Level
Cybersecurity Analytics and Operations (1)	Cyber Intelligence (1)	Cybersecurity (1)
Computing and Security Technology (1)	Cybersecurity (4)	Digital Privacy, Safety, and Security Studies (1)
Emergency Preparedness, Homeland Security, and Cybersecurity (1)	Information and Cybersecurity	(Information) Security and Privacy (3)
Security and Risk Analysis (1)	Information Assurance (2)	Information Assurance (1)
Information Security (1)	Information Security (4)	
Cybersecurity (1)	Information Security (Policy and) Management (3)	
Computing Security (1)	Identity Management and Security (1)	
	Secure Computing (1)	

and workforce development with shared lexicons between a multitude of stakeholders (Newhouse et al. 2017). The following concept map (Figure 4.2) depicts the hierarchical structure of the elements relevant to cybersecurity competency and their relationships in the NICE Framework. The NICE Framework contains seven workforce categories in which they are composed of 33 specialty areas and 52 work roles. Each specialty area or work role is associated with one or multiple KSAs and tasks. In total, there are 630 knowledge units (including 41 withdrawn), 374 skills (9 withdrawn), 176 abilities (1 withdrawn), and 1,007 tasks associated with 33 specialty areas and 7 categories in the NICE Framework.

A mapping of the NICE Framework components to a currently the most comprehensive cybersecurity curriculum among iSchools could lead to a better understanding of how cybersecurity education in a represented iSchool fits into the cybersecurity workforce. We extracted the keywords from the course titles and descriptions in the MICS curriculum, including three foundation courses: Beyond the Code: Cybersecurity in Context; Cryptography for Cyber and Network Security; and Software Security, and six advanced courses:

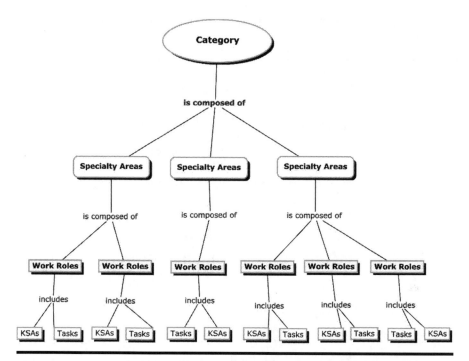

Figure 4.2   A Concept Map of the Elements in the NICE Cybersecurity Workforce Framework.

Source: Newhouse et al. (2017) National Initiative for Cybersecurity Education (NICE) Cybersecurity Workforce Framework. *NIST Special Publication 800-181.*

Network Security; Operating System Security; Usable Privacy and Security; Managing Cyber Risk; Government, National Security, and the Fifth Domain; and Privacy Engineering. Then, we utilized the extracted keywords and performed a search using the online keyword search system in the National Initiative for Cybersecurity Careers and Studies (NICCS) site (Figure 4.3) to identify three types of information: Work Roles, Specialty Areas, and Categories. The extracted keyword was matched to the Tasks and KSAs descriptions, which generate a list of result automatically in the field box. The match is recorded for each keyword and each description resulted in a list of work role ID, work roles, work role description, category, and specialty area(s). This mapping activity helped us to discover the keywords from the MICS curriculum in relevance to work roles, specialty areas, and categories in the NICE Framework.

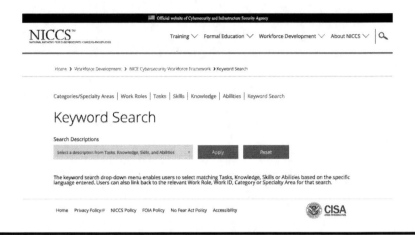

Figure 4.3   Screenshot of the NICE Keyword Search System.

Source: Department of Homeland Security, Office of Cybersecurity and Communications, NICCS (2019b).

Mapping to Workforce Categories

After conducting the keyword searches through the online system, the keywords were mapped between the UC Berkeley MICS curriculum to the NICE Framework – Workforce Categories, we calculated the count and presented the match between the two in Table 4.7. In general, the majority of KSAs and tasks identified in the curriculum fall into two primary workforce categories: Oversee and Govern (187 counts) and Securely Provision (180 counts). Interestingly, each of the three foundational courses covers a small number of KSAs and tasks, while the six advanced courses ended up with a lot more KSAs and tasks matches. The total counts of the match between the workforce category and the course title are presented in parenthesis in Table 4.8.

In addition, the topics in the first foundational course emphasize cybersecurity issues in various contexts like business, legal, behavioral economic, political, and ethical situations, spanning four workforce categories: Investigate; Protect and Defend; Oversee and Govern; and Securely Provision. The second foundational course concentrates on cryptography addressing the KSAs/tasks in three workforce categories: Protect and Defend; Operate and Maintain; and Securely Provision, whereas the third foundational course centers on software security, addressing the KSAs/tasks demands in two workforce categories: Oversee and Govern and Securely Provision.

As for advanced courses, a higher level of complexity becomes apparent in the engagement of diverse KSAs and tasks. Four out of six advanced courses matched cross seven workforce categories. Those courses are network security, operating

Table 4.7 Mapping UC Berkeley Cybersecurity Curriculum with NICE Cybersecurity Framework Workforce Categories

UC Berkeley MICS Curriculum	Workforce Category						
	Investigate (33)	Collect and Operate (44)	Analyze (42)	Protect and Defend (45)	Oversee and Govern (187)	Operate and Maintain (60)	Securely Provision (180)
Foundation Courses Beyond the Code: Cybersecurity in Context	X			X	X		X
Cryptography for Cyber and Network Security				X		X	X
Software Security					X		X
Advanced Courses Network Security	X	X	X	X	X	X	X
Operating System Security	X	X	X	X	X	X	X
Usable Privacy and Security	X	X	X			X	X
Managing Cyber Risk	X	X	X	X	X	X	X
Government, National Security, and the Fifth Domain	X	X		X	X		X
Privacy Engineering	X	X	X	X	X	X	X

Table 4.8   Mapping UC Berkeley Cybersecurity Curriculum with NICE Cybersecurity Framework Workforce Category and Work Role

Category	Work Role	Count[*]
Investigate	Cyber Crime Investigator	7
	Cyber Defense Forensics Analyst	14
	Law Enforcement/Counterintelligence Forensics Analyst	12
Collect and Operate	All Source-Collection Manager	9
	All Source-Collection Requirements Manager	8
	Cyber Intel Planner	5
	Cyber Operator	10
	Cyber Ops Planner	5
	Information Systems Security Manager	1
	Partner Integration Planner	6
Analyze	All-Source Analyst	5
	Exploitation Analyst	10
	Mission Assessment Specialist	5
	Multidisciplined Language Analyst	7
	Target Developer	5
	Target Network Analyst	5
	Threat/Warning Analyst	5
Protect and Defend	Cyber Defense Analyst	15
	Cyber Defense Incident Responder	8
	Cyber Defense Infrastructure Support Specialist	9
	Vulnerability Assessment Analyst	13
Oversee and Govern	Communications Security (COMSEC) Manager	8
	Cyber Instructional Curriculum Developer	6
	Cyber Instructor	7

(*Continued*)

Table 4.8   (Cont.)

Category	Work Role	Count*
	Cyber Legal Advisor	6
	Cyber Policy and Strategy Planner	5
	Cyber Workforce Developer and Manager	7
	Executive Cyber Leadership	7
	Information Systems Security Manager	16
	IT Investment/Portfolio Manager	9
	IT Program Auditor	10
	IT Project Manager	13
	Privacy Officer/Privacy Compliance Manager	**66**
	Product Support Manager	14
	Program Manager	13
Operate and Maintain	Data Analyst	5
	Database Administrator	6
	Knowledge Manager	9
	Network Operations Specialist	8
	System Administrator	11
	Systems Security Analyst	14
	Technical Support Specialist	7
Securely Provision	Authorizing Official/Designating Representative	14
	Enterprise Architect	17
	Information Systems Security Developer	**20**
	Research and Development Specialist	11
	Secure Software Assessor	14
	Security Architect	**20**
	Security Control Assessor	**20**

(*Continued*)

Table 4.8 (Cont.)

Category	Work Role	Count*
	Software Developer	15
	System Testing and Evaluation Specialist	12
	Systems Developer	**20**
	Systems Requirements Planner	17

Note: Any total count of equal and more than 20 are in bold.

* Note that total count is based on matched Tasks and KSAs through the keyword searches.

system security, managing cyber risks, and privacy engineering. The other two advanced courses incorporate five workforce categories. One course considers human usability factors that influence security and privacy; the other goes beyond the individual level to government, national and international level of cybersecurity issues. The KSAs and tasks mapped in both courses have three shared workforce categories, such as Investigate, Collect and Operate, and Securely Provision. The course "Usable Privacy and Security" deals with KSAs and tasks relevant to Analyze and Operate and Maintain to comprehend user needs, while the course "Government, National Security, and the Fifth Domain" comprises KSAs and tasks accounting for Protect and Defend as well as Oversee and Govern occurred at higher level.

Mapping to Work Roles

Similar to job titles, work roles indicate the major responsibilities of one's job. The work roles in the NICE Framework are identified by the category and specialty area. In this mapping activity, several work roles are identified along with the category in Table 4.8. The work role, Privacy Officer/Privacy Compliance Manager, in category Oversee and Govern, covers the most KSAs and tasks uncovered in the curriculum. Though Information Systems Security Manager is in the same category (Oversee and Govern), the number of KSAs and tasks covered is relatively lower. This is not surprising because privacy and ethical issues are more predominate among the different iSchools while security is typically identified as a technical concern. At the intersection between privacy and security is personal identification information (PII). The focus of security measure is to protect unauthorized access to PII and its confidentiality, integrity, and availability. On the other hand, the view on privacy is about managing the life cycle of PII and its alignment with the organization's governance framework, risk management, compliance, and regulations. The convergence of privacy and security in cybersecurity education will yield significant benefits to the workforce preparation and allow current professionals to develop collective solution at work.

The work roles with higher counts (counts >= 20) in Table 4.8 seem to fall into two categories: Oversee and Govern and Securely Provision. Other top five work roles include Information Systems Security Developer, Security Architect, Security Control Assessor, and Systems Developer, which are in the category of Securely Provision. Several work roles (with counts between 16 and 20) are also worth noting, such as Information Systems Security Manager in Oversee and Govern as well as Enterprise Architect and Systems Requirements Planner in Securely Provision. These identified work roles required both technical and nontechnical competency; particularly, the role of an enterprise architect involves aligning business processes with IT and system security, which require facilitation between the IT professionals and non-IT professionals.

Mapping to Specialty Areas

Cybersecurity is multifaceted, multidisciplinary, and interdisciplinary due to the nature of its complexity. Different disciplines focus their work in a specific area, but they also interact with other specialty area. Based on the keyword search activity, the results of matched Tasks and KSAs to the category and specialty areas are displayed in Table 4.9. The top three specialty areas by counts include Legal Advice and Advocacy, Program/Project Management and Acquisition, and Systems Development. Digital Forensics, Cybersecurity Management, Risk Management, and Security Architecture appear to be emerging specialty areas that are also a focus in the curriculum. The patterns unveiled in specialty areas with their categories are consistent with the previous findings in which Oversee and Govern and Securely Provision are the two major workforce categories. This observation reflects the iSchools UC Berkley School of Information current focus on competency development in specialty areas for cybersecurity.

Using the NICE Cybersecurity Workforce Framework to map the represented keywords and concepts from the MICS cybersecurity curriculum help us gain insight into what has been taught from one of the iSchools, UC Berkeley School of Information. The mapping activity of the keywords with the NICE Framework Tasks and KSAs supports the understanding of cybersecurity education emphases and possible direction for future development. Other specialty areas from the NICE framework (e.g., Training, Education, and Awareness in Oversee and Govern; Data Administration and Knowledge Management in Operate and Maintain; All-Source Analysis, Threat Analysis, Language Analysis in Analyze) can also provide work-related opportunities for information professionals who would like to develop their career horizontally or vertically.

Table 4.9 Mapping UC Berkeley Cybersecurity Curriculum with NICE Cybersecurity Framework Workforce Category and Specialty Area

Category	Specialty Area	Counts[*]
Investigate	Cyber Investigation	3
	Digital Forensics	**17**
Collect and Operate	Collection Operations	9
	Cyber Operational Planning	11
	Cyber Operations	5
Analyze	Threat Analysis	3
	Exploitation Analysis	6
	All-Source Analysis	4
	Targets	6
	Language Analysis	2
Protect and Defend	Cybersecurity Defense Analysis	10
	Cybersecurity Defense Infrastructure Support	4
	Incident Response	3
	Vulnerability Assessment and Management	6
Oversee and Govern	**Legal Advice and Advocacy**	**66**
	Training, Education, and Awareness	8
	Cybersecurity Management	**17**
	Strategic Planning and Policy	8
	Executive Cyber Leadership	6
	Program/Project Management and Acquisition	**45**
Operate and Maintain	Data Administration	6
	Knowledge Management	6
	Customer Service and Technical Support	3
	Network Services	5

(Continued)

Table 4.9   (Cont.)

Category	Specialty Area	Counts[*]
	Systems Administration	7
	Systems Analysis	9
Securely Provision	**Risk Management**	**18**
	Software Development	16
	Systems Architecture	**17**
	Technology R&D	7
	Systems Requirements Planning	7
	Test and Evaluation	7
	Systems Development	**23**

Note: The top seven areas by count are in bold.

* Note that total count is based on matched Tasks and KSAs through the keyword searches.

Potential Models for Cybersecurity Training

As smart innovations and the IoT are drastically growing and evolving, new technologies will inevitably outpace the current laws and regulations in protecting users from cybersecurity threats. The IoT Attack Surface Areas Project by the Open Web Application Security Project (OWASP) identified 18 surface areas that are vulnerable for IoT attacks. One of the IoT surface areas, privacy, primarily concerned with user data disclosure, device and user location disclosure, and differential privacy. This implied the need of cybersecurity education and awareness training for the users, especially to users from data breaches, and other cybersecurity threats (OWSAP 2015).

Cybersecurity threat is an ever-changing landscape. Identifying threats and addressing risks require ongoing training and education to meet both technical and social aspects of cybersecurity challenges. From the technical aspects of cybersecurity, the IT and cybersecurity professionals are often referred to the Open Systems Interconnection (OSI) model for the networked environment standard. The model was developed by the International Organization for Standardization (ISO) in 1984, as a protocol networking standards and information exchange between hardware, software, and applications. The last version was created in 1994 (ISO/IEC 7498-1:1994) and reviewed in 2000 (ISO 1994). The OSI model describes a seven-layers model of an information communication system where data is transferred across the layers in a networked environment. Each layer has its protocol to communicate and posed vulnerability as well because cyber threats and attacks can occur at any of these seven layers. The point of introducing the OSI model here is to demonstrate the extra layers added by Michael Gregg, Chief Technology Officer (CTO), and his colleague (2006), in comparison with the Pedagogic Cybersecurity Framework developed by Dr. Peter Swire (2018). The extended models provide a complementary perspective on the original technical OSI model. Table 4.10 shows a comparison of the OSI model and its extended models.

From the social/human perspective, the eighth layer, People, was added to the OSI model by Gregg and his colleague (2006) to address the human and social issues beyond technical functionality. Social engineering was identified as the biggest threat at that time, which is still a concern today. From a multidisciplinary perspective, Swire (2018) proposed a Pedagogic Cybersecurity Framework (PCF) which extends the horizon of our cybersecurity education to the organizational, government, and national levels. Although Swire (2018) did not list people in a separate layer, he placed users within the household, so-called people, in the Organization layer. The PCF delineates how the nontechnical areas of expertise, like cybersecurity management, policy, law, and international affairs, come into play. We usually think of cybersecurity threats from an individual or organizational level. However, cyber threats and impacts are also part

Table 4.10   A Comparison of the OSI Model and its Extended Models

		Models		
Layers		*OSI Model (ISO [1984] 1994)*	*Extended OSI Model (Gregg et al. 2006)*	*Pedagogic Cybersecurity Framework (PCF) (Swire 2018)*
Social and human factors	10	–	–	Nation
	9	–	–	Government
	8	–	People	Organization
Technical factors	7	Application	Application	Application
	6	Presentation	Presentation	Presentation
	5	Session	Session	Session
	4	Transport	Transport	Transport
	3	Network	Network	Network
	2	Data Link	Data Link	Data Link
	1	Physical	Physical	Physical

of the governmental concerns and national security. If we are taking a holistic view of cybersecurity education, the PCF helps in addressing the diverse stakeholders at the distinct levels of cybersecurity and locating potential vulnerabilities and threats at each layer. The PCF facilitates discussion across the different disciplines which reflect the nature of cybersecurity risks spreading at the individual, organizational, government, or national level.

Major Opportunities and Challenges for iSchools

Originally developed in the NIST Special Publication 800-16 Report: *Information Technology and Security Training Requirements: A Role- and Performance-Based Model*, the Learning Continuum (deZafra et al. 1998, Appendix A) provides a conceptual framework for IT security training. A further discussion of the learning continuum can be found in the NIST Special Publication 800-50 Report: *Building an Information Technology Security Awareness and Training Program* where the IT Security Learning Continuum (Wilson and Hash 2003, Figure 2.1) focuses on Awareness, Training, and Education as a continuum for learning with three levels embedded in each area: beginning, intermediate, and advanced. This report specifically identifies critical steps in the life cycle of IT security awareness and training program. This is a companion publication to NIST SP 800-16. The NIST special publications are also useful for nongovernment organizations to design their security awareness and training programs. As stated in the NIST SP 800-50, "Learning is a continuum; it starts with awareness, builds to training, and evolves into education" (Wilson and Hash 2003, 7).

From our examination of the North America iSchools program and curriculum, a majority of iSchools appear to recognize the importance of cybersecurity and have various levels of programs and courses for information security and cybersecurity. As presented in Figure 4.4, the Cybersecurity Learning Continuum illustrates the learning spectrum moving from security awareness to cybersecurity essentials, to role-based training, to education and/or experience. Thus, it shows the potential for iSchools cybersecurity program development to move from a single module or coursework to an integrated approach that addresses the various roles and responsibilities of cybersecurity professional in an interdisciplinary manner.

Figure 4.4 indicates that security awareness programs aim to raise the awareness of all users in the organization, while courses on cybersecurity essentials target at all users involved with IT systems. Cybersecurity awareness and cybersecurity essentials provide fundamental literacy for role-based training. Role-based training is required when the users' work roles intersect with IT systems or cybersecurity responsibilities. Role-based training and education phase mutually influence each other, which motivates the employees to consider cybersecurity as their profession through interdisciplinary or multidisciplinary education obtaining

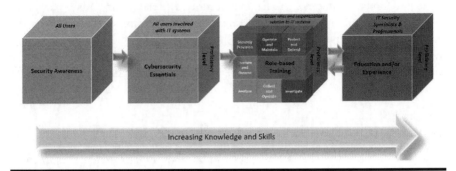

Figure 4.4   Cybersecurity Learning Continuum.
Adapted from Toth and Klein (2014, Figure 4.1).

a higher-education degree or certification to complement work experience (Toth and Klein 2014). Cybersecurity essentials, role-based training, and education and/ or experience involve different proficiency levels, namely, role competency levels spanning the range from basic to intermediate to expert. The knowledge gained from training or education and work experience accumulated over time will affect the role competency levels. Although most iSchools do not offer cybersecurity relevant degrees/programs further on the right of the cybersecurity learning spectrum, developing role-based training modules or courses for information professionals who would have taken IT/cybersecurity-related responsibilities could widen the iSchools representations in cybersecurity workforce.

In the ever-changing cybersecurity domain, it is crucial to help students keep pace with changing work roles and requirements in cybersecurity. CyberSeek (2018) surveyed job opening and created a heat map with visualizations to display their data collected between April 2017 and March 2018. In terms of understanding current job trends in cybersecurity workforce, we used the CyberSeek data grouped by the NICE Framework Cybersecurity Workforce Category to depict the proportion of job openings by categories (Figure 4.5). The breakdown of each category by percentages is dominated by Operate and Maintain (26%) and Securely Provision (24%), followed by Analyze (16%), Protect and Defend (16%), and Oversee and Govern (11%). Within the seven categories, Collect and Operate (6%) and Investigate (1%) have relatively few job openings to fill the various roles in cybersecurity workforce.

Based on Shoemaker, Kohnke, and Sigler (2016) analysis of the NICE workforce framework, it is suggested that the work roles in "Collect and Operate" and "Analyze" could be combined to complement the information lifecycle (i.e., collect, process, analyze, and present) and considered as intelligence work. They also mentioned the work roles in "Operate and Maintain" parallelly support both "Collect and Operate" and "Analyze" to interpret and manage the intelligence

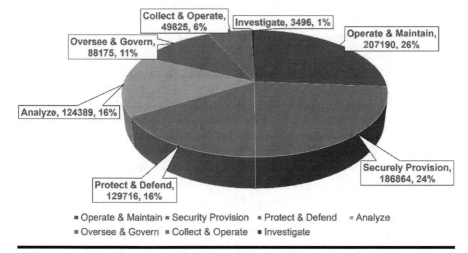

▪ Operate & Maintain ▪ Security Provision ▪ Protect & Defend ⁕ Analyze
▪ Oversee & Govern ▪ Collect & Operate ▪ Investigate

Figure 4.5 Cybersecurity Job Openings by NICE Framework Cybersecurity Workforce Category.
Source: Data from CyberSeek Cybersecurity Supply/Demand Heat Map – Job Openings by NICE Cybersecurity Workforce Framework Category (2018).

information and knowledge. In the preceding section, through mapping the cybersecurity curriculum from Berkeley's iSchool, we discerned that "Oversee and Govern" seems to be the most addressed/attended category comparing to other categories. Additionally, "Oversee and Govern" serves as a higher-level guidance structure for the organizations, which is vital to drive cybersecurity policies and management. When four relevant categories ("Collect and Operate," "Analyze," "Operate and Maintain," and "Oversee and Govern") are taken into account together, the contours of cybersecurity workforce for information professionals seem clearer. Furthermore, four workforce categories account for nearly 60% of job demand in cybersecurity. This does not mean that the specialty areas and work roles in the four categories are all suitable for information professionals, but they exhibit alternative career opportunities with either informational or managerial focus rather than solely a technical center.

Conclusion

"Cybersecurity issues are complex, and there is no standard recipes for protecting information assets within organizations" (Knapp, Maurer, and Plachkinova 2017, 102). This statement is proven by the diverse roles and interconnected responsibilities of cybersecurity professionals at the local and national level represented by the various publications and reports we discussed in this chapter. The global cybersecurity workforce shortage has a major impact on national cybersecurity as well. To respond to the Executive Order 13800 of May 11, 2017, *Strengthening the Cybersecurity of Federal Networks and Critical Infrastructure* by the president, there were several working groups and deliverables in response to the intensifying cybersecurity threats to our nation (Department of Homeland Security CISA 2017). Specifically, to address the matter of American cybersecurity workforce, the Secretary of Commerce and the Secretary of Homeland Security (2017) compiled a report: *Supporting the Growth and Sustainment of the Nation's Cybersecurity Workforce: Building the Foundation for a more Secure American Future.* They recommended the private and public sectors working together to improve cybersecurity workforce development and suggested three aspects where the public and private sectors can collaborate including transforming the learning environment to grow a dynamic and diverse cybersecurity workforce, aligning education and training with cybersecurity workforce demands by applying the National Initiative for Cybersecurity Education (NICE) Cybersecurity Workforce Framework, and establishing robust metrics that evaluate the effectiveness and impact of cybersecurity workforce investments (Department of Homeland Security 2017). A recent Executive Order on *America's Cybersecurity Workforce* by the president reiterates that more than 300,000 cybersecurity job vacancies remain unfilled in the United States and promotes the wider adoption of NICE's cybersecurity workforce framework (The White House 2019).

When cybersecurity is considered a national issue, our country should rally all the support to build and strengthen the current workforce. Higher education is in a unique position to support this goal due to their traditional role of teaching and research with industrial partnership to produce highly skilled workforce beyond basic education. Higher education can address the cybersecurity skill gaps by recognizing areas to improve and keep up with technology advances in this profession. Cybersecurity education is relevant to every other profession who comes "online," utilize technology, and being a part of a network environment. This chapter provides a glimpse of the current status of cybersecurity education at the different levels, in particular, higher education institutes such as the iSchools. Through mapping of the NICE Cybersecurity Workforce Framework with an exemplar curriculum from one of the iSchools and outlining the different extended OSI models and the NIST's Cybersecurity Learning Continuum to inform the ongoing training or education to address both technical and social/human aspects of cybersecurity challenges, we believe information profession can play a critical role to meet the need of a diverse cybersecurity workforce.

References

AACU (American Association of Community Colleges). 2002. *Protecting Information: The Role of Community Colleges in Cybersecurity Education.* January 26–28, 2002.

Washington, DC: Community College Press. www.nationalcyberwatch.org/ncw-con tent/uploads/2016/03/Workshop_Rpt-Role_of_CCs_in_Cyber_Ed-2002.pdf.

Assante, Michael J., and David H. Tobey. 2011. "Enhancing the Cybersecurity Workforce." *IT Professional* 13 (1). IEEE: 12–15. doi: https://doi.org/10.1109/MITP.2011.6.

CyberSeek. 2018. "Cybersecurity Supply/Demand Heat Map." Job Openings by NICE Cybersecurity Workforce Framework Category. Accessed January 10, 2019. www.cyberseek.org/heatmap.html.

Department of Education, Office of Postsecondary Education. 2018. "Pilot Program for Cybersecurity Education Technological Upgrades for Community Colleges." Last modified July 31, 2018. www2.ed.gov/programs/ppcetucc/index.html.

Department of Homeland Security (DHS). 2017. "Publications Library." Supporting the Growth and Sustainment of the Nation's Cybersecurity Workforce: Building the Foundation for a More Secure American Future. www.dhs.gov/publication/support ing-growth-and-sustainment-nations-cybersecurity-workforce#.

Department of Homeland Security (DHS), Cybersecurity and Infrastructure Security Agency (CISA). 2017. "Executive Order on Strengthening the Cybersecurity Net-works and Critical Infrastructure." American Cybersecurity Workforce Development. www.dhs.gov/cisa/executive-order-strengthening-cybersecurity-federal-networks-and-critical-infrastructure.

Department of Homeland Security (DHS), National Initiative for Cybersecurity Careers and Studies (NICCS). 2019. "Integrating Cybersecurity into the Classroom." Accessed March 25, 2019. https://niccs.us-cert.gov/formal-education/integrating-cybersecurity-classroom.

Department of Homeland Security (DHS), Office of Cybersecurity and Communications, National Initiative for Cybersecurity Careers and Studies (NICCS). 2019b. "Workforce Development." Keyword Search. https://niccs.us-cert.gov/workforce-development/cyber-security-workforce-framework/search.

Department of Homeland Security (DHS), Office of Cybersecurity and Communications, National Initiative for Cybersecurity Careers and Studies (NICCS). 2019a. "Workforce Development." NICE Cybersecurity Workforce Framework. Last published May 9, 2019. https://niccs.us-cert.gov/workforce-development/cyber-security-workforce-framework.

deZafra, Dorothea, Sadie Pitcher, John Tressler, and John Ippolito. 1998. *Information Technology Security Training Requirements: A Role- and Performance-Based Model.* NIST (National Institute of Standards and Technology) Special Publication 800-16. April 1998. https://csrc.nist.gov/publications/detail/sp/800-16/final.

Drexel University Online. 2019. "Online Master's Degree in Cybersecurity." Accessed January 5, 2019. https://online.drexel.edu/online-degrees/engineering-degrees/ms-cybersecurity/index.aspx.

Furnell, Steven, and Liam Moore. 2014. "Security Literacy: The Missing Link in Today's Online Society?" *Computer Fraud & Security* 2014 (5): 12–18. doi: 10.1016/S1361-3723(14)70491-9.

Georgia Institute of Technology. 2019. "Institute for Information Security & Privacy." Master of Science in Cybersecurity. Accessed January 5, 2019. www.iisp.gatech.edu/masters-degree.

Gregg, Michael, Stephen Watkins, George Mays, Chris Ries, Ronald Bandes, and Brandon Franklin. 2006. *Hack the Stack: Using Snort and Ethereal to Master the 8 Layers of an Insecure Network*. Rockalnd, MA: Syngress Publishing, Inc.

ISACA (Information Systems Audit and Control Association). 2019. "State of Cybersecurity 2019." Part 1. https://cybersecurity.isaca.org/state-of-cybersecurity.

(ISC)[2] International Information Systems Security Certifications Consortium. 2018. "Cybersecurity Workforce Study, 2018." www.isc2.org/Research/Workforce-Study.

iSchools Organization. 2014. "Welcome!" iSchools Vision. https://ischools.org/.

ISO (International Organization for Standardization). 1984. "Standards Catalogue/Publications and Products." ISO/IEC 7498:1984. Withdrawn publication. www.iso.org/standard/14252.html.

ISO (International Organization for Standardization). 1994. "Standards Catalogue/Publications and Products." ISO/IEC 7498-1:1994. Last edited June, 2016. www.iso.org/standard/20269.html.

Jones, Keith S., Akbar Siami Namin, and Miriam E. Armstrong. 2018. "The Core Cyber-Defense Knowledge, Skills, and Abilities that Cybersecurity Students Should Learn in School: Results from Interviews with Cybersecurity Professionals." *ACM Transactions on Computing Education* 18 (3), no.11. doi: 10.1145/3152893.

Knapp, Kenneth J., Christopher Maurer, and Miloslava Plachkinova. 2017. "Maintaining a Cybersecurity Curriculum: Professional Certifications as Valuable Guidance." *Journal of Information Systems Education* 28 (2):101–114. http://jise.org/Volume28/n2/JISEv28n2p101.html.

Leaning, Marcus, and Udo Richard Averweg. 2019. "Developing the Social, Political, Economic, and Criminological Awareness of Cybersecurity Experts: A Proposal and Discussion of Non-Technical Topics for Inclusion in Cybersecurity Education", In *Global Cyber Security Labor Shortage and International Business Risk*, edited by Christiansen, Bryan and Agnieszka Piekarz, 77–94. Hershey, PA: IGI Global. doi: 10.4018/978-1-5225-5927-6.ch005.

National Cybersecurity Alliance. 2017. "Keeping up with Generation App 2017: NCSA Parent/Teen Online Safety Survey." https://staysafeonline.org/wp-content/uploads/2017/10/Generation-App-Survey-Report-2017.pdf.

National Cyberwatch Center. 2019. www.nationalcyberwatch.org/.

Newhouse, William, Stephanie Keith, Benjamin Scribner, and Gred Witte. 2017. *National Initiative for Cybersecurity Education (NICE) Cybersecurity Workforce Framework*. NIST (National Institute of Standards and Technology). Special Publication 800-181. August 2017. doi: 10.6028/NIST.SP.800-181.

NICE (National Initiatives for Cybersecurity Education). 2019. "NICE K12 Cybersecurity Education Conference." Accessed March 5, 2018. www.k12cybersecurityconference.org/.

NSA (National Security Agency), CSS (Central Security Service). n.d. "National Centers of Academic Excellence." Accessed March 7, 2019. www.nsa.gov/resources/students-educators/centers-academic-excellence/.

Olmstead, Kenneth, and Aaron Smith. 2017. *Americans and Cybersecurity*. January 26, 2017. Pew Research Center. www.pewinternet.org/2017/01/26/americans-and-cyber security/.

OWASP (Open Web Application Security Project). 2015. "IoT Attack Surface Areas." Last modified on November 29, 2015. www.owasp.org/index.php/IoT_Attack_Surface_Areas.

Partnership for Public Service, and Booz Allen Hamilton. 2015. *Cyber In-Security II: Closing the Federal Talent Gap.* https://ourpublicservice.org/wp-content/uploads/2018/09/Cyber_In-Security_II__Closing_the_Federal_Talent_Gap-2015.04.13.pdf.

Ponemon Institute. 2018. "2018 Study on Global Megatrends in Cybersecurity." February 2018. www.raytheon.com/sites/default/files/2018-02/2018_Global_Cyber_Mega trends.pdf.

Shoemaker, Dan, Anne Kohnke, and Ken Sigler. 2016. *A Guide to the National Initiative for Cybersecurity Education (NICE) Cybersecurity Workforce Framework (2.0).* CRC Press.

Swire, Peter. 2018. "A Pedagogic Cybersecurity Framework." *Communications of the ACM* 61 (10), 23–26. http://peterswire.net/wp-content/uploads/Pedagogic-cybersecurity-framework.pdf.

The White House. 2019. "Executive Order on America's Cybersecurity Workforce." May 2, 2019. www.whitehouse.gov/presidential-actions/executive-order-americas-cybersecurity-workforce/.

Toth, Patricia, and Penny Klein. 2014. "A Role-Based Model for Federal Information Technology/Cybersecurity Training (3rd Draft)." NIST (National Institute of Standards and Technology). Special Publication 800-16 Rev.1(Draft). March 2014. https://csrc.nist.gov/publications/detail/sp/800-16/rev-1/draft.

UC (University of California) Berkeley, School of Information. 2019. "Cybersecurity @ Berkeley." The Master of Information and Cybersecurity Delivered Online from UC Berkeley. Accessed January 5, 2019. https://cybersecurity.berkeley.edu/.

U.S. Bureau of Labor Statistics, Office of Occupational Statistics and Employment Projects. 2019. "Occupational Outlook Handbook: Information Security Analysts." https://www.bls.gov/ooh/computer-and-information-technology/information-security-ana lysts.htm

Wilson, Mark, and Joan Hash. 2003. *Building an Information Technology Security Awareness and Training Program.* NIST (National Institute of Standards and Technology). Special Publication 800-50. October 2003. https://csrc.nist.gov/publications/detail/sp/800-50/final.

Chapter 5

MetaMinecraft: Cybersecurity Education through Commercial Video Games

Chris Markman, MSLIS, MSIT
Senior Librarian, Palo Alto City Library
Palo Alto, CA, USA

Introduction

This chapter was written halfway through a yearlong project funded by the California State Library LSTA "pitch an idea" grant program and in partnership with a group of 40 libraries in the San Francisco Bay Area called the Pacific Library Partnership (PLP), of which the author's home library in Palo Alto, California, is a member. It contains the views and observations of the author and should not be confused as an endorsement by any affiliate organizations.

The project was designed to test the effectiveness of Minecraft as a platform that librarians, and especially public librarians, could use to engage with teens and tweens on the subject of cybersecurity. Readers may find it interesting to know that testing course material against different target demographics was not the original intent of the grant "pitch" process, but at the advice of multiple librarians and state library's grant advisor, we identified tweens as largest consumers of Minecraft media, but not without some debate. This change is reflected in the final grant proposal title which specifically mentions both groups as "youths" and ultimately, a redesign of the original course material to reflect the cybersecurity understanding and cybersecurity need of young people who may not yet even poses a personal email account. That said, because Minecraft uses individual usernames and passwords to login, and has a history of data breaches affecting the community at large, the need for cybersecurity awareness had already been set.

Project Goals

The first phase, which ran from July to November 2018, consisted of several informal email surveys sent out to the PLP community, followed by interviews with librarians and teen advisory boards or groups, and more formal research avenues including pitching the idea at library conferences and networking with Minecraft and security professionals. Above all else, multiple years of Minecraft gameplay experience were also extremely important, because the plan was to use the actual survival gameplay mechanics of Minecraft to demonstrate cybersecurity threat modeling as an analogy. This connection is not obvious unless a person has spent at least 2–3 hours playing the game itself and understands the various threats embedded in the game environment and their options for mitigating these threats.

As much as possible, the goal of this project was to create a series of Minecraft-based lesson plans that blended the self-preservation skill sets inherent to Minecraft's survival mode with the survival skills a person needs to cultivate for the purpose of protecting their own personal online security cybersecurity. The second, overarching goal as a product intended for public libraries was to document the cost effectiveness of different delivery methods for the same material, of which there were many, and third and finally to create content that was accessible to multiple audiences in terms of cybersecurity awareness and Minecraft gameplay skills levels.

It should be noted that the first iteration of this lesson plan is still in the early testing stages, and will continue to evolve through the second half of the project, but was built on an idea presented by a researcher at a BSides, an ever-growing community run security conference series (White 2016) and later documented on the EDUCAUSE Security Matters blog (Markman 2016). As mentioned previously, the "prototype" lesson plan uses a design-thinking process to help illustrate the rise and fall of specific patterns in cybersecurity risk management based on different environmental factors. Students are asked to create and modify their self-designed threat profiles based on their understanding of the Minecraft universe, then build and test their "digital fort" against their classmates. All of this occurs within the game universe of Minecraft itself. This not only allows for a great amount of creativity, but also successfully gets people thinking about and building their conceptual framework around what makes for a good contingency plans and edges toward other cyberskills like network design, security layering, and social engineering.

Breaking the Mold

But why do all this? One could argue that while the information security world has put increasing resources toward job teaching cybersecurity skills to young adults and teens, and indeed, librarian-to-librarian training through grassroots initiatives like the Library Freedom Project and closely affiliated Tor Project (Macrina 2015) has already proven to be highly effective. Tweens however are largely left out of this equation, and at an age where basic security knowledge and key concepts could have a lasting impact on their digital lives. This is even more true in the state of California, which has seen significant drops in funding for school librarians and library spaces within schools. The California School Library Association notes that "since the beginning of the 21st Century, California has decreased the number of teacher librarians who are officially administering required school library programs by over 60%" (CSLA 2019). This shortage of school librarians, coupled with the ever-increasing impact of Internet technology in our daily lives, means public libraries should play a bigger role in developing online privacy awareness and cybersecurity skills. At the same time, over the past decade the US federal government has shown even greater interest in cybersecurity as a means of work-force development and promoting national security. One needs to only browse the millions of dollars of grant funding for cybersecurity education floating on Grants.gov, a central database for multiple departments, to see this is true.

The software "edutainment" model, which attempts to create games with the primary goal of education, rather than demonstrate learning objectives through pre-existing (and inherently fun) videogames, also never attempts to wrestle with or even approach the fact that youth and their digital selves have completely different motivating factors in cyberspace. In surveying traditional online cybersecurity education material and textbooks, you will often find it is heavily focused on the financial implications for data loss or data leaks, often from a business perspective. This does not work so well for tweens. For example, how does one explain the need to keep their smartphone OS up to date if they do not own or use one? Many of the teens I spoke to use laptops provided by their school – as a result, they have an IT team working in the background doing maintenance. This is not the "real world" quite yet. The impact of an identity theft is completely different when you are 12. Youth, when compared to their older peers, have invested less time in maintaining or curating their online identity, and thus have much less to lose. This is why Minecraft became the subject of cybersecurity education, because while a game world may seem less "real" to adults, our hypothesis was that it would better demonstrate the benefit of threat modeling to young people.

Although experimental in nature, through this project I hoped to show that reframing cybersecurity as a set of survival skills anchored in the fundamental elements of Minecraft gameplay (exploration, design, and self-preservation) and expanded to also include the meta-framework of the Minecraft online community (specifically the multiplayer experience, but also game modifications) will mesh better with youth audiences not only because educators and students can suddenly use this shared language, but also because it leverages *organic* gameplay and Minecraft game mechanics they are already familiar with, which is very empowering. This concept of a MetaMinecraft, or the sum whole of the Minecraft fandom, is an attempted to create a flipped classroom where the students do not know they have been exercising their cybersecurity skills until their gameplay experiences are introduced as analogous to the process of threat profiling and risk management.

This experiment is also in contrast to current uses of Minecraft in education, of which I was very familiar with as a member of the Minecraft Education Edition Mentor team, a volunteer cohort of educators and technologists from across the globe. Traditionally, Minecraft has been used as an education tool to explore virtual worlds either related to a subject area (walking through an ancient city or scale model of Martian terrain, for example) or as a means of self-expression (build a small town) and teaching the very basics of software programming through the acclaimed Hour of Code (https://hourofcode.com) lesson series. MetaMinecraft is different because the desired result is less about what a student design in their game, and more about reframing the mental processes of Minecraft gameplay toward a specific goal. If successful, students will never play Minecraft the same way again, creating a positive feedback loop and reinforcing the lessons learned through the workshop experience.

Videogames and Security

Software companies like Google and others have recently started to create "edutainment" style game experiences that teach security concepts as well. For example, https://beinternetawesome.withgoogle.com/ teaches students how to behave online, but these game experiences are somewhat lackluster. Unlike Minecraft, this is not a game that youth lose sleep playing or blame failing grades on because they stayed up too late at night playing.

It is interesting to note that videogames, and especially the ones found in arcades before home consoles, predate the Internet as we know it today and for much of their history did not pose many security threats to consumers. In the earliest days, the physical security of an arcade cabinet and its ability to consume quarters from the pockets of eager bowling alley users in the 1970s was about as complicated as it got for most consumers.

Videogame consoles made popular in the 1980s by Nintendo and Atari were also relatively risk-free endeavors for everyone involved. After the initial point of sale, there was not much information flow between game and publisher. This is in stark contrast with modern online multiplayer videogame ecosystems which often not only include both an account and password to manage, but also credit card information for in-game purchases, weekly and sometimes daily content updates, security patches, game exploits, and complex social hierarchies and cultural norms to virtually navigate.

Post-World Wide Web

Even during the 1990s with its mass proliferation of consumer electronic and first and second generations of home PCs becoming more widely available, many popular video games continued to be single-player experiences. This was in part due to the fact that the Internet just was not fast enough to enjoy playing a multiplayer videogame – the lag was too great, and the pings were too high. To get around this many videogame fans and community organizers hosted LAN parties where they instead *built* a high-speed network for a short time to facilitate gameplay.

In the 2000s and beyond, the true potential of massively multiplayer online (MMO) games began to blossom with titles like *RuneScape* and *World of Warcraft* gaining broader media attention and awareness as Internet speeds improved and 3D graphics hardware became cheaper. These games are important to consider in terms of cybersecurity because they often introduced virtual economies which relied on real-world currency to function. For the first time malicious users had a real incentive to trick or swindle other players out of their hard-earned money – with auction sites like eBay and others becoming more popular, there was a growing black market for compromised accounts and items. Video games became a thing that not only consumed cash, but, if you "played" it the right way, could generate cash as well.

For all these reasons it should be clear to see why it took so long for commercial videogames to find a foothold in classrooms, but through the popularity of MMOs and sudden rise in mobile gaming in the post-iPhone era, the need to educate youth about the risks associated with their favorite activity cannot be ignored.

The Cyberpunk Genre

While the context of gameplay has changed dramatically over the past 50 years, it is worth mentioning that throughout this entire expansion of the industry there were games that touched on security topics indirectly through the science fiction subgenre known as cyberpunk. One of the earliest examples, *Neuromancer* (Interplay 1988) was based on a novel by the same name written by William Gibson in 1984. Later examples include *Deus Ex* (Eidos Interactive 2000), *Uplink* (Introversion Software 2001), and *Watch Dogs* (Ubisoft 2014) all of which feature a style of point and click "hacking" as a game element but do little to educate players about Internet security – these are essentially role playing games which take place in a fantasy realm where technology is *intentionally* indistinguishable from magic.

Careful planning and risk remediation are not on the agenda when players can simply reload a save file. In these instances, security is most often presented as a temporary puzzle to be solved in the shortest amount of time possible; they are not training manuals or textbooks.

Darknet (McNeill 2016) is one exception to the rule. This virtual reality game still uses puzzles as a main game element, but also uses a 3D network visualization overview as its primary interface. Players have the option of using viruses, worms, and 0-day exploits to traverse increasingly difficult network topologies in a way that is closer to reality than any other example previously mentioned. Like Minecraft, the game works well in an education setting because it is an attempt to *simulate* reality rather than *amplify* reality in a way typically of the cyberpunk genre.

Minecraft in Context

Minecraft's popularity and growth over the past decade, starting as a small side project by a single software developer and exploding into a cultural icon recognized by iPad wielding toddlers many years before they even poses the fine motor skills necessary to operate the game's mobile app touch-screen interface, has allowed it to become the subject of close study across multiple disciplines (Nebel et al. 2016).

Before we proceed, it is important to note that there are two important periods in Minecraft's history to consider before engaging with past research. While the overall landscape of Minecraft in academia as it existed before Minecraft was famously acquired by Microsoft in 2014 for $2.5 billion, and the following tectonic shift in the Minecraft community this caused as a result is not as easy to see at the first glance. It is important to understand that while many gamers consider this time period before acquisition to be the "peak" of Minecraft's popularity (reflected in search tools like Google Trends and more currently, in sappy nostalgia videos created by Minecraft fans on YouTube, more on that later), this was a turning point from a technical standpoint as well.

Post-Microsoft, there was a major "cooling" period in terms of game changes and new features. Much like The Walt Disney Company acquisition of the Star Wars franchise from Lucasfilm in 2012, superfans were a bit apprehensive about the future of their favorite thing, and left to wonder if their fandom would survive. This was also an interesting period for Minecraft fans because it marked the kickoff of several data breaches in Minecraft community forums and multi-player servers (Hunt 2019) and was also the end of a popular "fork" of Minecraft for education called MinecraftEDU developed by a third-party game studio (TeacherGaming 2019).

In place of MinecraftEDU, Microsoft launched Minecraft Education Edition (MCEE) and, with that, brought integration with Office 365 for Education (which, as of 2019, still offers a free tier or service for qualifying institutions) and an online portal for lesson plans and course material. MCEE not only disabled some features found in normal Minecraft, like the ability to host a dedicated server (users would now have to manage save files themselves) but also paved the way for new in-game abilities and features to make the game more user friendly for instructors.

Public Perception: The Key Ingredient

With all of these market forces in play, any analysis of Minecraft for education would be incomplete without diving deeper into the subject (Willett 2018) but because this is not the focus of the project, I will quickly summarize my findings based on the San Francisco Bay area teen library community. It is pretty simple: tweens do not care, they play Minecraft because it is fun and easy to learn; however, as tweens age into teens, they may "grow out" of Minecraft or, perhaps more accurately, "grow toward" other videogames which feature more violent themes (e.g., Fortnite) that were previously unavailable to them due to their age or former levels of parental supervision.

In my experience, because Microsoft has invested so much time and money focusing on the youth market through merchandizing and annual events like MINECON, many teens and young adults identify Minecraft as something for kids. To put it another way, the game becomes categorically "uncool" and something "a younger brother plays." At best, Minecraft may become part of an occasional nostalgia trip in the same way people enjoy rewatching old movies. At worst, a point of ridicule toward peers with opposite interest and an extension of cultural "gatekeeping" across the gaming community. Indeed, it seems the game has become a victim of its own popularity.

Finding the Right Game for Your Library

From this, it is incredibly important to consider the fact that Minecraft, as a gaming platform, may not be the right fit for your community. In my experience as a video game enthusiast since the early 1990s, first acquainted with the Atari 2600 game library and nearly every generation of video game consoles since then, gamers are oddly territorial about not just the type of media they consume (in terms of genre: strategy, role playing, action, puzzle, etc.) but also for what *reason* you are a fan and the peer group associated with it.

From a practical standpoint, this is easy to understand in the context of face-to-face multiplayer games experienced in a computer lab setting. If everyone is already playing Roblox (https://roblox.com), one of the closest competitors to Minecraft in terms of gameplay style and intended audiences, one would have an uphill battle in persuading your classmates to switch to Minecraft. In an education setting, this is relaxed a bit because students are happy to play a video game during regularly scheduled class time, period, but for libraries the biggest challenge is getting people in the door, or perhaps more accurately, persuading the parents of teens and tweens to schedule time to attend a cybersecurity-based workshop.

What We Talk about When We Talk about Cybersecurity

One of the biggest challenge educators face in the realm of cybersecurity boils down to the inflexibility of tech jargon. Often these words had historical underpinnings or were spun off from military terminology that seem out of place in a classroom of tweens. For example, in the early days of the web when most Internet-connected computers were bigger than your refrigerator, a network "port" was something you could point at and understand its functionality simply by looking at the array of empty network plugs in front of you. One could literally see how and why communicating through the same signal interface would create confusion for a machine, and the security benefit of closing ports that were not in use. Now, the idea of a network port is an abstract concept, thanks to the miniaturization of computer parts – half the battle is about making sense of this disjointed terminology. Similarly, addressing a room of tweens about the merits of intrusion detection systems (IDS) to stop a cyberattack in its tracks becomes much easier when the security practice is decoupled from the three- and/or four-letter federal agencies that coined a term. Citing NIST does not get us any closer to understanding layered security as a learning objective when your audiences' top priority is protecting their XBOX account.

On the other hand, the same students sure as hell get why you cannot just camp anywhere overnight in Minecraft without at least putting a door and four walls between you and the swarm of monsters than intend to blow up all their virtual stuff and ruin the work-in-progress sugar cane farm they are building on the other side of the in-game volcanic island that was just discovered. Presenting the same IDS scenario in this context might surprise you and, depending on their age, might result in a "Well, duh" reaction from seasoned Minecraft players – after hundreds of hours of seeing these security scenarios played out in a virtual environment, what they lack is not experience or insight, but the conceptual framework their older peers use to describe it.

To illustrate this further, a team of researchers from multiple universities recently asked college age students to "think through" several cybersecurity-related questions in an effort to identify common misconceptions and problematic reasoning (Thompson et al. 2018). What this team of researchers found through a series of interviews (mostly) undergrads fell into four themes: overgeneralization, conflated concepts, biases (physical, user, and personal), and incorrect assumptions. This is a highly recommended read because the research team also took tremendous care in how they set up their interviews and in explaining their methodology.

Overgeneralization

This is the misapplication of an idea from one context to another, commonly happened in reference to passwords. Teens understand you should not share passwords. This is generally true, but an expert might point out that password sharing in some cases is necessary and there are additional security layers that can be put in place to mitigate the risk of password leaks. For example, a shared password that can be remotely revoked, reset, or expired after a certain amount of time is better than a permanently shared password that cannot be reset. Where a novice sees only one option or output, an expert can cite many alternatives.

In the game of Minecraft, novice players tend to overgeneralize the strength or utility of different security elements, like walls, because their builds follow a formula. In some cases, a pit with a drawbridge is easier and more effective than high walls and doors. An expert is more context aware – they can read the total environment better.

Conflated Concepts

This is similar to overgeneralizations but demonstrating a lack of deep understanding between two ideas. To state this more plainly, during focus group meetings I almost always had teens who were confused by the difference between security and privacy, and when prompted would often get the same answers for both. An expert understands the nuanced meaning both terms have, and why it is possible to have one without the other. My go-to explanation for librarians uses social media as an example: Twitter offers a secure sign in process, but the privacy of the service is much more about how you use it and the information contained in a given tweet or combined through a series of tweets. Similarly, Tor offers an anonymized IP address and greater browsing privacy but can be set up in an insecure way on a host machine. Understanding the difference between these two concepts is critical.

Biases

In cybersecurity this is often related to specific brands or tech companies. Teens might identify specific phone manufacturers as more or less secure than others (e.g., Android vs iPhone operating systems) based purely on rumor or market speculation. Expert opinions have more weight because they attempt to remove bias, or at the very least acknowledge their bias upfront. Another great example came from the research article itself, in which these researchers interviewed a subject who distrusted Internet services originating from a particular country but could not cite exactly what was at risk compared to other nation states.

Incorrect Assumptions

These are basic mischaracterizations of an attackers' intent or capabilities. In the study conducted with college students, they were often "fixed" on a single solution to a security problem statement. For example, the goal of unauthorized access to an information system is not always about stealing or exfiltrating data, it could be as simple as data destruction. In my focus group sessions, this was less prevalent among teens than the previous four themes, but this is not unusual because individuals did not get a chance to explain their answers at length – assumptions, even when incorrect, are more likely to arise in detailed explanations.

Recommendations

Of the four themes, overgeneralization and conflation were most prevalent in my sample group, often demonstrating familiarity with certain terms and security practices, but limited experience on the outer boundaries of their practical applications. By the third or fourth teen advisory group, it became very clear to me that this might be a result of how we often phrase security advice, out of necessity, into "sounds bites" that are easy to remember and digest, but lack substance. For example, it is generally true that sharing your home address with anonymous strangers online is a bad idea, but a deeper discussion about the process an attacker might use to piece together data points from different sources toward revealing or verify the accuracy of a home address does not fit on a poster at 72-point font size every October (corresponding with "national cybersecurity awareness month" in the United States).

As a result of this research, Thompson et al. generally advised embracing complex security scenarios and real-world case studies. I think this is true for college age students, but not good advice for teaching teens and tweens about cybersecurity in a public library setting. At the undergrad level and above, instructors are interested in using complex scenarios to help students essential "break free" from their preconceived ideas about how security works, but this approach requires *some* familiarity with cybersecurity concepts. Tweens have only just begun to scratch the surface of these concepts in their early years of Internet use. Recall that in theory, the Children's Online Privacy Protection Act (COPPA) means administration of *any* online account is happening only under adult supervision before the age of 13. Secondly, library workshop audiences tend to have a much wider range of technical skill levels and socioeconomic factors to consider. Knowing your community becomes much more important in the context of cybersecurity education for the public because you can anticipate a much wider range of risk factors based on how they access the Internet, not just what they are doing online. Because the range of security experience is much wider with this group, it is to our advantage to equally address both sides of the spectrum and still provide a meaningful experience. Video games make this process much easier because we can all speak a common language.

Videogame Culture

In planning a cybersecurity workshop in public spaces, it is also important to consider not only what topics should be covered, but also who you expect will attend. Previously, I alluded to this by referencing the wider range of skill level typically present in free or low-cost technology workshops versus academic settings, but market strategy goes beyond knowing your audience and should also consider competition.

Pairing cybersecurity training with commercial video games is one way to "piggyback" on market trends and research of a billion-dollar industry. In the same way a hacker performing a spear phishing campaign on a high-level executive might do extensive research on their target before crafting the perfect fake email, you might also want to begin consuming the same marketing material produced by video game software developers, over time, to pick up on trends you can also use to promote your own content. For extremely popular games like Minecraft there are annual conventions, a wide range of official merchandise (and knock-off brands), plus hours and hours of community made YouTube videos to help you understand the mindset of your target audience. The Internet is big – even "niche" games coming from independent game companies have thriving online communities you can observe or, better yet, participate in. Rather than categorizing potential customers in classically defined, socioeconomic segments of the total population like "Gen Z" or "Gen X," it might be more beneficial to think about fans of a specific videogame in terms of the mental processes involved in gameplay itself.

Ethnographic Research

In Palo Alto our local Microsoft Store hosted a global livestream event in October which was full of inside jokes and humor aimed squarely at superfans. This was a solid gold research opportunity in terms of seeing first-hand how Microsoft communicates back to the Minecraft community, but also the emphasis on video streaming and minigames within Minecraft. It was also at such event when the realization that a large portion of our market strategy for this project should address the parent or guardian of a Minecraft fan, not just youth who play the game. No one was walking to this mall on a Saturday morning to watch a livestream event at

9 AM without some extra help. Clearly this target audience was getting a ride to and from the mall (with perhaps a pitstop at Starbucks to refuel mom or dad on the way). It also seemed pretty clear that, on some level, Minecraft was recognized as being a generally good influence. "Kid tested, mother approved" as the low-sugar cereal commercial once said.

While not every videogame is as popular as Minecraft, there is undoubtedly an online community or subforum somewhere full of hundreds if not thousands of players who are willing to answer questions about the game you are thinking about using for education outreach or provide general help. Go out and talk to them! Find out what they like and do not like about the game. Though controversial because of its association with highly targeted online marketing campaigns and presidential elections, the study of psychographics versus demographics is incredibly useful in this case because we are interested in matching an education experience for a specific *type of thinker* based on a specific type of game (Godin 2018).

What about *Fortnite*?

No discussion of Minecraft in 2019 would be complete without contrasting the sandbox game with its contemporary, Fortnite. So here we are. If 2014 was the year of Minecraft, thanks to the added media coverage due to Microsoft's billion-dollar acquisition of the game, then 2018 was the year of Fortnite, with countless news articles and blogs citing the popularity of the game, simultaneously amplifying its reach through mass media while at the same time warning parents about the impact it has had in classrooms. One could argue "Peak Fortnite Panic" occurred in early December 2018 with USA Today's click-bait-style article titled '*This game is like heroin*.' *Fortnite addiction sending kids to gaming rehab* in which behavioral specialist Lorrine Marer is quoted comparing videogame addiction to substance abuse (Haller 2018). This quote from a news article in late November published in Bloomberg was later picked up by several other news publications as well, including the *Washington Examiner*, *Chicago Tribune*, *Boston Globe*, and *Forbes Magazine* (Feeley & Palmeri 2018).

Throughout this project many librarians were well aware of the rising popularity of Fortnite and ongoing competition in the hearts and minds of youth who play Minecraft. Fortnite presents several challenges in the context of education and outreach in public libraries, but to be clear, it is not off the table entirely. From a marketing perspective, the "viral" popularity of the game cannot be ignored. From a cybersecurity training perspective, Fortnite lacks several key features, like the ability to host your own server or install your own game mods, which make the "meta" elements of MetaMinecraft more interesting because through this metadiscussion an instructor can talk not only about game itself but also the surrounding software development mod community and the risks therein.

The rapid pace of Fortnite gameplay and emphasis on player versus player battle also makes it difficult to examine closely. Accomplished players have fast reflexes and good instinct, unlike Minecraft's slower paced resource finding and block by block building dynamics. Yes, both games feature fort building, and there is where much of the comparison comes from, but Minecraft builds can take hours, days, or even weeks to complete and requires careful planning and resource management, whereas an entire Fortnite match can happen in 10 minutes. Up until early December 2018, Fortnite also lacked a game mode where users could build and play in peace for as long as they wished.

One thing both games have in common and is a topic I suggest readers might want to consider exploring further is the prevalence of illegal game "hacks" or cheats in both communities. These are typically programs which run in the background and promise to grant the player special in-game abilities, like seeing through walls or automatic aiming of a weapon giving them superhuman accuracy and control. What is concerning about these applications, apart from the risk of being permanently banned from either game if caught using them (or legal action from the game developers themselves), are the instructions often paired with these programs, which sometimes suggest disabling any antivirus applications before installing or making specific changes to system files in order for the "hack" to work. What percentage of users that instead of hacking the game they want to play are instead hacking themselves, we may never know, but anecdotal evidence suggests this is a widespread means of propagating keyloggers and remote access Trojans on unsuspecting victims who are also disinclined to report the source of their computer problem because in doing so they would also reveal their intent to cheat.

Flipping the Equation: Gamified Cybersecurity

Commercial video games which not only *support* an education experience but also work as standalone games have a competitive advantage over games designed to be educational experience because potential students are already familiar with how they work and are more likely to participate or simply gain more from the experience because their favorite game is represented. This is exactly what made Minecraft offshoots like MinecraftEDU and later Education Edition popular, and in recent years more and more educators have been leveraging the power over commercial video games for this purpose. Indeed, there is now a cottage industry of education professionals who review commercial games online for this explicit purpose and categorize their potential toward different learning objectives. One such project, iThrive Games (https://ithrivegames.org)), under the direction of social psychologist Susan Rivers, PhD, features an online Curated Games Catalog which anyone can freely browse online and includes reviews split into four learning categories: empathy, curiosity, growth mindset, and kindness.

This is all wonderful news, and perhaps serves as an indicator that mindset of what makes for an effective classroom is changing rapidly. However, unlike a traditional classroom, where students are already present and are perhaps motivated to play *any* video game if their alternative is lecture based or noninteractive, there is some additional effort to get people to attend afterschool programs or weekend workshops in a public library. As a result, we need to think carefully about intrinsic (e.g., "I signed up to be here because I wanted to be here or had a desire to learn") versus extrinsic (e.g., "I'm here because I must be here or need a grade to pass") motivation. As one librarian phrased it in a PLP survey response, "what would make them want to come?"

We can shed some light on this idea by looking at two examples of gamified education experiences that very successfully did the opposite: the creation of original games which are simultaneously entertaining and educational, rather than one following the other. The first example was very recently launched by the Electronic Frontier Foundation (https://eff.org) and combines the virtual reality 360-degree panoramic view of Google Street Maps with a *Where's Waldo* approach to understanding the prevalence of surveillance cameras in urban areas, simply titled *Spot the Surveillance* (Maass 2018). The second example documented by blogger Matthew Farber for the National Public Radio KQED Mind/Shift blog describes an ARG or "alternative reality game" developed by two high school teachers called *Blind Protocol* (Farber 2018). This one is particularly interesting in the way it uniquely uses student collaboration taking place *between* different media formats rather than fixed in a specific medium or game space.

What makes these projects successful from a commercial gaming perspective is less about the content delivered and more about their process-based, trial-and-error approach to learning. *Spot the Surveillance* players are given very little instruction in terms of how to identify cameras or incentive to "win" the game. The takeaway message is clear though: cameras are everywhere, and they are not just recording traffic. In *Blind Protocol* students find a series of clues which unlock a story line, sent to them by a mysterious hacker whose intentions are unclear. They may go off course in the process because it is unstructured, but this is a feature, not a bug. The learning process is about teamwork, critical thinking skills, and how to combine disparate pieces of information to form a larger whole. These are prime examples of "learning by doing" rather than learning by hearing or seeing.

Practical Applications and Lessons Learned

Before jumping in to game-enhanced learning, it is important to consider how this compares to other eLearning modes (video tutorials, learning management software, etc.) not only from the instructor perspective, but from the student's perspective as well. One of the biggest weaknesses in Minecraft-based lesson plans is the ability of the instructor to "speak Minecraft" with students and draw parallels between the game world and reality. This, unfortunately, does not happen overnight. It means spending at least a couple hours in-game to understand *why* threat modeling fits so well. This is no different than any other science discipline, where line drawings on chalk boards are often used to represent high-level ideas on a microscopic level. Without fluency in these concepts, or having run the same experiment before, it is difficult to further illustrate those ideas with words.

From the student perspective, it is difficult to say if your lesson plan has worked because the true test is not about answering an essay prompt or filling out multiple choice questions, but will they recall the security advice or begin to apply the concepts you introduced at a later date. Clark and Mayer suggest there are five "promising features" that computer games offer over traditional eLearning modes (2016). Fully utilizing these elements are in many ways the "secret sauce" of game-based instruction.

Coaching

The instructors' ability to provide "over the shoulder" advice before or after students (or a group of students) make a move or implement a design idea can create a positive feedback loop for the entire class. Highlighting novel approaches as they happen is a great way to spark new ideas and can generate discussion topics organically.

Coaching can also take place before gameplay even begins; in the same way sports team might review a video recording of their competitor, we can review fort designs posted by the Minecraft community at large, and begin to critique their security efforts through the lens of threat modeling.

Self-Explanation

This evidence-based principle suggests that memory retention is enhanced when the "flow" of gameplay is maintained. When compared to control groups that were asked to type out their reasoning while using a video game enhanced learning module, the test groups that minimized disruption during gameplay performed better than their peers.

In this specific example, students were selecting answers from on-screen menus while in-game, which is not currently possible in Minecraft, but by designing the course in such a way that the student is given total freedom to decide not only what they are protecting in their Minecraft base, but also the complexity of their threat model and design, we help preserve autonomy and flow and their ability to self-explain.

Pretraining

It is interesting to note that this principle is already used frequently by game developers in the form of tutorials or linear introduction levels where the game introduces or demonstrates a concept, for example, that the "A" button will make your game character jump, then players have to immediately use this new information to advance to the next level.

Similarly, the process of threat modeling is greatly enhanced by the introduction of core concepts like risk, impact, and probability *before* they being hands-on in game coursework.

Modality

Through controlled experiments using educational video games, researchers have found information presented by voice rather than on screen text shows strong evidence for increased memory retention. This, coupled with the coaching principle, is a great reason to keep your class sizes small and interactive. Speaking from personal experience, one of the best aspects of teaching in a video-game-enhanced classroom is the instructor's ability to *see, translate, and share* design ideas as they happen, without asking students to break away from the activity itself.

Personalization

One of the easiest ways to leverage video games within a security education context is to pick a game that is easy to customize and speak in a conversational style. Minecraft is a great example of this because it allows players to not only build large structures block by block, but also modify the exterior environment to fit their needs. Similarly, asking students, and especially youth, to directly apply security knowledge or practice new security techniques like threat modeling to their personal life is much easier than bringing forward case studies about historic data breaches affecting millions of users.

At the same time, be aware that casually mentioning cyberthreats like online bullying and harassment are an extremely sensitive subject for those who have experienced it, especially in library workshop settings where participants are less familiar with each other. It is possible to overpersonalize, and given the norms established in most library settings around privacy and anonymity, you may want to begin classes or training sessions with a reminder to respect these boundaries.

Future Development

While Minecraft is not a new player in the world of youth and youth education, there are still many avenues for cybersecurity education to take place within the meta-framework of the game universe both online and offline. My primary goal has been to introduce the concept of threat modeling to youth using a systems design thinking methodology where students are tasked with building a fort or home base in the game and then reflecting on weaknesses in their initial design. Quite literally, they are building out a mental model mimicking security layers like firewalls and encryption in 3D space to better understand how security layers interact with one another and identifying ways to mitigate the impact of security events at these different layers. But this is only one example of how Minecraft gameplay could be leveraged to introduce cybersecurity concepts to younger audiences. In the process of all this, I have found a few more areas that could be generalized to fit newer games.

Social engineering is extremely common in most MMO games, ranging from wide scale market manipulation effecting the game world economy to a wide range of scams that occur inside and outside of the game. With MMOs it is not uncommon for high-level accounts to be bought and sold on auction sites, which incentives phishing scams and other attacks meant to compromise an account.

Network security principles and common threats could be demonstrated through simulated attacks on a game server. Why care about distributed denial of services (DDOS)? This is very easy to understand when a DDOS attack is directed at a server the class is using or attempting to use. There are several teachable moments here.

Cryptography could be explained in the context of multiplayer chat and demonstrated through rating the various means of communication players have available. Is it more secure to send encrypted messages over in-game chat, or to sit next to the person you are playing the game with and talk in the open? The answer really depends on context, and often there is no perfect solution. Exploring the pros and cons of end-to-end encryption versus offline communication requires a nuanced understanding of both, which are easier to explore in the context of game play rather than spy thrillers and trade craft we often see in action adventure films and spy thrillers starring James Bond.

Risk management could also be expanded to go beyond just the process of threat modeling and more toward the real-world application of business resources to solve or mitigate security problems. There are many real-time strategy (RTS) games which task players with collecting, managing, and distributing resources based on environmental factors which could be utilized to illustrate these principles. RTS players already use these skills in rapid succession, and "anchoring" their skills in a different field like cybersecurity could open up whole new career pathways.

Man in the middle attacks (MITM) are similar to how many computer memory or game-data editor-based cheats function, but on the network scale versus hardware scale. Instead of intercepting communication between computers or end nodes, these cheats demonstrate the power granted by editing data in real time, tricking the system into doing something it was not intended to do by delivering tampered information.

Conclusion

While it is a bit awkward to expound on a work-in-progress effort, my hope is that librarians, educators, security researchers, and game designers will all benefit from reading about what was learned over the last 9 months of research and development for this project, and perhaps inspire more people to try new styles of game-based education and "intentional play" activities in a range of public spaces, and not just library computer labs.

In this chapter we have explored past and present examples of how Minecraft, a wildly open-ended and popular commercial videogame, has been used to support teachers and classrooms across the globe. We also looked at one proposed use of its survival mode game mechanics to bring cybersecurity knowledge and security concepts to new audiences. Through interviews with teen librarians and teen focus groups in libraries, validated by a similar "talk aloud" exercise and research projects across multiple college campuses, four common areas of confusion or misconception around cybersecurity were also identified.

Different marketing strategies to support commercially available video games as an extension of ongoing information literacy education and outreach already taking place in public libraries were also discussed; however, many of these theories remain untested as the experimental project which launched this research moves forward. We also saw two best practice examples of how unique game experiences, as part of a process-based mode of inquiry, can begin to engage with cybersecurity topics and themes through the *Blind Protocol* classroom experience and *Spot the Surveillance*.

While the Minecraft project in question focused on threat modeling as the primarily learning objective, five additional topic areas in cybersecurity were also identified as areas showing major overlap with the meta-framework of Minecraft and other online multiplayer games worthy of future consideration from information professionals. At the same time, five general features or enhancements provided by video game-based learning were reviewed, in addition to how they could be applied to other security-focused classrooms.

References

Clark, Ruth C., and Richard E. Mayer. *E-Learning and the Science of Instruction: Proven Guidelines for Consumers and Designers of Multimedia Learning*. 4 edition. Hoboken, NJ: Wiley, 2016.

CSLA. "Advocacy Talking Points." CSLA—California School Library Association. Accessed January 13, 2019. http://csla.net/csla-advocacy-information/legislation-and-advocacy/.

Deus Ex. *PC Video Game*. Eidos Interactive, 2000.

Farber, Matthew. "How Data Privacy Lessons in Alternative Reality Games Can Help Kids in Real Life | MindShift | KQED News." 2018. www.kqed.org/mindshift/51772/how-data-privacy-lessons-in-alternative-reality-games-can-help-kids-in-real-life.

Feeley, Jef, and Christopher Palmeri. "Fortnite Addiction Is Forcing Kids into Video-Game Rehab." Accessed November 27, 2018. www.bloomberg.com/news/articles/2018-11-27/fortnite-addiction-prompts-parents-to-turn-to-video-game-rehab.

Godin, Seth. *This Is Marketing*. New York: Portfolio/Penguin, 2018.

Haller, Sonja. "'This Game Is like Heroin:' Fortnite Addiction Sending Kids to Gaming Rehab." USA TODAY. Accessed March 4, 2019. www.usatoday.com/story/life/allthe moms/2018/12/09/fortnite-addiction-sending-kids-gaming-rehab/2221149002/.

Hunt, Troy. "Have I Been Pwned: Check if Your Email Has Been Compromised in a Data Breach." 2019. https://haveibeenpwned.com/.

Maass, Dave. *EFF's 'Spot the Surveillance' Virtual Reality Project at the Cato Surveillance Conference*. 2018. Accessed January 14, 2019. www.youtube.com/watch?v=Q0ueYcGDO-s.

Macrina, Alison. "Library Freedom Project—Making Real the Promise of Intellectual Freedom in Libraries." 2015. https://libraryfreedomproject.org/.

Markman, Christopher. "Fast-Forward: Minecraft for Security Education." 2016. https://er.educause.edu/blogs/2016/5/fast-forward-minecraft-for-security-education.

McNeill, E. *Darknet. Oculus Go Virtual Reality Video Game*. 2016.

Nebel, Steve, Sascha Schneider, and Günter Daniel Rey. "Mining Learning and Crafting Scientific Experiments: A Literature Review on the Use of Minecraft in Education and Research." *Educational Technology & Society* 19, no. 2 (2016): 355–66.

Neuromancer. *PC Video Game*. Interplay, 1988.

TeacherGaming. "TeacherGaming—Teach Anything with Games." TeacherGaming. 2019. www.teachergaming.com/.

Thompson, Julia, Geoffrey Herman, Travis Scheponik, Linda Oliva, Alan Sherman, Ennis Golaszewski, Dhananjay Phatak, and Kostantinos Patsourakos. "Student Misconceptions about Cybersecurity Concepts: Analysis of Think-Aloud Interviews." *Journal of Cybersecurity Education, Research and Practice* 2018, no. 1 (July 17, 2018). https://digitalcommons.kennesaw.edu/jcerp/vol2018/iss1/5.

Uplink. *PC Video Game*. Introversion Software, 2001.

Watch Dogs. *PC and Home Console Video Game*. Ubisoft, 2014.

White, Jarred. *B04 Threat Modeling the Minecraft Way Jarred White*. BSides Nashville 2016. 2016. www.youtube.com/watch?v=9qs2uG6weHw.

Willett, Rebekah. "Microsoft Bought Minecraft … Who Knows What's Going to Happen?!': A Sociocultural Analysis of 8–9-Year-Olds' Understanding of Commercial Online Gaming Industries." *Learning, Media and Technology* 43, no. 1 (2018): 101–16. doi:10.1080/17439884.2016.1194296.

Chapter 6

Information Governance and Cybersecurity: Framework for Securing and Managing Information Effectively and Ethically

Dr Elizabeth Lomas

Associate Professor in Information Governance, University College London

Introduction

Over the last 30 years the value of information/data has increased as new technologies have made it more accessible, enabled it to be reused, and to add new uses, for example, through linked data, aggregated data, or big data. As such data has become increasingly commoditized with greater recognition of a range of information value(s). New technologies have created new forms of digital assets. An example is blockchain which has underpinned developments in cryptocurrency such as bitcoin. Another example is personal data, which in 2011 the World Economic Forum defined as a new asset class that it predicted will increasingly spur a host of new personalized services and applications with incredible velocity and global reach.

In order to take advantage of new information possibilities provided by technology, including the capacity for workers to connect 24/7, and new communication channels with enhanced audience reach, organizations have moved from a world in which they have been able to control information within internal boundaries to one in which the organizational boundaries are permanently perforated. New forms of data storage distribute and manage data in different ways, for example, the Cloud. In addition, there are new demands and expectations for organizations to interface and actively interact or even cocreate information with external stakeholders. Moreover, the digital world connects to and manages the physical world creating the "Internet of Things." Information now acts as the latest form of oil driving economies and societal living requirements.

In tandem with the growing potential, value, and social significance of information, the online management of a wide range of data has opened up information/data to new forms of theft and attack. New channels and the proliferation of information have led to new types of misinformation and subversion. The threat and reality of cybercrime and cyber warfare has significantly increased (Arquilla, 2012). A 2018 report by the security company Norton reported that in the year of 2017 alone, 44% of consumers were impacted by cybercrime. At an international level we see increasing reports that cyber warfare is ongoing. Denardis (2014) highlights the alleged 2010 use of the USA/Israeli Government to undermine the Iranian nuclear programs through the deployment of the Stuxnet Worm and the Russian Denial of Service attacks on the Estonian Government in 2007. Such attacks can cause both national reputational damage and tangible impacts. It has moved the focus of Internet/World Wide Web, telecommunication infrastructures, and mobile usage into an arena of open international dispute. For example, in 2018 the USA National Defense Authorization Act resulted in the Chinese company Huawei being banned from the 5G networks due to concerns over spying. Huawei are bringing a legal case to attempt to overturn this decision. This further evidences the complexity of international information and infrastructure control.

To deal with these complex information world dynamics, it is the field of information governance (IG) which has emerged as the solution for individuals, organizations, and nations (Lomas 2010; MacLennan 2014; Smallwood 2014) to manage information. IG provides the solution to dealing with information challenges in terms of both opportunities and dangers that are implicit within information creation. It deals with the management of information assets legally and ethically as well as providing guarantees around information confidentiality, integrity, and availability through time. It raises information to government and board level as an area for regulation, oversight, and active strategic management. IG is multidisciplinary drawing on a range of expertise to deliver information agendas. Embedded within IG are other governance components which form smaller parts of the IG framework delivery. These include, but are not limited to, cybersecurity and governance, computer/IT governance, data governance, information assurance, and Internet governance. Importantly IG provides frameworks that align people, processes, and technology in accordance with the law and best practice.

The Information Governance Context and Its Relationship with Cybersecurity: A Historical Perspective

IG has grown out of corporate governance thinking which has been legislated and regulated for and applied across public and private sector settings. Governance provides for governing, controlling, and regulating good order to deliver societal values including protection. As such, the term governance has developed to require a system of leadership and management that balances societal goals and is in essence "ethical." The system of governance may be applied to a nation, business, charity, or some other body. In 2000, a leading proponent of corporate governance, Sir Adrian Cadbury, described the complex balance which corporate governance should deliver in terms of providing for the delivery of economic and social goals which consider individual and communal needs. He stated that the aim of governance in this context is to align the interests of individuals, corporations, and society as nearly as possible (Cadbury 2000). This therefore includes the organizational management of relationships across organizational boundaries to sustainably deliver employment and prosperity while promoting, integrity, openness, value, and diversity for a wide range of interests (Financial Reporting Council 2018).

As such, governance is not a fixed concept but is dependent upon the ethical values of society and the governments in place, which inform and dictate the format for leadership, societal accountability, and trust. Within this context of corporate governance, information plays a critical dynamic role. As noted by Willis, information delivers: transparency, accountability, due process, compliance, the delivery of statutory and common law requirements, stewardship, systems and processes, and security of personal and corporate information (Willis 2005, 86–87). As technologies have advanced, the role of information within governance agendas has expanded and over time becoming a distinct activity particularly as information as an asset has been better recognized (Lomas 2010). In 2010, Deborah Logan (2010) wrote a Gartner blog post defining IG as the specification of decision rights and an accountability framework to deliver "desirable behaviors." She wrote the blog under an article titled, "What is information governance and why is it so hard?" Clearly to provide a holistic framework is not simple and the scale of this endeavor is not to be underestimated in terms of the support and resources required. In 2012, Barclay

T. Blair, a founder member of the Information Governance Initiative think tank, defining IG, again emphasized this complexity in terms of the need to deal with information through comprehensive IG programs which deliver the value of information assets while minimizing risk and cost (Blair 2012).

In part, the complexity of the IG endeavor relates to the wide-ranging nature of information assets which may be marketable resources but in addition represent something more. As Desouza (2009, 35) states, resources can be traded being purchased and sold in the marketplace but in contrast assets are things that organizations care deeply about. Such assets have a more strategic and complex set of values. In essence, information can be a product or service; it can deliver influence, a competitive edge, education, enrichment, and entertainment. It can have a monetary sales value and/or a cost to recreate it. However, it can have other wider societal values in terms of national, organizational, personal, or cultural memory and identity. It can be something to share or keep private. It is important to note the complexity and multiple realities of information value to different stakeholders through time. Reliable authentic information, delivered by systems with integrity, develop trust, accountability, and the potential for democracy and/or open systems of government. In this context, cybersecurity helps provide protection and strategies for authorized access to information. IG provides a wider vision of information needs.

Today IG has evolved to straddle four key domains, each of which is underpinned by risk management processes: information economics recognizing the value(s) of information assets, information laws and ethics, information management, and information security (which extends to cybersecurity and other information security including the management of paper records). IG balances stakeholder needs to provide access and information use in addition to protection. In terms of managing information, IG is the framework of choice as the broadest and most comprehensive. As noted by Eugen and Petruț (2018), IG encompasses data governance and IT governance. ENISA set out that cybersecurity is one aspect of a bigger governance delivery picture (2015, 12), with cybersecurity focusing on the specific protection required for information rather than a wider information picture which is required for organizational and national level delivery more generally. Cybersecurity does consider national and international safety but does not address citizen requirements in terms of other information needs. IG requires holistic thinking, experts, and collaboration to ensure successful information delivery for society. As defined by Lomas et al. (2019, 4), IG provides a holistic ethical framework, "which takes into account a range of societal and individual stakeholder information needs. It enables a just process of information co-creation, sharing, management, ownership and rights." In line with social justice concepts, all can be invested in the IG system. It takes into consideration individual, family, community, organizational, and societal needs supported by practitioner experts but in addition citizens more generally.

National Laws and Ethical Expectations for Managing and Protecting Information

Key in managing IG is ethical delivery and compliance with law and local expectations. Where there is ethical consensus there is law. However, there is limited consensus and few international information rights laws exist, even though we live in a world where information delivered through technology transcends international boundaries. There are differing national expectations for information ownership, publication, defamation, libel, sedition, computer misuse, confidentiality, privacy, and personal data. In a seminal article, Mason (1986, 5) sets out four key areas of ethical dispute with key questions of contention as summarized in Table 6.1.

These issues have been widely debated. The complexities surrounding these issues exist because it is not always possible to balance these differing dimensions and there are very different national and cultural perspectives. Van Den Hoven (2008, 52–57) discusses the different approaches to understanding the practice of

Table 6.1   Mason's Four Ethical Issues of the Information Age with Associated Questions

Area of Contention	Questions
Accuracy	■ Who is responsible for the authenticity, fidelity, and accuracy of information? ■ Similarly, who is to be held accountable for errors in information and how is the injured party to be made whole?
Property	■ Who owns information? ■ What are the just and fair prices for its exchange? ■ Who owns the channels, especially the airways, through which information is transmitted? ■ How should access to this scarce resource be allocated?
Accessibility	■ What information does a person or an organization have a right or a privilege to obtain, under what conditions, and with what safeguards?
Privacy	■ What information should one be required to divulge about one's self to others? Under what conditions and with what safeguards? What information should one be able to keep strictly to one's self?

applying ethical principles. Generalists see the possibility for there to be agreed overarching ethical principles while particularists see the importance of specific contextual circumstances. Reflective equilibrium moves back and forth between these perspectives to reach a balanced perspective which is potentially more suited to a complex world.

The closest thing to an agreed global or "generalist" agenda on ethical IG is established in the *Universal Declaration of Human Rights* which was adopted by the United Nationals General Assembly in 1948 in order to set out a global agenda for fundamental human rights. While setting out a global moral agenda one can see the need for a reflective equilibrium approach to apply these into practice. Article 12 and 17 set out rights in terms of privacy and property. Information property rights have emerged and developed through patents, trademarks, and copyright laws which over the past century have evolved into relatively agreed international frameworks. The World Intellectual Property Organization has provided a focal point for such discussions. However, in 2019, the EU has moved to protect intellectual ownership to a far greater degree than other Western counterparts. The 2019 Copyright Directive places increased responsibilities on social media platforms to regulate their content and take responsibilities for copyright infringements. Article 13 has been termed the "meme ban," which makes online platforms responsible for removing copyrighted content. Article 11 delivers what has been termed a "link tax," preventing news outlets reproducing content. Article 12 of the Copyright Directive prevents filming and sharing at events such as sports venues.

In terms of the Human Rights Declaration, Article 12 sets, "No one shall be subjected to arbitrary interference with his privacy, family, home or correspondence, nor to attacks upon his honour and reputation. Everyone has the right to the protection of the law against such interference or attacks." This thus enshrines privacy principles. Building on human rights legislation in 1980, the Organisation for Economic Co-operation and Development (OECD) passed the *OECD Guidelines on the Protection of Privacy*. This established key principles for managing personal data or "data protection." Within the *Universal Declaration of Human Rights*, Article 20 asserts a further information related right that, "Everyone has the right to freedom of opinion and expression; this right includes the right to hold opinions without interference and to seek, receive and impart information and ideas through any media and regardless of frontiers."

While not necessarily contradictory, the boundaries between these rights have not been consistently interpreted at a global level. Within Europe, legislation has placed an emphasis on strengthening privacy and personal data rights. In 1995 the European Union (EU) passed a Data Protection Directive (Directive 95/46/EC). This Directive regulated for a minimum standard for managing personal data across the 28 EU member states and those additional nations within the

European Economic Area (Iceland, Liechtenstein, and Norway). In 2016, the EU passed the General Data Protection Regulation (GDPR) with even stronger requirements for managing personal data (European Parliament and Council of the European Union 2016). This came into force in May 2018. It provides strict requirements for managing European citizens data even if the service is provided by a non-EU entity. The fines for personal data failings are significant and can cost an organization up to 20 million Euros or 4% of turnover whichever is greater. There are six key principles at the heart of the Regulation which require organizations to build in "privacy by design" when developing any system with personal data elements. This concept was first advocated for in the 1990s by the Ontario Privacy Commissioner, Ann Cavoukian. As such organizations are encouraged to undertake privacy impact assessments. This approach allows for the legislation to remain relevant as the needs for managing cybersecurity evolve. Under the terms of the legislation, personal data must be protected. The full range of protections are not defined but the privacy impact assessment requires that threats are identified and reviewed as new security dangers emerge. The recommended standard to build compliance in this regard is the International Standards Organization's *ISO 27000* standard series which aligns IG and cybersecurity considerations. If personal data is breached, then, in the GDPR context, the relevant EU regulatory authorities must be notified within 72 hours and penalties may be applied. Those individuals impacted should also be informed and the risks of the breach explained.

This legislation has evolved out of a recognition that personal data has become both a valuable asset which provides revenue streams, not least for harvesting marketing information, as well as a resource capable of costing an organization where it is not properly managed. In 2018, a number of high-profile companies received fines for data breaches and data misuse, for example, the two USA corporations Facebook and Uber. While these laws are sensible and pragmatic in principle, they are not necessarily easy to deliver into practice when considering competing personal data rights and demands for wide ranging information use. Businesses have continued to push the boundaries. For example, Amazon's Alexa records all conversations within its range regardless of whether the device is being actively used. As discussed by Day et al. (2019), the justification for this is that then Amazon staff can use the data to improve speech recognition. However, the information on this functionality and the ability to change it within Alexa privacy settings is limited. These instances happen because of the increasing possibilities of technology and the reality that the emphasis on protecting personal data is not interpreted uniformly at a global level. Within the USA context there has been a greater emphasis on fundamental freedoms of speech in contrast to privacy. This aligns the Human *Universal Declaration of Human Rights* Article 20 with the USA Constitution wherein freedom of speech is the First Amendment.

While there may be a need to protect personal data, the balance of this is contested across nations. The USA enacted freedom of information legislation in 1966 and as such was an early proponent of providing citizen access to public sector information. Sweden has been a pioneer of freedom of speech and the press with censorship abolished as long ago as 1766. It is therefore not a coincidence that WikiLeaks, which campaigns for open data, is based in Sweden. Ironically, the United States has been a significant target for WikiLeaks attacks. More generally, open data campaigners have called for Open Government Manifestoes with data more automatically made publicly available.

Globally national governments have legislated for different approaches to privacy and freedom of information particularly in respect of the parameters of the work of the security services. In 2013, Edward Snowden, who had been a USA Government employee within the Central Intelligence Agency, leaked classified information from the National Security Agency (NSA). The leaked information revealed a significant level of surveillance across citizens' digital lives including their usage of cell phones, social media such as Facebook, and other software such as Skype. The intelligence gathered was deemed to provide a "pattern of life" which provided a detailed profile of individuals and their networks of association. The UK's intelligence service was implicated in the surveillance. The targets of the surveillance were not limited to USA and UK citizens. The United States argued that all surveillance was in accordance with USA law and certainly legislation such as the Patriot Act 2001 allows far reaching powers to be exercised to allegedly keep the USA safe from terrorist attack. Snowden is seen by some as a traitor, given that it is argued the intelligence services do need to operate in secrecy to be successful and keep the nation safe. However, others see Snowden as a valiant whistleblower and freedom fighter, as his leaks have been claimed to expose a significant level of snooping on all citizens. Following this incident, the German Chancellor Angela Merkel famously condemned the USA and UK surveillance stating in response to the incident that there should be "no spying among friends." The USA Patriot Act has been challenged in other ways. In 2005 the so-called Connecticut Four (four librarians) filed a lawsuit Doe v Gonzales to challenge the powers of the Federal Bureau Agency under the Act, which was claimed to provide for access to libraries' patron reading records. In essence, this debate was about whether an individual should be judged and in part tried based upon what they read. In 2006 the USA Government gave up this battle and in 2007 the "Connecticut Four" were honored for their stance by the American Library Association.

The expectations for state intervention in overseeing citizens' lives through monitoring of their digital data are highly contested; it is at the heart of the moral agenda delivered by IG in terms of ensuring balanced information delivery. Recently there has been international criticism of China by freedom campaigners regarding the so-called Sesame credit which scores citizens for their online behaviors including providing points for not only personal behavior but the behaviors of those within a citizen's digital networks. Good behaviors, such as the purchase of Chinese goods or online educational study, may be rewarded, while bad behaviors, such as online gaming, may be punished, for example, through either travel rewards or travel bans.

The Chinese Government does take a differing stance on some aspects of state control. As cited by Zeng et al. (2017), the Chinese President Xi Jinping's address to the Beijing sponsored World Internet Conference in Wuzhen in 2015 indicated China's position that the Internet should be governed according to the same principles as other fields of international relations whereby Internet sovereignty is provided for and respected. As such nations, would in accordance with this, control and regulate their own cyberspace. In a world where cyber warfare presents real challenges, the ability to manage boundaries in cyberspace may become more accepted. Cybersecurity relies on national values for determining the protections and processes put in place around different types of information as opposed to opening up and creating trust across boundaries.

As technologies have advance, there have been new areas of contention. The ability for humans to understand and account for new technologies is complex. With the advancement of robotics, autonomous vehicles, machine learning, and artificial intelligence (AI), who holds responsibility for the actions of technology is being gradually defined in law. In the context of these technologies the role, decisions, and accountability in terms of human interventions are being further worked through. As new forms of technology emerge with biological components and increasingly sophisticated systems, the boundaries between human rights and "personhood" are a further area of contest. In 2017, the EU Legal Affairs Committee argued for the potential for AI and robots to have the status of personhood. However, this concept is as much about assigning responsibility away from individuals. Limited companies have legal personhood with legal responsibility. Nevertheless, there are boards of people with culpability for decisions if not financial payment. We are still working out:

1. What is a person and what are the rights that assign to "personhood"?
2. What are the responsibilities that assign to "personhood"?
3. Where are the boundaries and laws required between human and machine?
4. How are robots and AI understood and accountable?
5. How can/should these technologies be deployed?

In the latter context we see already the debates around surveillance but also with regard to AI predictive technologies and what they should be allowed to calculate or assume.

The link between information rights/data law link to education, consumerism, and networks is a contested ground globally. IG seeks to underpin and enforce good ethical information behaviors but in part relies on legislation as the ultimate boundaries for delivering these frameworks. Where moral behaviors are agreed, IG provides for whistleblowing to call out bad behaviors, and in some instances, this has changed engrained national norms. However, the complexity of navigating information boundaries is not insignificant.

Information Governance Frameworks

There is no one single approach to delivering IG. There are a wide range of frameworks that have been developed to put in place systems for managing, protecting, and leveraging information value through IG frameworks. The ARMA International's *Information Governance Maturity Model* developed in 2010 established eight key principles against which to measure IG delivery within an organizational context which include accountability, transparency, integrity, protection, compliance, availability, retention, and disposition (ARMA International, 2010). These principles have been largely developed from records and information management paradigms and while providing a strong internal framework, they nevertheless potentially require some development to take account of managing information across complex boundaries. A critical component within the framework is the delivery of retention/disposition schedules. In a cybersecurity context it is important not to retain redundant information which might pose risks to an individual or organization if accessed by an unauthorized party. Equally key data, as established under a retention schedule, must be protected to ensure its continued availability and integrity through time to authorized parties. Similarly there is a need to ensure information remains available. Cyberattacks have sought new ways to cause damage and unethical profit. For example, ransomware attacks take control of a system and deny access to core data by organizations or individuals subject to the payment of a ransom. Bitcoin has enabled ransoms to be paid with minimal potential for the payment to be tracked. The proliferation of new forms of attack continues each year. In this regard it is the International Standard's Organization's family of information security standards *ISO 27000* (see www.27000.org/) that builds a strong framework aligning IG and cybersecurity.

ISO 27000 is legislated for as the recommended standard for information security best practice, for example, to deliver personal data security under the EU's GDPR and to meet the requirements of the USA's Federal Information Security Management Act. The key requirements of the standard series are to deliver an "Information Security Management System (ISMS)" which provides for information as an asset to be managed to deliver:

- protection of organizational assets, ensuring both their ongoing availability for business purposes and their protection against unauthorized access;
- privacy of personal information;
- maintenance of intellectual property rights.

Critically it delivers (ISO 2012, 2 and 5):

- Confidentiality: The property that information is not made available or disclosed to unauthorized individuals, entities, or processes.
- Integrity: The property of safeguarding the accuracy and completeness of assets.
- Availability: The property of being accessible and usable upon demand by an authorized entity.

These defined characteristics are not synonymous and may sometimes need to be balanced with choices made through risk management decision processes. However, importantly they provide for the delivery of a system which is not dealing with a narrow definition of security which locks down information but rather one which delivers information in accordance with organizational/national needs. The framework is established through (ISO 2013a):

- a policy, objectives, and activities that reflect business objectives and can include whistleblowing processes;
- asset classification and control;
- physical and environmental security;
- personnel security;
- an approach and framework to implementing, maintaining, monitoring (including incident reporting systems), improving systems consistent with the organizational culture;
- systems development and maintenance protocols;
- business continuity management;
- legal compliance frameworks;
- visible support and commitment from all levels of management;
- effective marketing of the requirements to all managers, employees, and other parties to achieve awareness;
- distribution of guidance on policy and standards to all managers;
- provisions to fund key activities;
- provision of appropriate awareness, training, and education;
- implementation of a measurement system to evaluate performance and feedback suggestions for improvement.

It aligns to the ARMA International IG principles but places the delivery of a potentially wider scoped risk management framework. Some of the most effective implementations of an *ISO 27000* system may be very simple systems as it is argued these can enable individuals to engage with, understand, remember, and implement the system requirements. Key to successful IG delivery is human engagement with the values delivered by the frameworks.

The standard requires than an organization understands what information assets it holds and then ascertains the value of these assets. The starting point for evolving the ISMS is the information asset register which records the asset, its value, location, and owner in terms of assigning responsibility for the asset. Information assets are wide ranging and include the knowledge that individuals hold if it is significantly valuable and the future potential lack of availability of that knowledge presents a risk to organizational processes. In essence, *ISO 27000* links knowledge management concepts (which focus on human knowledge of organizational value) and records and information management. Furthermore, any other key components that are part of the delivery of information assets value must be listed, including software suppliers, systems hardware, and other non-technological information components.

Risk Assessment and Treatment

Risk assessment and risk frameworks are a mandatory part of the *ISO 27000* framework. The risk methodology requires the development of the information asset register to ensure that the threats and vulnerabilities for each information asset's confidentiality, integrity, and availability are managed. Risk assessment should involve the identification of information opportunities in addition to potential negative consequences from system failures. Thus, Cloud computing services provided by a third party may result in some loss of organizational control. However, working with a Cloud service that operates at a larger scale may provide additional expertise to mitigate against new and emerging software threats/vulnerabilities. Key to decision making in terms of the whole infrastructure deployed is to understand the value(s) of information and the competing requirements and threats/vulnerabilities.

The overarching risk management process requires the understanding of strategic objectives, the establishment of risk appetite (i.e., the level of risk exposure which is acceptable), risk assessment, analysis and evaluation, risk reporting, decisions and treatment, and then ongoing monitoring and review. In the simplest of systems, the information asset is provided with a value which aligns to the scored impact should the information be compromised in any way. In the context of *ISO 27000* compromise may mean that the information is:

■ disclosed to unauthorized parties, for example, if the information is stolen for the purposes of identity theft or other forms of cybercrime;

- unavailable, for example, through loss, data corruption, or as occurs in the instance of a ransomware attack;
- no longer trustworthy, for example, if a system has been tampered with.

In terms of valuing information, it is important to note that the value of information may not be static. Some information relies on immediacy, for example, information relating to financial markets may be highly significant at a very particular point in time. Other information accrues value. In the context of big data, data gains value by virtue of the scale of information held. The same piece of information or data may have multiple significances to different stakeholders. Against each asset, any threats/vulnerabilities relating to the asset are identified, described, and estimated. A threat itself cannot be managed but the vulnerabilities the threat can exploit can be dealt with. As risk seeks to mitigate the "effect of uncertainty on objectives" (ISO 2018) organizations will then determine an approach to risk. The risk exposure is calculated by multiplying the asset value/impact by the likelihood of the vulnerability being exploited.

In addition, to assessing the organizational and legislative context, a number of models can help with understanding risks. One such model is the STEEPLE model. This has seven factors which provide domains for considering risks. The seven factors are (Lomas and McLeod 2017) sociocultural factors (S), technological factors (T), economic factors (E), environmental impacts (E), political factors (P), legal factors (L), and ethical factors (E). These enable information values to be considered in a diverse way and to take into account risks including social agendas more widely. For example, the factors take into account environmental considerations considering the power resource implications of managing information through time. One case is that of bitcoin. Alex de Vries, a bitcoin specialist at PwC, has estimated that the servers required to run bitcoin consume almost as much power as that taken to run Ireland (De Vries 2018). Equally many individuals now have multiple devices running and consuming power for a wide range of nonessential uses. As such IG does encompass environmental considerations which may score differently in terms of an impact as opposed to economic impacts.

While the approach of quantifying risk by multiplying the asset value/impact by the likelihood of the vulnerability being exploited is one of the more common approaches, it is not the only one. Another approach commonly used in an information and communication technology (ICT) context relates to analyzing the single loss expectancy (SLE) for a single event which is calculated by the asset value multiplied by the likelihood or exposure factor. An annualized loss expectancy can then be calculated by multiplying the annual rate of occurrence by the SLE. This can be a useful approach in an ICT context where, as an example, power outages may influence ICT service delivery on multiple occasions or with regard to cybersecurity where there may be numerous cyberattacks.

Where there is legislation in place, the organization must act to manage and mitigate vulnerabilities. However, in many instances there will be a balance to be struck in terms of the action required. There is often no one right response to the risk choices to be made, as to share information will have benefits but may mean opening up information to some additional security risks. An organization may have an agreed risk appetite that will be critical to evaluating and implementing the appropriate risk approaches. The risk appetite can simply be set at a defined level whereby the risk score, if above the defined risk appetite, must be dealt with. This is typically dispensed with by terminating the process and thus erasing the risk, treating the risk through applying controls, or transferring the risk to another party. In the latter context it is important to note that not all risks can be transferred. For example, an EU organization that captures personal data can outsource the management of that personal data, but nevertheless it retains data protection responsibilities under EU data protection laws. In considering risk in an opportunistic context there will be certain calculations where the risk must be taken to leverage an advantage. In addition, risk can be tolerated. It remains impossible to negate all information vulnerabilities if the range of information values are to be leveraged.

Societal expectations for risk in different national and organizational contexts do differ. The financial sector is highly regulated to ensure confidence and consumer protection within the financial system. The public sector contains large quantities of diverse personal data sometimes sensitive in nature. There is an expectation, in this context, that the information will be protected and that public-sector organizations will be accountable and transparent regarding their actions. For example, the retail sector now often has large amounts of customer data which need to be managed to provide customer assurance and maintain reputational confidence. Within this context information is key for managing supply chains.

Within the *ISO 27000* standard are a list of 114 controls divided between process controls such as policies and procedures, physical controls, technical controls, legal and regulatory controls, and human controls including HR processes, education, and training (ISO 2013b). Organizations can also adopt their own additional controls. Controls can be preventive, detective, or corrective. It is a requirement of the standard to create a "statement of applicability." This is defined as a "documented statement describing the control objectives and controls that are relevant and applicable to the organization's ISMS" (ISO 2013b). Where a control is not selected, then it is necessary to justify within the statement of applicability why the system does not require that control. This process helps put in place a structure of linked information security responsibilities, for example, ICT will be responsible for network access controls and operating system access controls, HR for all employment recruitment and contracts including vetting, undertakings of confidentiality, and so on. While individual controls may fall within pre-existing frameworks, some will require new partnerships and the program of reviewing these controls therefore builds an information management framework of responsibilities (Lomas 2010). This is critical for IG and cybersecurity successful delivery. For example, HR may decide on homeworking policies, but these will also rely on ICT to facilitate access to online working spaces that do not compromise other network security considerations. As cyberattacks increase, it is important that training occurs with appropriate penalties for noncompliance. A bank may dismiss an employee for clicking on a link in an email that may contain a virus or responding to a phishing attack whereas a university is not likely to take such severe action. ICT will identify the risks and work with HR on training employees but HR will determine any dismissal procedures.

Organizations must have approaches to handling incidents, including accidents, as well as full-scale crises including internal and external attacks. Business continuity planning provides for pre-empting such situations. One of the listed controls relates to change management. In this regard, *ISO 27000* requires that information systems continue to be continuously monitored, through the cycle of planning, checking, doing, acting, and monitoring to make sure all aspects remain up-to-date and appropriate. All information security incidents, including near misses, must be recorded, reviewed, assessed, and new processes established as appropriate.

Other codes also assist with defining specific data governance and ICT security and management requirements. Examples include *DoD 5015.2* (Department of Defense 2002) as a specific security sets of measures, *COBIT 5* for audit approaches (Information Systems Audit and Control Association 2012), and *ITIL* for maintenance purposes (Office of Government Commerce 2011). In a technology context, vulnerability disclosure is required to enable organizations to identify and patch weaknesses within ICT software and hardware that can be exploited by cybercriminals. A recent report by ENISA (2018) sets out the different actors within a vulnerability disclosure process and the significant role of economic considerations and incentives that may influence their behavior. The report concludes that it is often due to economic protection that some vulnerabilities are disclosed responsibly while others are not. As such it is important for governments to hold software and hardware suppliers to account for such notifications and for the complex network of information relationships to be understood to provide IG.

Too simplistic risk profiling has increasingly been called into question (Gilb 2005; Lomas 2010). In the wake of the 2008 banking crisis, risk profiling was shaken to its roots. The complexity of managing risk and better understanding networked risks was given greater recognition. This is further recognized in the context of modelling the management of pandemics. This involves more sophisticated approaches to planning and considering a range of scenarios and future outcomes.

Risk management needs to take on board new realities of information creation and thus management. As information growth and variety expands, the ethics and realities of information value, ownership, and placement across legislative regimes need to be negotiated and risk assessed. *ISO 27000*'s central focus on frameworks with risk management processes at their very heart provide one tool which can assist. Nonetheless and critically is international cooperation and collaboration at a government and citizen level.

Information Governance Subcomponents

As noted, IG provides an overarching framework but within it there are component parts of the "technical" governance delivery which are sometimes isolated to focus on a specific aspect of the IG delivery. These include, but are not limited to, data governance, information assurance, cybersecurity and governance, information security, Internet governance, and IT governance. Where governance is aligned, it implies board level oversight and a broader ethical consideration of the delivery.

Information must be reliable to deliver value and this process is delivered through data governance. One way of providing greater assurance of information value is to break down the information into its smallest component parts (i.e., data) in order to ensure that each piece of data is reliable. This is termed *data quality*. Sarsfield (2009, 38) defines data governance as guaranteeing that data can be trusted, and people made accountable for any adverse event that happens when the data quality is poor. In this context, this means assigning ownership and preventing issues with data quality including fixing any issues that occur. Reliable data governance can provide powerful intelligence to impact on decision making and direction.

Data governance is delivered by the provision of data quality linked to the process of data curation and stewardship including the provision of metadata. Metadata encompasses a wide number of elements. The Dublin Core Metadata Initiative defined 15 core metadata elements, which have now been incorporated into the international standard *ISO 15836* (ISO 2017a) and are wide ranging including, for example, metadata about the author, creation date, and data format. These elements can be separate, encapsulated, tagged, hidden or explicit, automated, or manually applied. The international standard *ISO 23081* (ISO 2017b) defines six types of metadata including metadata about records, agents, business activities or processes, records management processes, business rules or policies and mandates, and metadata about metadata.

Metadata is also a critical component of information assurance which seeks to ensure the evidential value of information delivery. The five information assurance pillars are availability, integrity, confidentiality, authentication, and nonrepudiation. The first elements traditionally align to information security delivery and are at the heart of IG frameworks. So too are the additional components of authentication and nonrepudiation which refer to the application of processes which will assure that an author's statement or documentation cannot be disputed in terms of its validity. As such it is often associated with legal processes, for example, the delivery of a contract. Aligned within this aspect of legal delivery is e-discovery which provides a framework for the discovery and production of information in a legal suit.

IT governance looks specifically at the technical delivery (Weill and Ross 2004), "the leadership and organizational structures and processes that ensure that the organization's IT sustains and extends the organization's strategies and objectives." Another IG component is cyber governance which deals with organizational cyber risks at a board level engaging key stakeholders. Von Solms (2016) sets out a maturity model for this one IG subset. This has been developed based the PwC's global security surveys (PwC 2015). This surveys the budget, roles, and responsibilities of the security within an organization, security policies, security technologies, overall security strategy, and finally the review of the current security and privacy risks. Based on this data set, Von Solms has developed a maturity model that sets out four categories against which maturity is measured:

- Category 1: Understanding the strategic role of cyber risk in the company.
- Category 2: Understanding and providing guidance on the cyber strategy of the company.
- Category 3: Understanding and reviewing the cybersecurity budget for the company.
- Category 4: Understanding and evaluating the cybersecurity policies of the company.

The categories have further subcategories and against each, the level of maturity is assessed through a scale of nothing existing at all, to a very basic position, to a progressed position, and to finally a stable position. The maturity levels evidence that, in this context, security is locked down but there is no greater ambition than stability. Strategic and opportunistic ambition for information more widely is set into the bigger IG framework.

Cybersecurity systems seek to keep information safe from a wide range of attacks within a cyber context. Information security provides for the management of security risks across all platforms and information formats including technology and people with assets including knowledge, paper documentation, and online data formats. To harness the value of information/data while providing protection, sophisticated approaches to information creation and management are required. It is an oft quoted maxim that a security system is only as strong as its weakest link. In this regard, Kooper et al. denote the limitations and inadequacies of relying on only smaller parts of governance delivery, such as IT governance (Kooper et al. 2011). To deliver on information value, opportunistic and negative risks must be balanced. Kooper et al. discuss the balance of actors and the wider dimensions of delivering on information value. To deal with this complexity, holistic systems are needed that manage a wide range of information considerations not least the human factor.

Furthermore, as technology is driven globally, these systems must take account of information rights legislation at a global scale and regional difference in law and citizen expectations must be navigated. As such it is IG, which has emerged as a multidisciplinary field that provides for holistic thinking and frameworks to protect and enhance the management of information. The 2017 survey delivered by the Information Governance Initiative and published in 2018, which claimed to have reached 100,000 practitioners globally, stated that 48% of practitioners saw IG as essential for successful cybersecurity delivery even though cybersecurity is narrower in delivery as it cannot exist without broader underpinning. IG provides for a wider and more comprehensive approach.

The Human Factor

As noted by Rubino et al. in 2017, key to good governance is to have leadership throughout an organization in order that others take on the significance of managing and protecting information ethically. This applies at government levels too. In addition, IG needs to be understood and engaged with by multiple parties if it is to be successfully implemented and maintained. The Information Governance Initiative's (IGI) 2018 report further defines areas of information delivery and aligns these to a requirement for people with professional expertise to provide IG frameworks including analytics, audit, big data, business intelligence, business operations and management, compliance, data curation and stewardship, data governance, data science, data storage and archiving, e-discovery, enterprise architecture, finance, informatics, information security and protection, IT management, knowledge management, legal, master data management, privacy, records, and information management and risk management (IGI 2018, 17). The report (IGI 2018, 37) defines the roles of the:

- accountable (the boss)
- responsible (the doers) including professionals which encompass records and information management professionals, information security, legal and compliance business operations and management, risk management, data storage and archiving, and privacy;
- consulted (the advisors) including expertise from the above professions and in addition audit;
- informed (the dependents).

In the latter context, this looks largely internally. However, in accordance with evolving governance principles, organizations must look outside their boundaries to consider wider engagements with all information stakeholders. As a social construct, the concept of systems with human design and societal needs at their center are critical to IG delivery. In addition, humans need to be provided with skill sets to better navigate new digital realities. The United Nations Educational, Scientific and Cultural Organization (UNESCO 2013) recognizes the human right and requirement for all citizens to receive a media and information/digital literacy education to navigate information in order to access reliable information/data and use it successfully. In a world of "fake news" and social media subversion, this has become more critical. These boundaries are becoming further complicated by technological advancement. AI and algorithms are delivering new decision making into society; it is important that these remain controlled, understandable, and relatable for human needs. The potential for robots to have legal personality and rights is becoming a new and complex space of ethical debate. Human and societal needs must remain at the center of IG as it expands and proliferates. Individuals do need wide ranging training and education to fully engage with the potential risks and opportunities associated with creating, sharing, and using information. As noted within the World Economic Forum, personal data has an economic value due to the new potential to connect it to and profile individual service and consumer needs. The complexity of human needs will need to be better framed in terms of ethical considerations that can then be enshrined into international law. To date the Human Rights Declaration has developed fundamental moral tenets pertaining to IG but their complex balancing and application in law is only partially evolved and not globally agreed. In a digital era, legal responsibilities and accountability are not properly agreed and accounted for. Setting human moral agendas in place is key for IG to have effect.

Conclusion

The pervasiveness of information and technology with interdependencies between digital and physical spaces has created a world in which information and its governance impact on all aspects of society.

Key components of IG delivering are:

- putting the human dimensions at the heart of IG processes to ensure an ethical delivery which is framed in accordance with societal needs;
- contextual understanding of national and organizational information needs;
- greater global agreement on information rights laws;
- developed and mature IG professionalism with interlinking expertise from across a wide range of professional domains;

- understanding of information assets, their (co)creation, ownership, sharing, and evolution;
- IG frameworks around assets with government/board oversight on policies, strategies, and risk management including audit and regulatory underpinning.

The threats to our world through cybercrime and warfare are increasing. IG with its capacity to deliver a multidisciplinary and holistic response to contested and complex challenges is essential for successful cybersecurity and the bigger issues of organizational management, government leadership, national security, and international cooperation. Locking down all information is not possible as it creates alternative risks. Navigating, managing, mitigating, and taking risk are essential for human evolution and survival. However, it important that in so doing the moral needs of society/humans remain at the heart of an information-governed world.

References

Adesemowo, Adjeniji, Rossouw von Solms, and Reinhard Botha. 2016. "Safeguarding Information as an Asset: Do We Need a Redefinition in the Knowledge Economy and Beyond?" *South African Journal of Information Management*, 18(1).

ARMA International. 2010. *Information Governance Maturity Model*. Kansas City: ARMA.

Arquilla, John. 2012. "Cyberwar Is Already Upon Us. But Can It Be Controlled?" *Foreign Policy*, 27 February 2012. Available from: https://foreignpolicy.com/2012/02/27/cyberwar-is-already-upon-us/. Accessed April 22 2019.

Blair, Barclay. 2012. "Advancing a Definition of Information Governance." *Essays in Information Governance*. Washington, DC: IGI.

Cadbury, Adrian. 2000. *Global Corporate Governance Forum*. Washington, DC: World Bank 2000.

Day, Matt, Giles Turner, and Natalia Drozdiak. 2019. "Amazon Workers Are Listening to What You Tell Alexa". *Bloomberg News*, 10 April 2019. Available from: www.bloomberg.com/news/articles/2019-04-10/is-anyone-listening-to-you-on-alexa-a-global-team-reviews-audio. Accessed April 22 2019.

De Vries, Alex. 2018. "Bitcoin's Growing Energy Problem." *Joule*, 2(5), 801–805.

Denardis, Laura. 2014. "Cybersecurity Governance." In *The Global War for Internet Governance*, 86–106. New Haven, CT: Yale University Press.

Department of Defense. 2002. *DoD 5015.2 Design Criteria Standard for Electronic Records Management Software Applications*. DoD. Available from: https://www.archives.gov/records-mgmt/initiatives/dod-standard-5015-2.html. Accessed April 22 2019.

Desouza, Kevin. 2009. "Securing Information Assets: The Great Information Game." *Business Information Review*, 26(1), 35–41.

ENISA. 2015. *Governance Framework for European Standardisation*. Available from: https://webcache.googleusercontent.com/search?q=cache:XfQEvGl2IAgJ:https://www.enisa.europa.eu/publications/policy-industry-research/at_download/fullReport+&cd=2&hl=en&ct=clnk&gl=uk. Accessed April 22 2019.

ENISA. 2018. *Economics of Vulnerability Disclosure*. Athens: ENISA. Available from: www.enisa. europa.eu/publications/economics-of-vulnerability-disclosure. Accessed April 22 2019.

Eugen, Petac, and Duma Petruţ. 2018. "Exploring the New Era of Cybersecurity Governance." *Ovidius University Annals, Economic Sciences Series*, 18(1), 358–363.

European Commission. 2008. *On Promoting Data Protection by Privacy Enhancing Technologies*. Brussels: European Commission.

European Parliament and Council of the European Union. 2016. *Regulation on the Protection of Natural Persons with Regard to the Processing of Personal Data and on the Free Movement of Such Data, and Repealing Directive 95/46/EC (Data Protection Directive). The General Data Protection Regulation. L119*. 1–88. Brussels: European Parliament and Council of the European Union.

Financial Reporting Council. 2018. *UK Corporate Governance Code*. London: Financial Reporting Council.

Fliegauf, Mark. 2016. "In Cyber Governance We Trust." *Global Policy*, 7(1), 79–82.

Franks, Patricia, and Nancy Kunde. 2006. "Why Metadata Matters." *The Information Management Journal*, Sep–Oct 2006 (55–61).

Gilb, Tom. 2005. *Competitive Engineering: A Handbook for Systems Engineering, Requirements Engineering, and Software Engineering Using Planguage*. Oxford: Elsevier.

Information Governance Initiative. 2018. *IGI State of the Industry Report: Volume III*. Washington, DC: IGI.

Information Systems Audit and Control Association. 2012. *COBIT 5 – Control Objectives for Information and Related Technologies*. IL: Information Systems Audit and Control Association.

ISO. 2012. *ISO/IEC 27001: 2012. Information Technology – Security Techniques – Information Security Management Systems – Overview and Vocabulary*. Chiswick: BSI.

ISO. 2013a. *ISO/IEC 27001: 2013. Information Technology – Security Techniques – Information Security Management Systems – Requirements*. Chiswick: BSI.

ISO. 2013b. *ISO/IEC 27002: 2013. Information Technology – Security Techniques – Code of Practice for Information Security Controls*. Chiswick: BSI.

ISO. 2017a. *ISO/IEC 15836-1:2017. Information and Documentation – The Dublin Core Metadata Element Set- Part 1: Core Elements*. Chiswick: BSI.

ISO. 2017b. *ISO/IEC 23081-1:2017. Information and Documentation – Records Management Processes – Metadata for Records Part 1: Principles*. Chiswick: BSI.

ISO. 2018. *ISO/IEC 31000:2018, Risk Management – Guideline*. Chiswick: BSI.

Kooper, Michiel, Rik Maes, and Edo Roos Lindgreen. 2011. "On the Governance of Information: Introducing a New Concept of Governance to Support the Management of Information." *Information Management Journal*, 31, 195–200.

Logan, Deborah. 2010. "What Is Information Governance and Why Is It so Hard?" *Gartner Blog*. Available from: http://blogs.gartner.com/debra_logan/2010/01/11/what-is-information-governance-and-why-is-it-so-hard/. Accessed April 22 2019.

Lomas, Elizabeth. 2010. "Information Governance: Information Security and Access within a UK Context." *Records Management Journal*, 20(2), 182–198.

Lomas, Elizabeth, and Julie McLeod. 2017. "Engaging with Change: Information and Communication Technology Professionals' Perspectives on Change in the Context of the 'Brexit' Vote." *PLoS ONE*, 12, 11.

Lomas, Elizabeth, Basma Makhlouf Shabou, and Arina Grazhenskaya. 2019. "Information Governance and Ethics – Information Opportunities and Challenges in a Shifting World: Setting the Scene." *Records Management Journal*, 29(1), 1–2.

MacLennan, Alan. 2014. *Information Governance and Assurance*. London: Facet.

Mason, Richard. 1986. "Four Ethical Issues of the Information Age." *MIS Quarterly*, 10(1), 5–12.

Office of Government Commerce. 2011. *ITIL*. London: Her Majesty's Stationery Office.

PwC. 2015. *Managing Cyber Risks in an Interconnected World: Key Findings from the Global State of Information Security Survey 2015*. London: PwC.

Rubino, Michelle, Fillipo Vitolla, and Antonello Garzoni. 2017. "The Impact of an IT Governance Framework on the Internal Control Environment." *Records Management Journal*, 27(1), 19–41.

Sarsfield, Steve. 2009. *The Data Governance Imperative*. Ely: IT Governance Publishing.

Smallwood, Robert. 2014. *Information Governance: Concepts, Strategies, and Best Practices*. Hoboken, NJ: Wiley.

United Nations Educational, Scientific and Cultural Organization. 2013.*Global Media and Information Literacy (MIL) Assessment Framework: Country Readiness and Competencies*. Paris: UNESCO. Available from: www.karsenti.ca/archives/UNE2013_01_MIL_FullLayout_FINAL.PDF. Accessed April 22 2019.

Van Den Hoven, Jeroen. 2008. "Chapter 3: Moral Methodology and Information Technology." In Himma, Kenneth and Herman Tavani (Eds.) *The Handbook of Computer Ethics*, 49–69. Chiswick: John Wiley & Sons.

Von Solms, Basie. 2016. "Towards a Cyber Governance Maturity Model for Boards of Directors." *International Journal of Business and Cyber Security*, 1(1), 1–9.

Weill, Peter, and Jeanne Ross. 2004. *IT Governance—How Top Performers Manage IT Decision Rights for Superior Results*. Boston, MA: Harvard Business School Press.

Willis, Anthony. 2005. "Corporate Governance and Management of Information and Records." *Records Management Journal*, 15(2), 86–97.

Zeng, Jinghan, Tim Stevens, and Yaru Chen. 2017. "China's Solution to Global Cyber Governance: Unpacking the Domestic Discourse of 'Internet Sovereignty'." *Politics and Policy*, 45(3), 432–464.

Chapter 7

Providing Open Access to Heterogeneous Information Resources without Compromising Privacy and Data Confidentiality

Daniel G. Alemneh and Kris S. Helge

University of North Texas
USA
Texas Woman's University
USA

Introduction

It is a well-accepted fact that emerging trends in technology (such as the rapid growth of mobile devices and cloud computing solutions) are changing the landscape and the way business is conducted in many organizations, including in cultural heritage institutions. Digital technologies provide scholars with access to diverse and previously unavailable contents that span various formats and myriad technologies across institutions and nations. As noted by Janes (2018), the digital shift has been upon us all for some time now, and the issues and realities are getting deeper and more complex as library service continues to be transformed by the multifaceted changes already in place and others on the horizon. Although technologies such as automation, artificial intelligence, and machine learning already help to facilitate access and interactions with big data, such innovations can also increase risks. Data growth has reached explosive levels. As the legacy system simply cannot keep up with the pace of the digital transformation. Some of the concerns with big data applications relate to:

- system security (e.g., protecting digital preservation and networked systems/ services from exposure to external/internal threats);
- collection security (e.g., protecting content from loss or change, the authorization and audit of repository processes);
- legal and regulatory aspects (e.g., personal or confidential information in the digital material, secure access, redaction).

This chapter will discuss challenges raised by concerns about ensuring long-term access to digital resources verses data confidentiality and balancing the right level of data security that addresses compliance requirements in the context of libraries and cultural heritage institutions.

Background of Open Access Movement

The digital shift has challenged the status quo and existing values, and as a result data security is a critical imperative for all institutions. According to the 2018 Thales Global Data Security and Threat Report, the rate of enterprises that are encountering data breaches grew from 21% in 2016 to 26% in 2017 and now to 36% in 2018. Digital transformation requires new data security approaches. In fact, increasingly many nations are articulating and releasing their national cyber strategies. Accordingly, the White House published a comprehensive National Cyber Strategy in September 2018 detailing how the United States current administration aims to improve cybersecurity in government, critical infrastructure and the private sector, as well as tackling cybercrime and international issues.

The open access (OA) movement is part of the broader "open knowledge" or "open content" movement that transforms scholarly communication. In reviewing the literature of the past few years, there is no shortage of views on the role of digital libraries and open access in facilitating digital access to knowledge by reducing barriers. Many researchers articulate a vision of a digital library environment that resonates with possibilities to create a knowledge management system that will enable scholars to navigate through these resources in a standard, intuitive, and consistent way. Many researchers including Alemneh and Hastings

(2006), and Verma (2018) agree that the new scholarly communication systems will inevitably be based on capabilities of interoperable network technology.

As cultural heritage institutions embrace such digital environments, they are facing unprecedented pressures to ensure privacy and reduce the exposure of their institutions to all kinds of data-related risks. Escalating cyberattacks, together with the insider threats for data breaches, make balancing the open access aspirations of cultural heritage institutions without compromising privacy and data confidentiality challenging. In July 2018, the US National Academies of Sciences (NAS) released a consensus report titled *Open Science by Design: Realizing a Vision for 21*[st] *Century Research*, which lays out a vision for a fully open global science environment, and provides the following five specific recommendations for moving from vision to implementation:

1. Research institutions and funders to work to create a culture that actively supports open science by better rewarding and supporting researchers engaged in open science.
2. Research institutions and other entities to support the development of educational and training programs to support students and researchers in adopting open science practices.
3. Research funders and institutions to develop policies/procedures to identify research outputs for long-term preservation and public access, and funding to be made available to support these activities.
4. Funders and institutions to ensure that research archives are designed and implemented according to the FAIR (findable, accessible, interoperable, and reusable) principles.
5. The research community to work together to advance open science by design in order to advance science and help science better serve the needs of society.

The Promise and Security Challenges of Open Access Big Data

The term "big data" increasingly refers to the use of advanced data analytics methods that extract value from data. According to the 2018 Thales Data Threat Report, compared to traditional relational databases, the data generated and stored within big data environments can be orders of magnitude larger, less homogeneous, and change rapidly. There are a number of concepts associated with big data, including the three top attributes what are often referred to as the "Three 'V's: Volume, Variety and Velocity." Some experts (including Jain 2019; Van Rijmenam 2018) go on to add two more Vs to the list, variability and value.

It would be difficult to define what these 5Vs mean in ways that can work in various contexts. When it comes to handling big data, different disciplines or organizations might use the same tools for collecting and manipulating the data at their disposal, but there are significant differences in how they use technologies to organize, analyze, interpret, and put the output data to work in general. The following brief description provides some points about the five Vs and their impacts on information professionals:

1. Volume: Data is being produced at astronomical rates, and size in this case is measured as volume. As Cano (2014) noted, with the Internet of Things (IOT) and all kinds of smart devices that feed smart living, the sheer volume of the data continues to grow every second. No wonder 90% of all data ever created was created in the past two years.

2. Velocity: In the context of big data, velocity refers to the speed at which huge amounts of new data are being created, collected, and analyzed in near real-time using various technological tools. Big data technology helps to cope with the enormous speed the data is created and used in near real time.

3. Variety: With increasing volume and velocity comes increasing variety. Big data technology allows structured and unstructured diverse data to be harvested, stored, and used simultaneously (George 2017).

4. Variability: It refers to the inconsistency, which is the quality or trustworthiness of the data. According to Van Rijmenam (2018), variability is the variance in meaning, or the meaning is changing (rapidly). In indexing the same term or word can have a different meaning. In the same way, to perform proper sentiment analysis, algorithms need to be able to understand the context and be able to decipher the exact meaning of a word in that context.

5. Value: This refers to the worth of the data being extracted. Big data can create enormous value for the global economy, driving innovation, productivity, efficiency, and growth. Despite the size, unless big data can be turned into value, it is useless (cost-benefit). In other words, the value is in the transformation and how the data is turned into information and then into knowledge.

Firican (2019) emphasized the importance of understanding the characteristics and properties of big data to prepare for both the challenges and advantages of big data initiatives. Some used the term complexity to refer to the complex process in which large volumes of data from multiple sources is collected, linked, connected, and correlated to be reliable in order to grasp the information that is supposed to be conveyed by in original data.

Unintended Consequences of Open Access

Good intentions of open access may results in deleterious consequences. As mentioned earlier, most modern information institutions attempt to offer hosted data, information, and knowledge as openly as possible. Yet, such open access can result in data and information descending into the possession of sinister individuals. For example, copious libraries now offer data repositories for their researchers, faculty, and students (University Libraries, Data Repository Services 2018). Faculty place their raw data, both quantitative and qualitative, into such a repository. Uploading their data benefits faculty by assuring that it will be accessible to colleagues who may comment, utilize, question, and otherwise implement their raw data; their data will be preserved and safe from corrupt jump drives or personal drives; and their data will be harvested and visible globally. However, since these researchers' data is globally accessible, danger of a data parasite obtaining and misusing this data is also possible.

Data parasites are individuals who through little or no achievement of their own obtain other people's data and use it maliciously to ultimately publish articles or other written documents, and fail to give attribution to the original data gatherers or creators. For example, data parasites will troll several different data repositories from universities, colleges, and other research institutions in hopes of gathering specific data about new technologies that could promote cleaner forms of energy for automobiles. They will then piece this data together and attempt to publish a paper or offer a conference presentation using the fragmentary data, while offering no credit to the original data gatherers (Longo and Drazen 2016). Thus, they take credit for proposing some form of this new technology and convey it as their idea and research – a form of plagiarism (Helge and McKinnon 2013). Such parasitic pseudo-research harms the original creators of the data and scientific research as a whole.

Plagiarism of others' data, information, and knowledge occurs frequently and for various reasons. Sometimes, researchers accidentally use another's research and data without giving proper attribution. Other times, such as with data parasites, plagiarism is intentional. Such malevolent intent can occur because a student researcher simply believes he or she will not get caught in such a malicious act. Other reasons for plagiarism include not taking an academic course seriously; not understanding self-plagiarism, improper conceptualization of what common knowledge is; and not knowing how to accurately cite scholarship, research, and data (Helge 2017). Dissertations and theses also often become the target of cybercriminals preying on academic informational institutions.

Intellectual Property Rights and Pirated Theses and Dissertations

Many benefits arise when students and faculty place their dissertations or theses into an open access scholarship repository. Their research is instantly accessible to anyone around the globe; sharing their research globally results in personal and professional benefits; they have a permanent and convenient hyperlink with which to refer prospective employers, research collaborators, and other research entities; they may receive invaluable constructive criticism from many researchers globally; and other altruistic researchers have perpetual and efficient access to this invaluable scholarship (Abrizah et al. 2015). Despite these benefits, as with open data, negative ramifications may manifest with open access to dissertations and theses as well. Serving as a scholarly communications librarian at the University of North Texas, I was approached by a faculty member who had just obtained her Ph.D. from North American university. She deposited her recently completed dissertation into her university's digital scholarship repository and was excited about the potential benefits of such a deposit. However, she discovered her dissertation had been pirated and was being sold in China. She queried whether anything could be done to stop the scholarship bootlegging. The response given to her explained that, unfortunately, legally not much could be offered. In the United States, one may be sued for copyright infringement and other intellectual property crimes when a dissertation is improperly reproduced, distributed, displayed, or when illegal derivatives are created within the borders of the United States of America (17 U.S.C. 2018). However, when such intellectual property crimes occur outside of the United States of America, such as in China, the US courts do not have legal personal jurisdiction to allow a prosecution to proceed, without proper extradition. Pennoyer v. Neff, 95 U.S. 714 (1878). Obtaining proper extradition from China is very cumbersome, especially for a stolen dissertation or thesis. So, at best for faculty or students whose dissertations or theses are stolen and sold, they should be happy someone is actually reading their scholarship.

Internet Crimes Complaint Center (IC3) Roles in Protecting IP Rights

Besides being elated someone is actually reading and paying money for their dissertation or thesis, victims of scholarship piracy may also file a complaint with the Internet Crimes Complaint Center (IC3) www.ic3.gov/default.aspx. IC3 is a branch of the Federal Bureau of Investigation and examines Internet-facilitated criminal activity (Federal Bureau of Investigation, Internet Crime Complaint Center 2018). There is no guarantee victims of scholarship theft will receive any equitable or monetary relief from filing a complaint with the IC3; however, filing a complaint with this federal entity could help in such recovery. Although it is difficult to legally punish cybercriminals who steal and misuse intellectual property outside of specific legal jurisdictions, some countries such as the European Union (EU) formed alliances and passed legislation that protects the use of certain individual data.

Current Data and Information Laws

To help assuage individuals' fears of their data being misused in sinister guises, the EU recently passed legislation that directly addresses data misuse. In 2018, the EU passed into law the Regulation (EU) 2016/679 of the European Parliament and of the Council of 27 April 2016 on the protection of natural persons with regard to the processing of personal data and on the free movement of such data, and repealing Directive 95/46/EC (General Data Protection Regulation (GDPR)) (Intersoft Consulting 2018). The GDPR regulates how specific individuals' (e.g., student, medical patient, etc.) data (e.g., grade point average, medical diagnosis, etc.) is processed or utilized by an individual, or an organization in a professional or commercial guise (European Commission 2018). For example, the GDPR does not apply to a private individual utilizing home addresses and phone numbers of other individuals who live in the same neighborhood in order to organize a block party. However, a commercial organization that is collecting that same data along with data about the shirt sizes of all who live on that block, and that plans on selling that data to another commercial company, is regulated by the GDPR. The difference is that the individuals or entity whose motivation is to utilize the personal data for monetary purposes is regulated by the GDPR, whereas the private individuals using the data for personal noncommercial purposes is not regulated by the GDPR. Newly passed GDPR also grants to EU citizens many protections and opportunities to become aware of how their data is being utilized.

Impact of GDPR in Protecting European Union Citizens

The GDPR grants to EU citizens many rights, which include to discover and have access to personal data other entities hold, be aware of the processing of one's personal data, have incorrect data about an individual be corrected, have obsolete personal data deleted, object to the processing of one's data for commercial purposes, request the restriction of some personal data, and to obtain personal data in a machine-readable format (European Commission 2018). Such regulation allows EU denizens the opportunity to be more aware of where their data is being utilized, how it is used, who is using it, what it is being used for, and for EU citizens to rightfully object, correct, and have more control over the use of their personal data. Ultimately, in information warehouses, such as libraries, this could help researchers, students, and other patrons become more aware of how their research data is being utilized, by whom, where, and also allot them more legal protections to reverse the use of data usage when their data is used for malevolent purposes. In fact, GDPR impact companies beyond user privacy. Article 33 of GDPR specifies that organizations must report a breach to the supervisory authority within 72 hours of detection, detailing the nature of the breach, the approximate number of data subjects and personal data records impacted, the likely consequences of the breach, and measures taken or proposed to address the breach and its negative effects (Woods 2019). In the past few decades, not many countries, including the United States, have passed similar broad sweeping legislation that offers as much macrolevel protection.

Proposed US Model Statute for Data and Information Privacy

In the United States, legislation passed offers more directed, microlevel protections, targeting specific industries. Laws such as the 1996 Health Insurance Portability and Accountability Act (HIPPA), the 1999 Gramm-Leach-Bliley Act, and the 2002 Homeland Security Act direct healthcare entities, financial institutions, and federal agencies to ensure data and information systems are protected with a reasonable level of security. Such reasonable levels of security are usually satisfied via an entity tangibly displaying it has created and documented specific protocols, policies, principles, standards, and guidelines that reasonably protect and secure healthcare, financial, and/or other data, information, and knowledge. Such vague statutory language does not afford private citizens an opportunity to edit incorrect data, to know how one's data is being utilized, by whom it is being utilized or to whom it is being sold, where one's data is being transferred, and other microlevel uses of one's data. Perhaps it is time for the US Congress to follow the example of the EU and create legislation that better empowers the citizens of the United States to have more control over their personal data, and that fosters them more knowledge about who, what, when, and how their personal data is being utilized.

A proposed model statute that could offer denizens of the United States more protection of the use of and more awareness of how their data is being utilized is conveyed as follows.

1. Citizens may request and be given information and access to information regarding who, what, when, why, and where their personal data is being utilized. They further may be given access to when, where, and why their data was transferred, sold, or otherwise disseminated.
2. Any organization using data for monetary purposes (whether a for-profit or not-for-profit entity such as a school, doctors office, law firm, charity, church, etc.) must deliver a tangible response to a request for data within 30 days of the request for personal data, and the use of personal data. The response may detail how much of the personal data was sold, to whom it was sold, how the sold data may be utilized, the date the data was sold, and other possible pertinent requested information.
3. Misuse of personal data occurs when a for-profit or not-for-profit entity collects a person's personal data, and then uses it in a deceitful manner to create false digital personas, false digital likenesses, or otherwise uses the data in a malicious manner.

4. Any individual may have the right to force a for-profit or not-for-profit entity to cease the use of his or her personal data if he or she was not properly notified of such sale and use of data in a timely manner.
5. Any individual may force any for-profit or not-for-profit entity to cease the use of an individual's data for commercial purposes, to cease using an individual's data in an incorrect or deceitful manner, to cease the use of obsolete data, or to cease the use of a deceased person's data.
6. Penalties for using a person's data in a malicious manner, or for noncompliance or disclosure of delivery of data, correction of inaccurate data, delivery of the location, specific use, and identification of the persons or function of such use of data may result in a fine of at least $250,000 and two years in prison.

Due to the ubiquitous threat of cybersecurity breaches and the misuse of a person's data, information, or knowledge; such a statute needs to be proposed and enrolled into law at the state and federal levels. Such a statute could provide more legal guidance and general protection to information entity users' privacy, which is a cornerstone value to libraries and other information entities. By ensuring such privacy and data protection, patrons of these information entities can confidently and comfortably use various information warehouses without fear of having their data, information, and knowledge compromised. Another consideration, beyond proposed legislation, is horizon technologies that will be adopted by cultural heritage institutions, such as libraries.

Effect of Horizon Technologies on Data Privacy and Confidentiality

Many horizon technologies will soon affect library patron data privacy and confidentiality. One such technology currently being experimented with is termed sixth-sense technology. This technology interacts with digital world phenomena as a person gestures with his or her arms, hands, or other parts of the body in the physical world (Nuistry 2009). Such a gesture could involve motioning with one's hand and fingers to swipe from right to left to turn a page, pointing to an object to retrieve more information about it, or setting an object on a tablet to discover where similar items may be located. For example, a person may place a soccer ball on a tablet, and utilizing the sixth-sense technology, he or she could then theoretically locate all nearby physical locations where a soccer match was being played within the next two weeks.

Another burgeoning technology that will be adopted by cultural heritage institutions is the IoT. IoT digitizes the physical world, and allows an information warehouse to digitally connect all of its electronic items and simultaneously collect, share, analyze, and project data and information (Geng 2017). Therefore, a library could digitally connect its data visualization screen with student check-out records, student grade and attendance data in the registrar's office, and student financial aid records to determine whether any of these factors correlates to students' academic success, and ultimately to high student retention and enrollment. Both of these above-mentioned horizon technologies allow for efficient access to various types of information beneficial to students, library staff, faculty, and possibly members of other cultural heritage institutions. They also can assist in generating synergized data that can help predict what types of behavior, financial assistance, information retrieval, and study habits may correlate highly with student success, re-enrollment, and retention. However, each of these technologies also potentially expose students, staff, faculty, and other members of cultural heritage institutions to breaches of confidential data and information.

Each of these above-mentioned technologies gathers sensitive and private information pertinent to users. If cybercriminals implement one of the types of cyberattacks mentioned at the beginning of this chapter (e.g., unleashing ransom-ware, hacking into the servers holding such private information and data, or some other type of cyberattack), then student, faculty, staff, or other uses of these new technologies may place their financial, medical, academic, or other personal data and information at risk of being stolen by cybercriminals. Such misappropriated data and information may eventually be sold or used in another malevolent manner against the will of the original data owner.

Data Confidentiality

Digital tools hold a lot of promise in terms of empowering individuals to take control over their personal data. However, there is a significant gap in terms of practices around different groups. For example, collecting data about vulnerable populations by humanitarian organizations may not adequately address the possible implications of collecting and using data about such populations (Vannini et al. 2019). Depending upon the importance and sensitivity of the data being shared, this may be especially critical for marginalized individuals, such as students and undocumented or irregular immigrant. Similarly, the use of the digital tools increases the probability that a patron's data confidentiality may be violated as well. Anytime a patron utilizes any technology in a cultural heritage institution, he or she also should feel confident his or her data will also be kept confidential. Such confidentiality is vital so that patrons medical history, financial research, and religious preference are not exposed and coupled with his or her name, lest the public discover personally sensitive data about specific patrons. If such confidentiality is breached, basic tenants of all cultural heritage research institutions are eroded and patrons are likely not to return and utilize technology that helps them learn, discover, and synergize new information and data. Unfortunately, because cybercriminals persist, some degree of risk of breach of confidentiality is ever present. Thus, information warehouses must remain cognizant of such risks and perform every possible action to assure patron confidentiality.

Although a full-proof manner of preventing all types of cyberattacks will probably never exist, state and federal legislatures should exercise due diligence in ensuring laws are current to address the expeditious changes in technology and the sinister ways in which cybercriminals exploit such abrupt changes. Further, staff of cultural heritage institutions should ensure they utilize the most current cyber securities, both in software and hardware, to protect their patrons' privacy and confidentiality in data, information, and knowledge.

Data, Information, and Knowledge Privacy

Horizon technologies will affect data, information, and knowledge privacy. Although the tools to manipulate big data (including capturing data, data storage, data analysis, search, sharing, transfer, visualization, querying, updating, preserving) are improving by the hour, the top big data challenges include information privacy. Jail (2016) noted that as more and more medical devices are designed to monitor patients and collect data, there is great demand to be able to analyze that data and then to transmit it back to clinicians and others. With increasing adoption of population health and big data analytics, we are seeing greater variety of data by combining traditional clinical and administrative data with unstructured notes, socioeconomic data, and even social media data. All these will only lead to increasing velocity of big data and presents data security challenges, as sensitive data can be anywhere – and therefore everywhere.

Tene and Polonetsky (2019) among other privacy advocates and data regulators call for the development of a model where the benefits of data for businesses and researchers are balanced against individual privacy rights. This presents us with what Janes (2018) calls a classic balancing act: for instance, taking advantage of what libraries could learn from their communities' habits, tastes, and activities without crossing the line into what would be perceived as misuse of personal data.

A cornerstone value that library personnel perpetually promote is patron privacy. This value stems from the Association for Research and College Libraries (ACRL) and the American Library Association (ALA) – promoted fundamental right to have one's research, checkouts, search history, and other library conduct remain confidential, unless otherwise consented to by a patron. ACRL and ALA's privacy perspective supports current US law opined in the US Supreme Court case, Olmstead v. U.S., 277 U.S. 438 (1928), which conveys one of the most comprehensive rights free people have is the right to be left alone. The ALA Bill of Rights (ALA Library Bill of Rights 2018) additionally suggests a lack of privacy for library patrons chills patrons' research choices and access to information, and further it undermines basic tenets of a democratic society (ALA Privacy and Confidentiality 2018). Thus, a library patron's right to privacy is a preeminent value for all information institutions. This right to privacy extends to all types of patrons' data such as checkout records, home addresses, phone numbers, gender, digital research trials, cached images on utilized library computers, and other digital trails. Ensuring privacy of such data is increasingly a challenge in information warehouses such as libraries as more and more data, information, and knowledge is offered globally via open access.

According to Verma (2018) and Raisaroa (2018), the top choices to secure big data were stronger authentication and access controls, monitoring, and encryption. In most modern libraries, privacy is balanced with an effort to ensure global access to information to as many individuals as possible. Such global access, which relies upon intentions of open access, often necessitates the uploading of digital data, information, and knowledge which may overtly or covertly convey health records, academic records, financial records, or other personal information or data to a university or college server, institutional repository, or other digital storage medium. However, some digital data, information, or knowledge that library or other information warehouses may store might not be intended to be accessed openly by the public. This data or information may concern student grades, patron research history, or other data or information normally considered to be private. Yet such data or information may still be uploaded to a library server. Uploading this data and information, whether it is intended to be open to the public or not, may result in privacy breaches via cybercriminals' malevolent conduct, data parasites, or other inadvertent breaches.

Case Studies of Security and Privacy Issues

Ransomware in St. Louis

Some case studies may further convey the possible pitfalls of digital access to data, information, and knowledge. Some data, information, and knowledge stored in information institutions, such as libraries, may be susceptible to cyber breaches, even though it is not proffered to the public via open access. In 2017, cybercriminals breached the information systems of 17 separate public libraries in St. Louis, Missouri. As a result of this digital trespass, all 17 libraries' computer systems were infected with ransomware that encrypted most of the library system's digital files, and to unencrypt these files the cybercriminals demanded a $35,000 payment in the electronic currency bitcoin. The cybercriminals effectively shut down the entire St Louis library system and destroyed the library staff's email system in a matter of minutes (Pagliery 2017). Until the sabotage was rectified, this caused patron and library staff angst due to them not having access to their normal information retrieval channels, and due to not knowing if confidential information regarding their search history, personal data, and checkout histories would be made public.

Malware in Singaporean Health Records Database

In a separate unexpected attack in 2018, cybercriminals broke into a Singaporean health records database, SingHealth, and took the names and addresses of approximately 1.5 million medical patients, and the names of medicines dispensed to some of these patients. This breach occurred via malware through which the cybercriminals gained access to the personal health data. Just this year alone, numerous other nation-states claimed some of their governmental agencies have been hacked by cybercriminals such as Germany's government IT network and the UK's National Health Service (BBC Asia 2018).

These unanticipated criminal breaches of information entities exemplify that despite the best efforts of any security protocol, no information depository is impervious to cyber hackings. As a result, for libraries in the digital age, no patron can ever be completely confident that his or her research history, checkouts, social media posts, grades, and even university digital health records are inpentrateable. For university library patrons in particular, student grades, courses taken, information literacy classes attended, entrance into a library, and other metrics are often collected by libraries to show correlations between library usage and student success in college or graduate school (LeMaistre et al. 2018). Due to these metrics being utilized and stored on library servers, cybercriminals could breach security protocols and obtain student health records, student grades, student or faculty research history, and utilize this data in malevolent guises. If such private student information and data is stolen, irreparable damage could be caused to students. Along with uploading students' data to information warehouse servers, promoting some data and information via open access initiatives may also lead to malicious uses of data and information. These cybercriminals also seem to troll and assail universities and other related research institutions.

Cybersecurity Breaches Affect Universities and the Military

Cyberattacks also affect universities and the military. In February 2019, numerous universities' information technology systems were allegedly breached via Chinese hackers known as Temp.Periscope, Leviathan, and Mudcarp. These hackers targeted military defense information housed at various universities such as the University of Hawaii, University of Washington, Duke University, Massachusetts Institute of Technology, Penn State University, and other universities and colleges around the United States, Canada, and Southeast Asia. During this cyberattack, hackers aimed to locate and steal United States, Canadian, and

Southeast Asian military and economic secrets. The universities and colleges assailed in this cyberattack house key research institutes focused on undersea technology. Thus, it appears this is the focus of this cyberattack. Temp.Periscope, Leviathan, and Mudcarp are linked to previous cyberattacks where the hackers were seeking highly secure military information and data related to submarine missile creation and ship maintenance data (Volz 2019).

Allegedly, Temp.Periscope has targeted many US universities in the past because they tend to partner with military branches and usually have the digital infrastructure to house and preserve valuable military research. Hackers leverage the natural trust and desire to share information that most researchers display and promote while working at a university or college. Hackers know because of this desire to share and build upon stored information and data, some researchers at colleges and universities may be more likely to click on a well-produced spear phishing email. The universities and colleges hacked in this particular case probably had their information technology infrastructure hacked due to researchers clicking on spear phishing emails. Hackers have become well aware that researchers are willing to share their research, and they use this as a conduit in which to hack, steal, and illegally utilize sensitive military information (Volz 2019).

Academic libraries are often the location in which such sensitive data, information, and knowledge is digitally preserved. Scholarly communication departments often maintain digital scholarship repositories, digital data repositories, and other digital repositories that may house some of this highly classified military information. Since libraries personnel also naturally gravitate toward sharing information, this can be problematic when housing highly sensitive military data and information. Any data, information, or knowledge stored in a type of digital repository may easily be safely stored behind a dark archive perpetually. Such a dark archive theoretically ensures that only specific entities or individuals from those entities are able to access any deposited data, information, or knowledge located in that dark archive. That is partially why military institutions choose libraries in which to store sensitive information. However, as is mentioned in this chapter, cyber hackers constantly create new means to breach information technology security safeguards. Thus, no digital repositories are impenetrable. Knowing this, cybercriminals often target academic libraries to hack into, steal, and maliciously utilized highly sensitive military data, information, and knowledge. Despite such technological vulnerabilities, academic libraries should continue to keep abreast of the best security measures that may prevent any data, information, and knowledge from being pirated, and then sold on the dark web, used for malicious purposes in other countries, or utilized in some other malevolent manner. Along with breaking into the digital infrastructure of academic libraries, cybercriminals also attempt to pillage data from small businesses too.

Small Businesses Are Easy Targets for Cyber Assailers

Small businesses might assume they are nontargets from cybercriminals because larger corporations might have more data for criminals to quickly pillage and sell. This is an incorrect assumption however. Small businesses are frequently targets of sinister hackers because most cybercriminals are cognizant that most small businesses either cannot afford adequate antivirus protection, or simply do not have the awareness to keep such security software current.

One example of how easily cybercriminal can infiltrate small business is from a case study of Quaint Bakeries. A couple of years ago, two recent graduates of California Polytechnic State University commenced a vegan bakery that took orders online. They even hired a third-party vendor to install and maintain what they thought was adequate software to protect their online assets. This third party also set up a virtual personal network to keep the bakery's IP addresses confidential and encrypt various Internet connections. However, this was not quite enough protection to prevent all cyberattacks (Strauss 2018).

While participating in a demo with a Microsoft store (lucky for the bakery, they learned their site was not full-proof via a demo rather than really losing valuable data to malicious hackers), the two bakers learned that their website could be easily spoofed. In other words, Microsoft successfully created a derivative of their site with one slight change in the sites URL. The bakers did not notice this small derivation and logged into the fictitious site thinking it was their actual website. Luckily this was just a demo and the bakers learned a valuable lesson (Strauss 2018).

However, many real cybercriminals create fictitious websites and embed links with malware on these sites. When one clicks on one of these links the malware can secretly install key-logging software on the user's computer, and then cyber hackers can discover exactly what a user is typing (Strauss 2018). Or worse, what the user is doing via video, or saying via various sensors.

Small businesses need to remember they are just as vulnerable as any other entity to cybercrime. They are often a more desirable target due to cyber hooligans' knowledge that owners of small businesses might not have the financial means or appropriate digital security knowledge to protect themselves. These small business entrepreneurs should educate themselves as did the Bakers at Quaint Bakery about valid and reliable products and services that may protect their digital assets from cybercriminals. Cybercriminals also target the secure data and information located in larger companies as well.

Cyber Breaches Occur in the Larger Corporate Arena as Well

Cybercriminals attack corporate information technology venues as well. In January 2014, Ukrainian cyber hackers broke into Target's information technology system and stole up to 110 million customers private data, including credit card and debit card accounts, names, phone numbers, and email address. A month prior to this attack, similar cybercriminals hacked into Neiman Marcus's information infrastructure and pirated customers' credit card data. One huge concern for Target is that when a customer at this store purchases alcohol, a store clerk scans his or her driver's license (Jayakumar 2014). Thus, during this hack, millions of customer's driver's license data was stolen. Today, with a person's driver's license number, much other data can quickly be garnered online such as residence and business addresses and phone numbers, public civil and criminal records may be quickly tracked, and other personal data.

At least some of the perpetrators of this crime have been arrested and prosecuted. Rusland Bondars, a Latvian citizen, was arrested, found guilty, and sentenced to 14 years of prison for designing a program that helped hackers improve malware. This malware was used by hackers later to breach Target's information infrastructure. Those criminals who breached Target's system first used their developed Scan4You malware to determine whether an antivirus program would recognize their software as malicious. Cybersecurity officers believe the other hacker responsible for this hacking of Target and possibly Neiman Marcus is "Profile 958," who is likely a Ukrainian named Andrey Hodirevski (Weiner 2018). What do hackers do when they steal personal and private data from corporate, educational, governmental, and public sector entities?

Cybercriminals Quickly Make Money Off of Stolen Data

After breaking into a business, educational, governmental, or other public sector digital database, cybercriminals quickly sell stolen private and personal data on the dark web. The dark web is a part of the Internet not discoverable by traditional search engines such as Google of Firefox, and is only discoverable via special web browsers such as Tor. The dark web accounts for approximately less than 0.01% of the Internet. The entire Internet itself is known to be broken into three parts: the surface web, the deep web, and the dark web. The surface web makes up about 10% of the Internet. Information, data, and knowledge located in the surface web may easily be retrieved via simple search from most web browsers such as Google and Internet Explorer. An example of surface web content is a gaming company selling legally downloadable video games via their commercial website. The deep web consists of 90% of the Internet, and offers data, information, and knowledge to users via search engines such as scholarly books or articles located in databases protected by a paywall. An example of this would be when a student searches in EbscoHost and downloads a scholarly article via his or her university or college username and password. The dark web is a part of the deep web, consisting of about 0.01% of the Internet, where much illicit trading of black market items and services occurs. Using primarily the web browser called Tor, stolen data, information, and knowledge is often illegally traded and sold (Ablon et al. 2014).

Cybercriminals have a complete logistics system setup on the dark web on which stolen data can be marketed, sold, delivered and lightning speed to buyer via a sophisticated logistics system, and implemented via buyers for their malicious purposes (Ablon et al. 2014). Thus, when cybercriminals are able to successfully break into and private millions of customers' private data, they are highly likely to have a quick return of possibly hundreds of thousands of dollars to possibly millions of dollars by selling this data within minutes on the dark web.

What Recourse Do Victims of Cybercrime Have?

Of course, when one falls victim to cybercrime, the natural response, other than feeling violated and somewhat helpless is, is there any legal recourse, or anywhere one may turn to for help. While receiving legal help from cybercrime is somewhat cumbersome due to cybercriminals usually residing in legal jurisdiction from where the victim's data or information is stolen, there are some interjurisdictional agencies to where one may attempt to seek help. For example, one may contact the IC3 www.ic3.gov/default.aspx and file a complaint regarding cybercrime. The complaint should include the victim's name, address, telephone, and email; financial transaction information; specific details regarding how one was victimized; and any other relevant information.

If one falls victim to a phishing email scam, or if one simply receives an email phishing scam attempt, one may forward the attempt to reportphishing@antiphis ing.org (the Anti-Phishing Working Group (APWG)) and to phishing-report @us-ert.gov where it will be received by the United States Computer Emergency Readiness Team (US-CERT). The APWG is a private, international work group that tracks various types of cybercrime, and attempts to deter such future cybercrime. The US-CERT group is an arm of the US Department of Homeland Security, and attempts to garner information about cybercrimes and subsequently better educate the its denizens regarding awareness and potential dangers of certain Internet activity (Smith 2016).

If an individual finds him or herself in one of the mass data/information breaches similar to the ones mentioned above (e.g., Target), he or she may enroll in the online account monitoring software called WebWatcher. This system offers reimbursement for monetary loss due to cyber breaches and fraud up to one million dollars. When one enrolls in these services, he or she also obtains fraud counseling services and reimbursement coverage for free (Smith 2016). So, though sometimes a labyrinth to navigate, some sites are prepared to attempt to help victims of cybercrime when it affects people.

Lessons Learned

It is apparent that cybercriminals target any and all type of organization when seeking to quickly breach an information technology security barrier and steal private data and information. All sectors are at risk, government, academic, small business, large corporations, nonprofit charities, and so on. The motivation behind pirating private data and information is that cybercriminals can quickly sell such stolen goods and make a large profit on the dark web. Cybersecurity simply does not have the technology nor adequate person power yet to effectively patrol the dark web. There are too many discrete locations and ways for cybercriminals to complete illegal sales on the dark web. There are government and private agencies that can lend some help to victims of cybercrime on the dark web. However, all cultural heritage institutions need to remain cognizant of the possible malware and cyberattacks to which their information infrastructure and their patrons may be susceptible. Such awareness needs to prompt information professionals to ensure the utilization of the most valid and reliable antimalware software and that its patrons are properly notified of all possible risk when using digital materials in their institution. Information professionals should also actively advocate for the passage of updated laws that offer strict punitive consequences for those who commit cybercrimes, and advocate for adequate funding with which to invest in proper safety measures to prevent cybercrimes.

Summary and Conclusion

Emerging trends and horizon technologies are allowing cultural heritage institutions to develop new ways of gathering, preserving, analyzing, and synergizing data, information, and knowledge. These new technological endeavors offer great benefits to global humanity, but also, as with any new development, open new opportunities to individuals with malicious intent. As many data, information, and knowledge warehouses adhere to the global open access movement; opportunities for data breaches are further apparent. Dichotomously opposed to such data, information, and knowledge breaches, most cultural heritage institutions, such as libraries, adhere to valuing the utmost guarantees of privacy and confidentiality for their patrons. Such strong dedication to privacy and confidentiality is exemplified in international library policy and is further reflected in international law. Despite such efforts of cultural policy makers and legislators, cybercriminals continue to implement malevolent tactics such as ransomware, viruses, worms, spyware, Trojans, phishing, pharming, and other malicious endeavors.

Some of these types of cyberattacks result from well-meant initiatives to globally share data, information, and knowledge. Dissertations, theses, e-journals, data, art, and other types of scholarship are increasingly deposited into open access digital repositories to allow and promote universal access. Promoting such access, unfortunately, sometimes leads to data parasites stealing researchers' data, or other cybercriminals pirating scholarly works such as theses or dissertations. These cybercriminals then claim such stolen information and data as their own, plagiarize it, offer no credit to the original creator(s), and often utilize this pilfered information, data, and knowledge for illegal commercial purposes.

The current state of the dark web also presents a challenge to all types of cultural heritage institutions based in government, academic, corporate, and not-for-profit sectors. Cybercriminals may quickly pirate private information and data from these entities and turn a quick profit by selling these goods on the dark web. To help reduce the illegal behavior being carried out on the dark web, more research needs to be completed about how the dark web works, more effective technology needs to be implemented to help reduce dark web crime, and new laws need to be passed that can enable security to better monitor the dark web and prevent and reduce the percentage of cybercrime occurring on the dark web.

Increasingly, organizations are reassessing their operational readiness to detect and respond to a breach. The European Union has addressed some data theft and misuse via the General Data Protection Regulation (GDPR). In addition to privacy, thanks to GDPR's stringent breach notification regulation, organizations have revamped their incident response programs over the past year to meet the requirements. With the recent release of the National Cyber Strategy (White House 2018), the United States now has its first fully articulated cyber strategy. However, the United States has not recently passed proposed legislation into law to address modern data and information misuse. A proposed model statute in this chapter could offer insight into how the US federal government, and the 51 states and district legislative bodies could provide legal guidance to this issue.

In the current data-intensive environments, all global standard organizations and legislative bodies will be precipitously challenged in perpetuity to continuously update standards and legislation as horizon technologies arise that alter the way in which data is collected, preserved, utilized, shared, and synergized. No wonder most ISO27k standards, which include many aspects of information technologies, are under review on an ongoing basis, and the publication of ISO/IEC 27045 (the specific standard for big data security and privacy) may not even be expected until the year 2022.

In today's cybersecurity landscape, realizing the promise of open access and big data may well depend on our ability to continue our quest to maintain our cyber readiness in the face of ever evolving threats. With appropriate legal policies, guidelines, and a combination of right technology, right people and right processes in place, organizations and nations at large will be able to contain damage and minimize risk when (yes, it's a matter of "when," not "if") they are breached and digitally attacked.

References

Ablon, L., M. C. Libicki, and A. A. Golay, *Markets for Cybercrime Tools and Stolen Data: Hackers' Bazaar* (Washington, DC: Rand, 2014).

Abrizah, A., M. Hilmi, and N. A. Kassim, "Resource Sharing through an Inter-Institutional Repository: Motivations and Resistance to Library and Information Science Scholars." *The Electronic Library* 33, no. 4 (2015) 730–748. doi: https://doi.org/10.1108/EL-02-2014-0040.

Alemneh, D. G., and S. K. Hastings. "Developing the ICT Infrastructure for Africa: Overview of Barriers to Harnessing the Full Power of the Internet." *Journal of Education for Library and Information Science* 47, no. 1 (2006). https://digital.library.unt.edu/ark:/67531/metadc38890/.

"American Library Association Library Bill of Rights," 2020, http://www.ala.org/advocacy/intfreedom/librarybill.

"American Library Association: Privacy: An Interpretation of the Library Bill of Rights," 2020, http://www.ala.org/reedom/librarybill/interpretations/privacy.

"BBC Asia," "Singapore Personal Data Hack Hits 1.5m, Health Authority Says," 2018, last modified January 22, 2019, www.bbc.com/news/world-asia-44900507.

"European Commission," "Policies, Information and Service," last modified July 22, 2018. https://ec.europa.eu/commission/priorities/justice-and-fundamental-rights/data-protection/2018-reform-eu-data-protection-rules_en.

"Federal Bureau of Investigation," Internet Crimes Complaint Center, last modified January 22, 2018, www.ic3.gov/about/default.aspx.

Firican, G., "The 10 Vs of Big Data. A Publication in Transforming Data with Intelligence (TDWI)," (2019), https://tdwi.org/articles/2017/02/08/10-vs-of-big-data.aspx.

Geng, H., "Internet of Things and Data Analytics in the Cloud with Innovation and Sustainability." In *Internet of Things and Data Analytics Handbook*, ed. H. Geng (Hoboken, NJ: John Wiley & Sons, 2017), 3–128.

Helge, K., "Law Student Information Seeking, and Understanding of Citation, Common Knowledge, and Plagiarism." In *Knowledge Discovery and Data Design Innovation*, eds. G. A. Daniel, J. Allen, and S. Hawamdeh (London, England: World Scientific, 2017), 249–263.

Helge, K. and L. McKinnon, *The Teaching Librarian: Web 2.0, Technology, and Legal Aspects* (Amsterdam, The Netherlands: Elsevier, 2013).

"Intersoft Consulting," "General Data Protection Regulation (GDPR)," last modified June 2, 2018, https://gdpr-info.eu/.

Jain, A., "The 5 Vs of Big Data," last modified January 12, 2019, www.ibm.com/blogs/watson-health/the-5-vs-of-big-data/.

Janes, J., "Balancing Privacy & Innovation | Reinventing Libraries." *The Digital Shift*, last modified November 15, 2018, www.thedigitalshift.com/2013/08/uncategorized/balancing-privacy-innovation-reinventing-libraries/.

Jayakumar, A., "Target Tries to Reassure Customers after Data Breach Revelations." *The Washington Post Online Edition*, last modified January 2014, www.washingtonpost.com/business/economy/target-tries-to-reassure-customers-after-data-breach-revelations/2014/01/13/3c0323e0-7c7e-11e3-95c6-0a7aa80874bc_story.html?utm_term=.009b25a8e9fa.

LeMaistre, T., S. Qingmin, and S. Thanki, "Connecting Library Use to Student Success." *Libraries and the Academy* 18, no. 1 (2018) 117–140. https://muse.jhu.edu/.

Longo, D. and J. Drazen, "Data Sharing." *New England Journal of Medicine* 374 (2016) 276–277. doi: 10.1056/NEJMe1516564.

Mistry, P., "The Thrilling Potential of Sixth Senses Technology," Ted Talk, 2009, http://www.ted.com/talks/pranay_mistry_the_thrilling_potential_of_sixthsense_technology?language=en.

Olmstead v. U.S., 277 U.S. 438 (1928).

Pagliery, J., "St. Louis' Public Library Computers Hacked For Ransom," last modified February 1, 2017, https://money.cnn.com/2017/01/19/technology/st-louis-public-library-hack/index.html.

Pennoyer v. Neff, 95 U.S. 714 (1878). 17 U.S.C. 106 (2018).

Raisaroa, J. L., "Are Privacy-Enhancing Technologies for Genomic Data Ready for the Clinic? A Survey of Medical Experts of the Swiss HIV Cohort Study." *Journal of Biomedical Informatics* 79, March (2018) 1–6. doi: https://doi.org/10.1016/j.jbi.2017.12.013.

Smith, B. R., *Don't Step in the Trap: How to Recognize and Avoid Email Phishing Scams* (Seattle, WA: CreateSpace Independent Publishing Platform, 2016).

Strauss, S., "How Do Cyber Criminals Hack Small Business Startups? Here's What We Learned from Microsoft," *USA Today Online Edition*, last modified October 17, 2018.

Tene, O. and J. Polonetsky, "Privacy in the Age of Big Data: A Time for Big Decisions," last modified January 22, 2019, www.stanfordlawreview.org/online/privacy-paradox-privacy-and-big-data/.

"University Libraries Data Repository Services," last modified August 1, 2018, www.library.unt.edu/datamanagement/data-repository-services.

Van Rijmenam, M., "Why the 3V's Are Not Sufficient to Describe Big Data," last modified June 14, 2018, https://datafloq.com/read/3vs-sufficient-describe-big-data/166.

Vannini, S., R. Gomez, and B. C. Newell, "Documenting the Undocumented: Privacy and Security Guidelines for Humanitarian Work with Irregular Migrants." *Lecture Notes in Computer Science: 14th International Conference, iConference 2019*, Washington, DC, (March 31–April 3, 2019), last modified March 8, 2019, www.conftool.com/iConference2019/index.php?page=browseSessions&form_session=365.

Verma, S., "Data Confidentiality in Public Contracts: Why Typical 'What Me Worry' Attitudes May Not Really Be an Acceptable Position for Government Contracting Professionals Anymore." *The 8th International Public Procurement Conference (IPPC)*, Arusha, Tanzania, (August 2018), http://dx.doi.org/10.2139/ssrn.3159135.

Volz, D., "Chinese Hackers Target Universities in Pursuit of Maritime Military Secrets: University of Hawaii, University of Washington. And MIT are among Schools Hit by Cyberattacks." *The Wall Street Journal Online* Politics, National Security, (March 2019), www.wsj.com/articles/chinese-hackers-target-universities-in-pursuit-of-maritime-military-secrets-11551781800.

Weiner, R., "Hacker Linked to Target Data Breach Gets 14 Years in Prison," *The Washington Post Online Edition*, last modified September 21, 2018, www.washington

post.com/local/public-safety/hacker-linked-to-target-data-breach-gets-14-years-in-prison/2018/09/21/839fd6b0-bd17-11e8-b7d2-0773aa1e33da_story.html?noredirect=on&utm_term=.e3b5fd6574df.

"White House National Cyber-Strategy of the United States of America." last modified September 1, 2018, www.whitehouse.gov/wp-content/uploads/2018/09/National-Cyber-Strategy.pdf.

Woods, T., "GDPR's Impact on Incident Response," last modified April 25, 2019. https://securitytoday.com/articles/2019/04/24/gdprs-impact-on-incident-response.aspx.

Chapter 8

Cybersecurity and Social Media

Hassan Zamir

School of Information Studies, Dominican University

Introduction

Social media is a tool for building networks, interact, and practice democracy. Across the globe, users share contents, emotions, and engage in conversations effortlessly. People of all walks of life can communicate with each other in an unprecedented way. Individuals, businesses, governments, and charities can get connected with ease. However, with greater technological affordances come greater risks. Nefarious sources can sabotage the systems and steal individual and organizational information. Cybersecurity risks are higher on social media platforms. Moreover, there is citizen distrust on both government and social media companies with regard to data protection. Recent data and security breaches have escalated the concerns even more.

In the United States, 64% Americans do not have faith on government and social media sites over cybersecurity risks and information privacy protection (Pew Research 2017). The US citizens are also divided over data encryption, particularly during crime investigation. However, more and more people are shifting toward storing their information in digital tools. In the case of social media, user data can be harnessed easily at this current status. Hackers steal information, breach the data, and manipulate in any way they wish. They intend to leak data and take control over the systems by slowing it down and hijacking the accounts. Their illegal access to information causes physical interferences and service unavailability.

Recently, Federal Communications Commission ruled that Internet Service Providers (ISPs) can use and sell customer data (Consumer Reports 2018). The ISPs have customer data divided into two categories – sensitive personal data and less-sensitive personal data. Sensitive data includes "geographic location, children's information, health information, financial information, social security numbers, web browsing history, app usage history, the content of communications." The less-sensitive data includes user's name, address, IP address, subscription levels, and anything else not in the "opt in" category. Under the new rule, ISPs can sell the second category or less-sensitive data.

In the age of big data, social media platforms generate exabytes of data in seconds. While the technology companies capitalize on big data, online predators weaponize sophisticated data science and machine learning techniques for security concerns like social engineering. The threat actors can be anyone including individuals, inside agents, disgruntled employees, third parties, government-backed officials, or careless workers. They can sabotage the system and cause major security breaches out of financial gains, espionage, entertainment, revenge, convenience, fear, and ideology.

This chapter discusses specifically about cybersecurity risks associated with social media. It highlights frequently observed security risks on social media platforms. The chapter rigorously stresses on the most common forms of cyberattacks on social media including phishing and social engineering. In this regard, it describes related security concerns and mitigation techniques. It also describes recommended social media cybersecurity implications for information professionals.

Social Media Cybersecurity Risks

Social media platforms generate millions of user data every second. Abundant information provides enough indicators to the attackers. Numerous cybersecurity risks are associated with social media usage. Security experts and scholars mention similar kinds of cyberattack approaches on social media based on recent and well-known security incidents. Some of the common social media security risks are phishing, social engineering, malicious shorten URLs, information leakage, impersonations, account hijacking, identity theft, scamming, Denial of Services (DoS), Distributed Denial of Services (DDoS), and bots (Table 8.1). Some risks are connected with cause and effects of social media use such as cyber-extremism, hate crime, cyberbullying, disinformation, and social media policing. These risks are commonly related either to information or systems. The attackers set missions to capture victim's information or sabotage the system. Facebook assessed information operations executed on its platform and identified three major areas

Table 8.1 Commonly observed security risks across social media including Facebook and Twitter

Common Social Media Security Risks	Cyber Security Risks Observed on Facebook	Platform Manipulation: Twitter
Phishing, social engineering, malicious shorten URLs, information leakage, impersonations, account hijacking, identity theft, scamming, DoS, DDoS, social media bots, cyber-extremism, hate crime, cyberbullying, disinformation, and social media policing	Targeted data collection ▪ Information collection/ reconnaissance ▪ Cyberattacks against organizations or individuals ▪ Spear phishing ▪ Data theft Content creation ▪ Meme, stories, and fake profiles False amplification ▪ Dissemination of fake contents and memes ▪ Organized activities for or against political discourses	Misleading or disrupting users via deceptive techniques including spams, malicious automation or malicious use of bots, inauthentic account abuse or fake accounts

of ploys – targeted data collection, content creation, and false amplification. Malicious actors steal data from all kinds of users including individuals, government, nonprofit and for-profit organizations, and media outlets. They use phishing infected with malware and steal user data. Stolen information typically is used for creation of fake contents, meme that shared and interacted on Facebook. Twitter reports that its platform can experience manipulation from malicious sources. Users on Twitter can be victim of bulk or aggressive disruptions through deceptive activities that include but are not limited to spams, bots, and fake accounts.

Social Media Phishing

Phishing is the most common form of cyberattack even on social media platforms. Although phishing attacks are conventionally conducted by sending malicious links through emails, in modern days this attack has tremendously grown over social media sites. Phishing messages sent over social media sites are highly likely to be opened by users than those sent via regular emails (Seymour and Tully n.d., 3; Frenkel 2017). Phishing comes in forms of spear or whale attacks. In spear phishing, specific individuals or organizations are strategically victimized with recognition of enough trustworthy background information about them. In whale phishing attacks, the phishers target top executives and higher officials with intentions of large form of fraud and embezzlement. Spear phishing is the fastest growing attack on social media primarily due to the networked architecture of the platform and lack of security trainings of users. The attackers send malicious links in harmless messages to the users in spear phishing attack (Figure 8.1). Even if one user in the entire network clicks the link the entire system gets affected by it. This practice is not limited within general hackers, but state-hackers also take the benefit of it. For example, Russian hackers entered Pentagon computer network through one of its official's Twitter account (Frenkel 2017). The hackers sent infected summer vacation messages to the Pentagon official's spouse's Twitter account, which was later shared with the Pentagon official. Therefore, the breadth of phishing attack on social media is far-fetched. Government social media accounts are usual targets of spear phishing. For example, Iranian hackers attempt to attack US Department of State social media accounts (Frenkel 2017; Sanger and Perlroth 2017), Russian-led hacking efforts of 10,000 US Defense Department Twitter accounts are classic spear phishing attacks (Calabresi 2017; Frenkel 2017).

Social Engineering

Individuals and organizations can be victim of social engineering. Commonly, victims lose personal and sensitive information because of deceits. Some popular approaches of social engineering are phishing, quid pro quo or vishing, tailgating, impersonation, and baiting. While its attacks are quick, recovery from one incident is lengthy. Individuals need awareness about cybersecurity policies and guidelines in order to avoid social engineering risks. The organizations must train its employees with cybersecurity skills so that they can avert away any potential security risk for the company. Social engineering categorically can be executed in couple of ways – hunting and framing. Hunting is

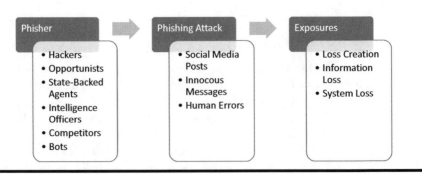

Figure 8.1   Social media phishing phases.

more popular approach in comparison to farming method. In hunting process, victims notice negligible attentions and in farming process, the victim gets hooked for longer terms.

The malicious attacks intend to extract confidential information from individuals and organizations. Social engineering is even more dangerous than other forms of cybersecurity risks because it grows on the basis of trust. The victims are lured to believe information provided to them. The levels of deception are stronger with regard to its breadth, depth, and intention. These complex tasks are often performed with the help of regular tools and applications. Extensive coding skills are not always required for conducting social engineering.

The core purpose of social engineering is to exploit victims and control access to their confidential information and resources (Conteh and Schmick 2016, 32; Breda et al. 2017, 2). In cybersecurity context, human targets are tempted to release sensitive information knowingly or unknowingly being affected by security breaches. It can be defined as the practice of provoking individuals or organizations for disclosing confidential information and network resources through the means of deceitful techniques either controlled by human skills or socio-technical tools. The attackers capitalize human trust and the tendency to make errors.

Social Engineering Attack Types

The attackers use different types of resources including human and technical tools for the purposes of system exploitations. Cybersecurity risks on social media platforms take many forms. Those are accomplished either by reliance on entirely human involvements or by utilization of socio-technical tools. Based on these two different sources, social media cybersecurity risks are divided into two areas – in-person attacks and socio-technical attacks (Table 8.2).

In-Person Attacks

Cybersecurity attacks that involve human skills largely are termed as in-person attacks. Popular in-person cybersecurity attacks are impersonation, tailgating,

Table 8.2 Social engineering attack types

In-Person Attacks	Socio-Technical Attacks
■ Impersonation ■ Tailgating ■ Dumpster diving ■ Shoulder surfing ■ Reverse social engineering	■ Phishing ■ Vishing ■ Vishing or quid pro quo ■ Eavesdropping

dumpster diving, shoulder surfing, and reverse social engineering (Breda et al. 2017, 2). There are other forms of human-involved cybersecurity risks but essentially are associated with the risks listed here.

Impersonation

The practice of deceiving victims by representing as another person with aims to gain access to information resources and networks is known as impersonation (Conteh and Schmick 2016, 32; Breda et al. 2017, 2). The attacker pretends or pretexts by playing a role of someone close to the target. It is one of the most common conventional methods of social engineering. Persons are critical victims of such practices and they may lose control over their information networks to the social engineers. Intentions of impersonation involve committing fraud, identity theft, industrial espionage, and so on. The impersonators collect information about the victim granularly over the time and connect those for building a coherent and credible source of information. They fulfill their aim through stalking, researching company websites and social media accounts, and pretexting over phones. For example, impersonators play roles of IT help desk and provide related services like change of passwords and so on. This opportunity gains them power to obtain confidential information and sabotage the targets. In the context of social media, fake profiles of celebrities and their fan group accounts are common impersonation trap. Technically impersonators spread rumors and increase followers' counts with aim to distribute spams whenever needed.

Tailgating

The attackers plan to enter the network through restricted access points. They convincingly request legitimate employees for availing unauthorized access in a company (Bourne 2014, 242–67). They portray plausible scenarios to the legitimate employees such as loss of keys or forgetting keys at home. Their false identities help them to be in areas of a company that otherwise will be impossible to enter. Their investigations on target company and its employees bring expected opportunities for them. This attack sounds innocent. It requires fewer computing skills and more human skills. Often companies neglect this criminal activity and mostly rely on security logs for mitigating specific cases.

Dumpster Diving

This risk is associated with cybersecurity indirectly. However, it has direct connections with physical security risks. Old company resources like forms, credit cards, invoices, memos, and notepads contain valuable information about the organization and its human resources. The attackers can produce credible information from these weeded sources (Liu et al. 2009, 87–108). The dumpster diver generates databases with target company and employee information. Prior to disposal of old materials, it is important to shred the printed resources and erase any kind of storage media.

Shoulder Surfing

The attackers often look over the shoulders of the victims to gain access to their login information. Shoulder surfing is a low-tech strategy of social engineering. In modern days, employees work in open environments and log in workstations locally for administration purposes. Shoulder surfers take the advantage of stealing PIN codes and passwords just by looking at the authorized users' activities (Liu et al. 2009, 87–108). Although it has low-tech requirements, this technique proves effective for the social engineers because they can easily modify their strategies on the spot.

Reverse Social Engineering

Social engineering techniques are gradually becoming well-recognized cybersecurity knowledge. The attackers are also perfecting their skills even more and placing them one step ahead in strategy deployment. They are now reversing social engineering attacks. They place them among authorized users as trustworthy allies in attempts to provide technical supports (Rountree 2011, 135–59). But their fundamental target lies in organizational disruption by the means of gaining trusts of official users.

Socio-Technical Attacks

Socio-technical attacks heavily rely on digital tools and computing skills (Breda et al. 2017, 5). Some techniques require moderate to some degree of coding skills. Others can be achieved based on significant technical knowledge. Most common types of socio-technical cyberattacks on social media platforms are phishing, baiting, vishing, and eavesdropping.

Phishing for Social Engineering

Phishing is the most common form of email scams that targets to access confidential information (Flick and Morehouse 2011, 109–42; Conteh and Schmick 2016, 32; Breda et al. 2017, 2). Simple clicks on suspicious links overthrow the entire system. Its contents look believable and legitimate to users, which is why they often fall into the trap and lose their confidential information to the attackers. Phishing can easily transport sensitive information and computer network access to the devices of the attackers. This technique is adopted by the hackers most commonly due to its simplicity and efficacy. With simple HTML and CSS codes, building a copycat website for a well-known website is easier. Once the link is sent through emails in the target networks, the built-in worms and malware can effectively collect confidential information from victim's devices. The phishers commonly choose email networks, social networks, and text messages to spread the spams. They circulate spams on social media like Facebook, Twitter, Instagram, LinkedIn, and other such platforms for acquiring background information about victims. They build phishing operations around large-scale social and political events like national holidays, celebratory moments, elections, and so on. These campaigns are often generated in the form of breaking news. Their ultimate target is persuading the victim to click on the malicious links. In social phishing cases, the phishers scrape contents from social media accounts of the victim and use those for constructing user-specific phishing emails (Jagatic et al. 2007).

Baiting

Baiting works with concepts similar to phishing. It provides backdoor entrance of IT infrastructure to the attackers (Conteh and Schmick 2016, 32; Breda et al. 2017, 6). They use either digital tools or physical media like USB devices and place those in tactical spots. Once these are used in any organizational workstation, effects of chain reactions from the installed malware compromise devices in the entire network. The attackers on social media, for example, on Twitter, detect the trending topics. They locate key messages with the trending topic and hashtags. They replace the original links from the Twitter posts with their desired shortened links and URLs associated with malicious sites. It is hard for regular users to distinguish the original contents from malicious contents out of millions of comments and retweets. The security flaws start to affect devices when these deceptive links are clicked.

Vishing

Vishing, also known as quid pro quo, is another widespread type of social engineering model. It is widely known as "voice phishing" (SocialEngineering. Org, n.d.). The deceiver tricks the users with legitimate information by representing well-known companies or even government agencies. They make random phone calls and convince users to exchange personal information in relation to IT services, IRS tax records, utility services, and more. The perpetrators also attract users to click on certain online advertisements and contents that make users' personal devices compromised and provide a phone number. While some users call to that phone number by providing their personal information, which ultimately land in wrong hands with ill purposes.

Eavesdropping

Unapproved interference of users' real-time conversations over social media, mobile phones, and online applications is known as eavesdropping (Breda et al. 2017, 5). Its physical form, that is, interception of phone conversations, is conventionally called as wiretapping. Internet-based calls without VoIP and encryption technology become playground for digital trespassers. With rootkit applications, the intruders can easily listen to users' conversations on their laptops via the installed microphones. Eavesdropping is treacherous in nature because it does not malfunction the network data transmission and leaves little room for detection. Public WiFi networks are the most vulnerable targets for eavesdropping.

Social Engineering Attack Phases

A careful attention to the overall process of any social engineering can be broken down to various steps. It involves stages of investigations, data collection, manipulation, and termination. These phases can be sub-grouped into three categories – pre-attack, the attack, and post-attack (Figure 8.2). At first, an extensive amount of research is conducted on victim's profile. Owing to the advancement of information systems, personal and professional networks are becoming easier to harness to gather information. The more information collected about the victim, the better the chances for entrapment. In the second stage, victim's work information, geographical attachments, and social information are valuable resources for the perpetrators to conduct criminal masquerade. The attacker plots tricks to hook the victim to the stories without increasing suspicions. Once the desired information is collected, in the third stage, the attacker closes the interactions with victim. The consequences of such practice are beyond imagination and its longevity is difficult to predict.

Cyber-Extremism and Online Hate Crimes

Religious extremist groups use social media platforms to recruit members. They use multimedia contents like images and videos, chatrooms, website links, retweets,

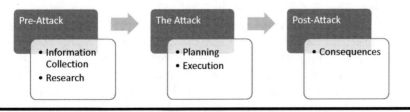

Figure 8.2   Social engineering attack phases.

hashtags, social media interaction features, and mobile applications to distribute propaganda and radicalize new generations. Cyber-extremists have seven key characteristics in their online behaviors – cyber mobs, isolates, fantasists, thrill seekers, moral crusaders, narcissists, and attention seekers (Awan 2017, 143). Ignorant groups use social media like Facebook with negative and discriminatory attitudes for sharing religious slurs such as Islamophobia (Awan 2016, 1). Scholars categorized hate speech by collecting data from two social media platforms – Whisper and Twitter, and found those are commonly related to race, behavior, physical, sexual orientation, class, gender, ethnicity, disability, and religion (Silva et al. 2016, 1–4).

Cyberbullying

In a survey of college-going students, Whittaker and Kowalski (2015) found they were victims of cyberbullying via social media posts. Although texting was most common method, the students were targets for cyberbullying, social media platforms like Twitter, Facebook, Instagram, and YouTube were pre-dominant media where they were affected as well. Cyber aggressors use comments and forum replies as venues to post their comments by protecting their anonymity.

Disinformation

Social media is a cost effective and easily accessible tool for promotion of low-quality news mixed with false information, that is, fake news. Malicious accounts such as trolls, social bots, and automated accounts use social media for dissemination of propaganda. Social media is an effective tool for creating groups of like-minded people. They share similar ideas and eventually believe homogenous thoughts. This can create social echo chambers where users only share and consume same types of information (Shu et al. 2017). Propagation of fake news in social media echo chambers is faster, which can create polarized communities (Vicario et al. 2016, 554–9).

Social Media Bots

Social media bots generate automatic interactions with users and interfere in public dialogues, civic discourses, political agenda, policy discussions, elections, social movements, and fake news distribution. The bots can easily be trained with sophisticated computing techniques (Mønsted et al. 2017). Those can be customized and controlled for when and how many social media posts will be created on various social and political events. Evidence of bot usage can be traced during global issues like US elections (Allcott and Gentzkow 2017, 211–36; Bessi and Ferrara 2016; Shao et al. 2018), Brexit (Howard and Kollanyi 2016), the 2017 French Presidential Election (Ferrara 2017), stock market manipulation, and vaccination policy (Ferrara et al. 2016). Ferrara (2017) mentions the existence of potential underground black-market for "reusable political-disinformation bots."

Social Media Policing

Identifying criminal information on social media and build investigation on it is an attractive practice to the IT forensic and cybercrime departments in police forces (Denef et al. 2012). They use tools developed by IBM, SAS, SAP, Oracle, Radian6, Attensity, Kapow, and Palantir for information storage, analysis, and visualization. The law enforcement departments can track IP addresses used to send private messages over social media, alongside email addresses and phone numbers. European laws and orders provide supports for police to harness criminal data available from social media. Data collection from social media sites protected under international laws is complicated for local police departments. For example, it requires advanced protocols for European police forces to gather data from Facebook, as it is an US-based company. Facebook, however, has teams for collaborations with international police operations, especially, to work in areas like child abuse, child pornography, and so on. Facebook can suspend social media accounts and freeze users' data that the company can share with police departments when appropriate legal requests have been placed. In addition to criminal investigations, the police forces generally use social media platforms for communications and building relationships with citizens, distributions of emergency news, and community policing.

Social Media Cybersecurity Risk Mitigation Techniques

Use of social media is manifold. From keeping in touch with close networks to build new friendships, planning events to sharing experiences through rich multimedia contents like photos, videos, and interactions with network – all can be done in single hyper-connected media. While this brings the world closer, the consequential prices are often higher.

Browser Updates

To prevent malware attacks, loss of personal information by clicking links, it is safer to disregard social media login invitations through links sent by unknown parties. Bookmarks and opening social media accounts by accessing through browsers is wiser. Sometimes browsers get affected by virus and loss of personal stored information becomes obvious. Therefore, personal information storage on browsers for social media accounts is unsafe. It is acutely dangerous when the accounts are accessed on public computer systems. Public WiFi connections with unsecured network are risky for social media usage for various reasons. Hackers can access to these untrusted networks easily and track all the general activities online.

Passwords

Strong passwords are highly useful to protect social media accounts. Security experts advise the use of stronger and newer passwords regularly. Use of one password for all of the social media platforms is never a good practice. Passwords related to contents shared on social media like favorite animals or places are easy to crack. Random words in strings make stronger passwords. Use of password managers are often highly recommended in case of fear of losing and forgetting passwords. Although highly sophisticated password manager applications are growing, Americans are still relying on good-old fashioned strategies such as memorization and writing down in notes (Olmstead and Smith 2017).

Security Updates

Security updates are crucial for social media account protection. Right version of the application is safer to use. Otherwise, security glitches and errors are prone to exploitation by hackers. Unrecognized network requests without proper details should be treated cautiously. Social media can be a playground for scammers to lure users with false information and contents. Users become easy targets of social engineering with the expense of sensitive personal information, financial resources etc.

Company-specific security protection initiatives are on the rise. Facebook address spear phishing attacks by providing specialized notifications, developing detection systems, and promoting phishing counteractive measures among the users. Most of the social media companies practice dual-factor authentication as standard. In addition to passwords, users get links and codes sent to their emails and phones for account verification purposes. Many use third-party applications in connection with social media accounts for the purposes of playing games, productivity, and so on. It is highly important to review third-party applications on regular basis.

Privacy

Privacy risks are widely known concept among social media users. Although many do not trust social media company with data protection and privacy, they seem to ignore privacy risks and use the platforms anyway. Change of privacy settings to desired theme is important. Business users tend to have greater concerns over privacy protections. Because they believe their information can get jeopardized for not paying proper attention to it.

Social Media Policy

Importance of social media policy is highly recognized. It can work as the guidelines for everyone in the company and offer solutions to legal procedures, if necessary. From actions to preventions, individual and group responsibilities, confidentiality, and copyright – all areas can be clearly defined in policy documents.

Training and Security Education

Promotion of social media security education is immensely necessary. Security education can make individual users aware of how to protect and avoid any risks on the platform. In the business cases, employees must go through security trainings, which can prove very effective. They can be better equipped with right knowledge on how to mitigate and tackle company security challenges. In this regard, appointment of individuals and teams based on size of the company can be useful. The designated official can take charge and update everyone time-to-time with security alerts.

Security Technologies

We safeguard our computers and internet services by installing firewalls and anti-virus software. Similarly, investment in security technologies for social media site protection is a timely decision. The security technologies regularly alert and update with security concerns on the system. It can track, monitor phishing and suspicious activities, and report for actions.

Laws and Regulations

With the brink of number of cybersecurity attacks and users' data exploitations, governments around the world are closing the regulation gaps for social media companies. Governments of North American and European countries are clamoring on social media companies like Facebook, Twitter, Google, and others for their liabilities in data breaches. The Online Safety Act, in Australia, for example, in response to horrendous New Zealand mosque mass shooting, aims to hold accountable social media platforms and the online predators for any illegal and abusive contents. The United States has enacted several laws and regulations for the protection of online safety. Major data breaches have escalated the privacy concerns. Securing safer online environment is a monumental task as sources of security risks are always dynamic and unpredictable. Although overarching cybersecurity laws in the United States is absent, various area-specific federal laws such as Health Insurance Portability and Accountability Act (HIPAA) for health data protection and Federal Information Security Management Act (FISMA) are widely accepted. Moreover, state laws such as California Consumer Privacy Act and New York State Department of Financial Services cybersecurity regulation are highly regarded. The most expansive online data protection law is, however, Europe's General Data Protection Regulation (GDPR).

Implications for Information Professionals

Information professionals in libraries and cultural institutions strongly practice privacy protection of user information for long. According to the American Library Association (ALA) Code of Ethics, it protects "each library's user's right to privacy and confidentiality with respect to information sought or received and resources consulted, borrowed, acquired or transmitted" (American Library Association 2006). The International Federation of Library Associations and Institutions (IFLA) Code of Ethics express similar message as it believes that "librarians and other information workers respect personal privacy, and the protection of personal data, necessarily shared between individuals and institutions" (IFLA 2012). The ALA 2003 Policy Manual added more specific direction about privacy based on the Library Bill of Rights. It reads

> in a library (physical or virtual), the right to privacy is the right to open inquiry without having the subject of one's interest examined or scrutinized by others. Confidentiality exists when a library is in possession of personally identifiable information about users and keeps that information private on their behalf.
>
> (American Library Association 2014)

Therefore, it is highly important for information professionals to protect all kinds of digital resources and user information from breaching through disruptive technologies like social media. Libraries frequently use social media platforms like Facebook, Twitter, blogs, Instagram, Pinterest, and Flickr (Sieck 2014). Libraries around the world use social media typically for the purposes of promotions of events, library services, new acquisitions, and collection of items, sharing news and updates, customer services, and development of connections. Libraries can adopt various steps for better digital privacy including updating privacy policies, encryption technologies, requiring vendors to follow library privacy rules, and training information professionals (Caldwell-Stone 2017). On the basis of social media security, best practices designed by social media security providers such as ZeroFOX (2017) following recommendations are suggested for information professionals.

Current Status

Information professional can identify current security status by checking if the social media accounts are breached or not. Some platforms such as havei beenpwned.com provide information if any email address associated with social media accounts has been disclosed from any data breach.

Multi-Factor Authentication

Uses of dual-authentication or multi-factor authentication are now a security practice standard. Enabling multi-factor authentication can strengthen security protection. With this practice, users now require to log-in their accounts by providing a security code sent to their smartphones or other devices. Without matching codes, any unknown access will be difficult.

Strong Passwords

Avoidance of using simple passwords for social media accounts is always a good practice. Usernames, date of births, and location can be easily traced and those can put the library social media accounts in cybersecurity risks. Passwords with complex and longer characters are difficult to crack for hackers. Use of combination of symbols, numbers, and upper and lowercase letters can generate strong passwords.

Security Setting Updates

Use of the most updated applications is an essential task to avoid security risks. Web browsers, operating systems, device settings, and social media applications provide notifications for updates and security glitches. To mitigate any security risks, it is important to install the updated versions of digital systems and applications.

Connection Curation

Library social media accounts will have many followers and connections. Monitoring the types of followers and friends on the social media accounts is a good practice to identify any malicious profiles. Phishers may post spams and false information in conversations of library social media accounts. Weeding these spams and malicious accounts can help avoid cybersecurity risks both for the libraries and its users.

Designated Officers

One or more officers need to be appointed for identification and protection of social media cybersecurity cases in libraries. The designated officers can monitor and track any suspicious activities on the accounts. They can generate security protection steps and procedures for everyone.

Training and Awareness

Cybersecurity risks are complex and commonly use modern technologies. In order to cope up with it, regular security trainings are required for information professionals. Knowledgeable professionals know how to handle complex social media security cases. Offering correct cybersecurity education in the academic programs can help the information professionals to learn the nuances of security risks. Workshops, webinars, conferences can be valuable avenues for increasing knowledge in cybersecurity.

Social media are often unnoticed when it comes to cybersecurity. However, libraries and information professionals regularly use social media for various professional purposes. It contains user communication and digital resources. Anyone with wrong motifs can threaten the overall library systems through its social media platforms. Protection of privacy and confidentiality has been an utmost priority for libraries and information professionals. They should uphold this noble tradition by providing and offering secure and safer social media platforms for its users.

Conclusion

Online security is often compromised due to lack of knowledge about it. Three components – process, technology, and people are significant in overall online security. The core of which, however, is people. It is common sense how people should develop a more secure online environment. Carefulness on friendship network development is crucial. Avoidance of unknown web territories is wiser and safer. The CIA (confidentiality, integrity, and availability) model ensures online security protection goals. Security experts commonly advice for the use of secured and updated web browser, look for popups, pay attention to warning messages, certificate errors, and suspicious links.

As technological advances proceed, the capabilities of attackers are renewed in the same rate. Their expertise to filtrate systems for stealing sensitive information becomes greater, too. However, human psychology is central to their attention. Therefore, right security education is necessary for all kinds of users. Correct knowledge about social media cybersecurity risks can prove effective. Moreover, users need to pay attention to what they share on social media. Their interactions become a data point for machine learning. What users share on social media, sometimes is held against them and they are judged for their interactions and behaviors. Users have their own intentions behind every post and machine learning-based knowledge may expose quite different meanings, which is often times terrifying.

Online security awareness is mandatory in every sector. Users may think of quitting social media platforms. However, their voices against social media malpractices can bring better solutions. At the same time, state and governments need to place appropriate regulations to protect user information. Adequate laws and orders can hold offenders and platforms accountable for deceptive activities. It is the duty of social media companies as well to ensure safety of its users. Their business model should not be to produce profits at the expense of user data. The users deserve a safer and protected online environment that can nurture true democracy.

References

Allcott, Hunt, and Matthew Gentzkow. 2017. "Social Media and Fake News in the 2016 Election." *Journal of Economic Perspectives* 31 (2): 211–36. doi:10.1257/jep.31.2.211.

American Library Association. 2006. "Privacy." *Advocacy, Legislation and Issues.* July 7, 2006. www.ala.org/advocacy/intfreedom/librarybill/interpretations/privacy.

American Library Association. 2014. "Privacy and Confidentiality: Library Core Values." *Advocacy, Legislation and Issues.* April 25, 2014. www.ala.org/advocacy/privacy/toolkit/corevalues.

Awan, Imran. 2016. "Islamophobia on Social Media: A Qualitative Analysis of the Facebook's Walls of Hate." *Zenodo,* July. doi:10.5281/zenodo.58517.

Awan, Imran. 2017. "Cyber-Extremism: Isis and the Power of Social Media." *Society* 54 (2): 138–49. doi:10.1007/s12115-017-0114-0.

Bessi, Alessandro, and Emilio Ferrara. 2016. "Social Bots Distort the 2016 U.S. Presidential Election Online Discussion." *First Monday* 21: 11 doi:10.5210/fm.v21i11.7090.

Bourne, Kelly C. 2014. "Security." In *Application Administrators Handbook*, edited by Kelly C. Bourne, 242–67. Boston, MA: Morgan Kaufmann. doi:10.1016/B978-0-12-398545-3.00015-7.

Breda, Filipe, Hugo Barbosa, and Telmo Morais. 2017. "Social Engineering and Cybersecurity." 4204–11. doi:10.21125/inted.2017.1008.

Calabresi, Massimo. 2017. "Inside Russia's Social Media War on America." *Time.* http://time.com/4783932/inside-russia-social-media-war-america/.

Caldwell-Stone, Deborah. 2017. "Why Is Privacy so Important in Libraries?" 64.

Conteh, Nabie Y., and Paul J. Schmick. 2016. "Cybersecurity: Risks, Vulnerabilities and Countermeasures to Prevent Social Engineering Attacks." *International Journal of Advanced Computer Research* 6 (23): 31.

Denef, Sebastian, Nico Kaptein, Saskia Bayerl, and Leonardo Ramirez. 2012. "Best Practice in Police Social Media Adaptation." https://repub.eur.nl/pub/40562/.

Ferrara, Emilio. 2017. "Disinformation and Social Bot Operations in the Run Up to the 2017 French Presidential Election." *First Monday* 22: 8. doi:10.5210/fm.v22i8.8005.

Ferrara, Emilio, Onur Varol, Clayton Davis, Filippo Menczer, and Alessandro Flammini. 2016. "The Rise of Social Bots." *Communications of the ACM* 59 (7): 96–104. doi:10.1145/2818717.

Flick, Tony, and Justin Morehouse. 2011. "Attacking the Utility Companies." In *Securing the Smart Grid*, edited by Tony Flick and Justin Morehouse, 109–42. Boston, MA: Syngress. doi:10.1016/B978-1-59749-570-7.00007-8.

Frenkel, Sheera. 2017. "Hackers Hide Cyberattacks in Social Media Posts." *The New York Times*, December 22, 2017, sec. Technology. www.nytimes.com/2017/05/28/technol ogy/hackers-hide-cyberattacks-in-social-media-posts.html.

"Have I Been Pwned: Check if Your Email Has Been Compromised in a Data Breach." 2019. Accessed May 1, 2019. https://haveibeenpwned.com/.

Howard, Philip N., and Bence Kollanyi. 2016. "Bots, #StrongerIn, and #Brexit: Computational Propaganda during the UK-EU Referendum." ArXiv:1606.06356 [Physics], June. http://arxiv.org/abs/1606.06356.

"IFLA – IFLA Code of Ethics for Librarians and Other Information Workers (Full Version)." 2012. Accessed May 2, 2019. www.ifla.org/publications/node/11092.

Jagatic, T. N., N. A. Johnson, M. Jakobsson, and F. Menczer. 2007. "Social Phishing." *Communications of the ACM* 50 (10): 94–100.

Liu, Dale, Max Caceres, Tim Robichaux, Dario V. Forte, Eric S. Seagren, Devin L. Ganger, Brad Smith, Wipul Jayawickrama, Christopher Stokes, and Jan Kanclirz, eds. 2009. "Chapter 5 – SSH Shortcomings." In *Next Generation SSH2 Implementation*, 87–108. Burlington: Syngress. doi:10.1016/B978-1-59749-283-6.00005-2.

Mittal, Sudip, Prajit Kumar Das, Varish Mulwad, Anupam Joshi, and Tim Finin. 2016. "CyberTwitter: Using Twitter to Generate Alerts for Cybersecurity Threats and Vulnerabilities." In *2016 IEEE/ACM International Conference on Advances in Social Networks Analysis and Mining (ASONAM)*, 860–7, San Francisco, CA: IEEE. doi:10.1109/ASONAM.2016.7752338.

Mønsted, Bjarke, Piotr Sapieżyński, Emilio Ferrara, and Sune Lehmann. 2017. "Evidence of Complex Contagion of Information in Social Media: An Experiment Using Twitter Bots." *PLoS One* 12 (9): e0184148. doi:10.1371/journal.pone.0184148.

Morran, C. 2017. "House Votes To Allow Internet Service Providers To Sell, Share Your Personal Information." https://www.consumerreports.org/consumerist/house-votes-to-allow-internet-service-providers-to-sell-share-your-personal-information/.

Olmstead, Kenneth, and Aaron Smith. 2017. "Americans and Cybersecurity." *Pew Research Center* 26: 311–27. https://www.pewresearch.org/internet/2017/01/26/americans-and-cybersecurity/.

Rountree, Derrick. 2011. "Organizational and Operational Security." In *Security for Microsoft Windows System Administrators*, edited by Derrick Rountree, 135–59. Boston, MA: Syngress. doi:10.1016/B978-1-59749-594-3.00005-3.

Sanger, David E., and Nicole Perlroth. 2017. "Iranian Hackers Attack State Dept. Via Social Media Accounts." *The New York Times*, December 21, 2017, sec. World. www.nytimes.com/2015/11/25/world/middleeast/iran-hackers-cyberespionage-state-department-social-media.html.

Seymour, John, and Philip Tully. n.d. "Weaponizing Data Science for Social Engineering: Automated E2E Spear Phishing on Twitter." 8.

Shao, Chengcheng, Giovanni Luca Ciampaglia, Onur Varol, Kaicheng Yang, Alessandro Flammini, and Filippo Menczer. 2018. "The Spread of Low-Credibility Content by Social Bots." *Nature Communications* 9 (1): 4787. doi:10.1038/s41467-018-06930-7.

Shu, Kai, Amy Sliva, Suhang Wang, Jiliang Tang, and Huan Liu. 2017. "Fake News Detection on Social Media: A Data Mining Perspective." ArXiv:1708.01967 [Cs], August. http://arxiv.org/abs/1708.01967.

Silva, Leandro, Mainack Mondal, Denzil Correa, Fabrício Benevenuto, and Ingmar Weber. 2016. "Analyzing the Targets of Hate in Online Social Media." In *Tenth International AAAI Conference on Web and Social Media.* www.aaai.org/ocs/index.php/ICWSM/ICWSM16/paper/view/13147.

SocialEngineering.Org. n.d. "Vishing." *Security Through Education* (blog). Accessed May 2, 2019. www.social-engineer.org/framework/attack-vectors/vishing/.

Vicario, Michela Del, Alessandro Bessi, Fabiana Zollo, Fabio Petroni, Antonio Scala, Guido Caldarelli, H. Eugene Stanley, and Walter Quattrociocchi. 2016. "The Spreading of Misinformation Online." *Proceedings of the National Academy of Sciences* 113 (3): 554–9. doi:10.1073/pnas.1517441113.

Whittaker, Elizabeth, and Robin M. Kowalski. 2015. "Cyberbullying Via Social Media." *Journal of School Violence* 14 (1): 11–29. doi:10.1080/15388220.2014.949377.

ZeroFOX. 2017. "7 Social Media Security Best Practices." *ZeroFOX.* November 14, 2017. www.zerofox.com/blog/social-media-security-best-practices/.

Chapter 9

Healthcare Regulations, Threats, and their Impact on Cybersecurity

Mitchell Parker

Executive Director, Information Security and Compliance
Indiana University Health

Introduction

Cybersecurity in Healthcare is important because it touches the patient, the provider, and the facilities that store and process patient data. Everyone is affected by the loss of access to electronic systems such as Electronic Medical Records used in the care process. From data breaches, ransomware, to improper medical records access, healthcare information security is pervasive. The risks present are unlike every other industry because of the interconnected nature of medical devices, electronic health records, interconnectivity between organizations, and patient access.

One of the most targeted industries is Healthcare. With the number of large data breaches such as Anthem's 2015 90 million record data breach caused by foreign intrusion into their network (Teichert 2018), Community Health Systems' 2014 4.5 million record data breach caused by foreign network intrusion (Bosshart n.d.), and MD Anderson's 2012 and 2013 loss of two flash drives and a laptop containing Protected Health Information for over 30,000 patients (Flahive n.d.), it is easy to see that attackers are not only targeting healthcare, they are succeeding. The threats, such as Ransomware, continue to evolve.

This chapter will delve into four categories to explain the impact of regulations and threats on a healthcare organization's cybersecurity posture. The reason why is to examine what they are and their multidimensional impact on an organization.

We will look at the Security and Privacy components as they relate to cybersecurity for the Health Information Portability and Accountability Act (HIPAA), Health Information Technology for Economic and Clinical Health Act (HITECH), the EU General Data Protection Regulation (GDPR), Payment Card Industry – Data Security Standards (PCI-DSS), and 21st Century CURES Act and Trusted Exchange Framework.

We will then examine the key programs and processes an organization needs to proactively defend themselves against cyberattacks, which are the establishment and maintenance of a Security program, establishment and maintenance of a Privacy program, and implementation of a Data Management, Classification, and Monitoring program. We will also discuss the requirements to implement an Asset Management and Tracking program, conduct comprehensive risk assessments that cover both Information Security and Physical Security, and develop and implement a risk management plan to address open issues.

From this, we will discuss ten emergent and current threats to the healthcare environment, and fifteen ways healthcare organizations can address these threats under the described programs. The intent of this is to educate the reader as to how to build their own comprehensive program, and address risks in their own organization.

Healthcare Privacy and Security Legislation and Requirements

HIPAA and Its Applicability to Privacy and Security

The Health Information Portability and Accountability Act, which was signed on August 21, 1996, was originally designed to "improve the portability and accountability of health insurance coverage" for employees transitioning between jobs, combat fraud, waste and abuse in the industry, promote the use of medical savings accounts, provide coverage for pre-existing conditions, and simplifying the administration of health insurance (HIPAA Journal n.d.). The Privacy Rule, which had an effective date for compliance of April 14, 2003, defined Protected Health Information (PHI) as "any information held by a covered entity which concerns health status, the provision of healthcare, or payment of healthcare that can be linked to an individual" (HHS n.d.a). It also defined the intended purposes of access to PHI, which are for purposes of coordinating payment, treatment, or business operations (HIPAA Journal n.d.).

It also defined the terms Business Associate (BA) and Covered Entity. Covered Entities are Health Plans, Health Care Clearinghouses, or Health Care Providers who electronically transmit any health information in connection with transactions for which the Department of Health and Human Services (HHS) has adopted standards (HIPAA Journal n.d.). They can be institutions, organizations, or persons. Researchers are also covered entities if they are also providers that electronically transmit PHI in connection with any transaction for which HHS has adopted a standard (HIPAA Journal n.d.).

BAs are a person or entity, acting on behalf of a Covered Entity, that perform certain functions involving the storage or processing of PHI on behalf of them who are not workforce members. They require a Business Associate Agreement, which is a legal contract that ensures that the BA will adequately and appropriately safeguard the information they store and/or process on behalf of the Covered Entity. The Business Associate Agreement must establish the permitted and required uses and disclosures of PHI by the BA. It must also provide that the BA will not use or further disclose the information other than as permitted or required by the contract or law. It requires the BA to implement reasonable and appropriate safeguards against unauthorized use or disclosure. It also requires the BA to disclose PHI to individuals' requests for copies and make the information available for amendments and accountings. This also requires BAs to make available their internal practices, books, and records relating to the usage and disclosure of PHI to the Department of Health and Human Services (HHS). At the termination of the contract, require the BA to return or destroy all PHI received or created by or on behalf of the CE. BAs have to ensure that all subcontractors meet the same requirements as they have to. Finally, there needs to be a termination clause authorizing termination should the BA violate material term(s) of the contract. The Privacy Rule also provides instructions on how PHI should be disclosed. It gives rules for getting consent from patients before using their information for marketing, fundraising, or research purposes. It also gives patients the right to withhold information about their care from private insurers if they fund their own treatment. It empowers patients to get copies of their Medical Records, ideally in an electronic format (HHS n.d.e).

The most important and critical concept is that of Minimum Necessary. This means that organizations need to use only the minimum amount of PHI needed to complete a task. This requires organizations to implicitly inventory their PHI and sensitive data, classify it, and understand where the information is (HHS n.d.h).

The HIPAA Security Rule, which has an effective compliance date of April 21, 2005, deals specifically with electronically stored Protected Health Information (ePHI). It has a number of controls and safeguards organized into three categories. These controls and safeguards have to be abided by and followed. In addition, a Security Officer has to be named for the organization. This person does not have to be dedicated, but they have to have the assigned responsibility.

The Administrative safeguards require organizations to create and maintain policies and procedures which will illustrate how organizations will comply with the Security Rule. The Physical Safeguards are a set of controls that organizations have to follow to protect physical access to both printed and electronic PHI from unauthorized access. Technical Controls are to protect ePHI during its storage, transmission, and processing (HHS n.d.a).

There has been a major misconception with the Security Rule in terms of Cybersecurity. There are two types of controls in the Security Rule, which are Addressable and Required. The Addressable controls were interpreted as that you needed to have a compensating control to address the issue, rather than directly meeting the requirement of the control. Specific requirements for Encryption in the Security Rule, 45CFR 164.312(a)(2)(iv), which require organizations to implement mechanisms to encrypt and decrypt ePHI, were bypassed in favor of controls that did not encrypt data and required the Breach Notification Rule and HITECH Act to clear up (HHS n.d.d). A number of data breaches were caused by organizations not encrypting data to the requirements of the Security Rule.

A number of breaches were also caused by organizations not understanding the concept of Minimum Necessary, not inventorying their data and assets, and not understanding where their data flows were to be able to protect the data (HHS n.d. a). A major misconception of companies that has led to data breaches has been that HIPAA only applies to communication with outside entities, and that internal communications or storage do not need to be protected with the same rigor. This is not true.

The Security Rule also requires organizations to conduct a periodic risk assessment against it to identify any potential gaps, and to reasonably and appropriately address them with a security plan. A common misconception of HIPAA is that it requires organizations to use only supported operating systems, hardware, and software. What the intent is to reasonably and appropriately protect those hardware, software, and devices (HHS n.d.c). While no periodic frequency has been specified in the Security Rule for either the Risk Assessment or Security Plan, the HITECH Act does mandate that organizations who apply for Meaningful Use funding are supposed to have a Risk Assessment for each reporting period. The Office For Civil Rights, the division of the Center for Medicare and Medicaid Services responsible for enforcement of HIPAA, has made statements that they expect the Risk Assessment and Risk Management Plan to be done annually at each reporting period (HHS n.d.a). As part of this risk assessment, organizations are expected to know what assets they own, what risks they pose, and plan to address those risks through a combination of Administrative, Technical, and Physical controls.

Finally, and most critically, the Security Rule requires Data Backup plans, Disaster Recovery plans, Emergency Mode Operations Plans, their testing and revision procedures, and an application and data criticality analysis (HHS n.d.c).

This means that organizations need to make sure that they back up their data, test the backups periodically, and test and update their disaster recovery and emergency mode operations plans regularly. In the case of a cybersecurity incident, knowing what to do when the systems are down is critical. This is mandated by Federal law. However, many organizations have not implemented these or adjusted them, as evidenced by the ransomware attacks that several hospitals have paid the ransom for. The prime example of this is Hollywood Presbyterian Hospital, who paid $17,000 in Bitcoin to get their data back after a 2016 attack (Winton 2016). The Online Trust Alliance, in their Cyber Incident and Breach Trends Report, cites that out of 159,700 total incidents analyzed by them in 2017, 93% were avoidable and could have been prevented with basic cyber hygiene (The Internet Society 2018). A report from the Department of Health and Human Services Office of Inspector General published in July 2016 reveals that only two-thirds of the 400 hospitals they surveyed had contingency plans that met the four HIPAA requirements (Levinson n.d.).

The emphasis of the Privacy and Security Rules is to provide a framework that organizations can follow and periodically review to check their compliance. The controls in the framework emphasize the need to have good Administrative controls in place to protect ePHI.

The HITECH Act and Its Applicability to Privacy and Security

The Health Information Technology for Economic and Clinical Health Act, also known as the HITECH Act, was enacted as part of the American Recovery and Reinvestment Act of 2009 and was signed into law on February 17, 2009 (McGraw n.d.) with an effective date of February 18, 2010 for the majority of the Act (HHS n. d.a). The main aim of the Act is promoting the adoption and meaningful use of Health Information Technology. However, Subtitle D addresses many of the privacy and security concerns of the Privacy and Security rule by doing the following:

(1) Clarifying who a BA is and providing explicit definitions. Before the HITECH Act, organizations had to enter into explicit contracts to define their responsibilities under HIPAA. Under the HITECH Act, BAs are required to comply with most provisions of the Security Rule. Entities that store, transmit, or process data on behalf of Covered Entities are also considered BAs, as are entities that contract with Covered Entities to allow them to offer a personal health record to patients (McGraw n.d.). This is important as many smaller hospitals and providers do not have the resources to offer patients electronic copies of their records, and this sets the minimum cybersecurity standards that they have to follow.

(2) Requiring organizations to notify individuals of breaches of their PHI within 60 days of discovery of a breach and defining what a breach is. A breach, according to the Department of Health and Human Services (HHS), is considered an impermissible use or disclosure under the Privacy Rule that compromises the security or privacy of the PHI.

    (a)   However, if encryption technology is used to protect the data, and is consistent with guidance given in NIST Special Publication 800–111, Guide to Storage Encryption Technology for End User Devices, or NIST Special Publications 800–52, Guidelines for the Selection and Use of Transport Layer Security (TLS) Implementations; 800–77, Guide to IPsec VPNs; or 800–113, Guide to SSL VPNs, or others which are Federal Information Processing Standards (FIPS) 140–2 validated, the data is not considered breached. In effect, this closes the encryption loop discussed earlier and establishes minimum standards for the encryption technologies to be used.

    (b)   If an unauthorized person who received it cannot be reasonably expected to have retained it, it is not a breach.

    (c)   If it's unintentional, access is within the scope of employment or a professional relationship, and not further disclosed, it is not a breach.

    (d)   If it's an inadvertent disclosure that occurs within a facility and the information does not go any further, it is not a breach (HHS n.d.a).

(3) Introducing four categories of violations that reflect increasing levels of culpability.

(4) Introducing four corresponding tiers of penalty amounts that increase the minimum penalty amount for each violation.

(5) Setting a maximum penalty amount of $1.5 million for all violations of an identical provision (HHS n.d.d)

(6) Striking the previous bar on the imposition of penalties if the covered entity did not know and with the exercise of reasonable diligence would not have known of the violation (such violations are now punishable under the lowest tier of penalties) (HHS n.d.a).

(7) Providing a prohibition on the imposition of penalties for any violation that is corrected within a 30-day time period, as long as the violation was not due to willful neglect. (HHS n.d.a).

(8) When organizations attest to Meaningful Use, requiring a Security Risk Assessment and Risk Management Plan as part of the attestation process for the reporting period (HHS n.d.a).

What this does in terms of Cybersecurity is provide enhancements to the Security Rule that put hard time limits on completion of Risk Assessments and Risk Management Plans, implicitly requires the usage of approved encryption technologies to protect PHI and requires organizations to consistently assess and address risk for their assets. This means that organizations that ignore updating vulnerable components of web server applications, or that do not update their operating systems will be penalized for willful neglect. In addition, they will need to come up with risk management plans that address identified risks (HHS n.d.a). These will need to be kept consistent for the device, hardware, and software in the environment or under their control. The HITECH Act, when properly interpreted, clears up the ambiguity of the Security Rule.

The European Union GDPR and Its Applicability to Healthcare

The EU General Data Protection Regulation, better known as GDPR, is a replacement for the existing EU Data Protection 95/46/EC (Intersoft n.d.b). It is designed to provide EU-wide data privacy laws. Due to the evolution of data usage, the laws needed to be updated. Due to the effects of both fascism and communism in the European Union, there has been a long history of organizations such as the Stasi, Nazis, or KGB using personal data to oppress citizens and dissidents, which provides the basis for such strong legislation.

The scope of GDPR, according to Article 3, includes two groups of entities:

(1) Firms located in the European Union that store and process Personal Data there.
(2) Firms not located in the European Union that directly and specifically offer goods and services to EU residents, or monitor the behavior and/or activity of EU residents (GDPREU.org n.d.).

If your organization specifically markets to or offers services to EU citizens, then GDPR will apply to your organization. The reason for inclusion here is because a significant amount of academic healthcare organizations openly solicit international patients.

GDPR is based around the concept of three entities. The Subject is any natural person whose personal data is being stored, processed, or transmitted. Controllers, which can be a natural or legal person, corporation, or government authority, establish the purpose, method, and reason for data processing either by themselves or in concert with others. Processors process data on behalf of Controllers or other Processors (Intersoft n.d.a). GDPR is designed to empower and protect EU Subjects from governmental or organizational overreach through enhanced privacy laws, provide transparency into Personal Data usage by data controllers and data processors, require data security by default when processing data, and allow citizens to control what information data controllers and processors have about them and require explicit consent to do so. This is a major change for many companies. However, one of the major issues is that organizations only focus on one of the main tenets, the Right to Be Forgotten, as opposed to the numerous others. GDPR has 99 articles, and an additional 171 requirements. It is a regulation, as opposed to a directive, and therefore is enforceable with fines and penalties.

Its scope is to all Data Controllers and Data Processors based in the European Union, and organizations that access the personal data of EU citizens (Intersoft n. d.c). This has a major impact because corporations that access data and explicitly provide and market services to EU citizens outside of the EU need to abide by the GDPR as well. This includes companies such as Facebook, Microsoft, and Google. Processors can include third parties.

According to Article 37 of the GDPR, these organizations need to have Data Protection Officer(s) assigned, who can be assigned to more than one organization (Intersoft n.d.d). They can be a contractor, employee, or company assigned to the role. Their contact information needs to be made available to the appropriate supervisory authorities. They also need to be professionally qualified to complete the duties assigned under Article 39. They need to inform the Controller and Processor(s) under their purview of their obligations under the GDPR, monitor compliance with GDPR and any local superseding provisions with regard to the protection of Personal Data, including the assignment of responsibilities, awareness, and training. They are also responsible for conducting and monitoring the required audits (Intersoft n.d.e).

They are also responsible for monitoring the Data Protection Impact Assessment (DPIA). A DPIA needs to be conducted when an organization is planning on implementing new processes or technologies that present a risk to Personal Data which can present a high risk to the rights and freedoms of natural persons. This means that if the processing of data can result in harms or damages to the Subject, DPIA needs to be done. A DPIA, under Article 35 of the GDPR (Intersoft n.d.f), is an assessment of the following:

(1) The intended usage of Personal Data by the Controller and/or Processor.
(2) The intended data elements to be used in processing operations.
(3) A description of the processing operations used to store, process, or manipulate the data.
(4) A description of the responsibilities of the Controller(s) and Processor(s) in the operations.
(5) A needs analysis of why these data elements are required.
(6) An assessment to the risks of rights and freedoms of Subjects.
(7) Intended countermeasures, including safeguards, security measures, and mechanisms to protect Personal Data and regulatory compliance in harmony with the GDPR, specifically Article 25, Data protection by design and default (Intersoft n.d.g)
(8) Intended countermeasures to protect the rights and individual interests of Data Subjects and other concerned parties.

(9) Consultation with the appropriate supervisory authorities during the process, as defined under Article 36, Prior Consultation (Intersoft n.d.d), which involves allowing them to review the DPIA, supporting processing documents, risk assessments, and intended countermeasures. The supervisory authorities are allowed up to eight weeks to provide an initial response, and can, under Article 58, order a deeper investigation, fines, reprimands, withdrawal of certification, or erasure of Personal Data if an organization is noncompliant (Intersoft n.d.h).

Organizations also need to prove that they are accountable by not only following Articles 37, 35, and 25, but also by demonstrating a culture of monitoring, reviewing, and assessing procedures regularly. They need to minimize data processing and retention to only what is needed by processes identified by the DPIA. They need to have the appropriate safeguards and countermeasures as defined by Article 25. They also need to take the Article 36, Prior Consultation process, seriously. Most critical is that Processors and Controllers assess, and address risk continually, as opposed to once.

The other important four important tenets are that of Explicit Consent, the Right to Be Forgotten, the Right for Data Portability, and the Right to Object to Profiling (Heimes n.d.; Intersoft n.d.h, n.d.i; Irwin n.d.). Subjects must freely give consent for their Personal Data to be used for explicit consent (Irwin n.d.). A number of US-based websites that use third-party ad networks for revenue generation have stopped offering services to EU citizens because they resell information without affirmative consent (Heimes n.d.).

The Right to Be Forgotten, Article 17 of the GDPR, allows Subjects to request that Personal Data held by Processors or Controllers concerning them be erased using reasonable and practical steps if the data is no longer necessary for the purpose for which it was collected or processed, they withdraw affirmative consent and there is no grounds for processing based on that, there is objection to the processing by the Subject and there is no legitimate need to do so, it's illegal, it has to be erased for compliance purposes, or it was collected on a person under the age of 16 without parental consent, or the child is now 16 and objects. However, there are five caveats to the Right to Be Forgotten, which are that it does not apply for exercising the right of freedom and information, does not take precedence over legal obligations or tasks done in the public interest, does not take precedence over legal obligations or tasks done in the public interest for public health, or for archiving purposes in the public interest, scientific or historical research purposes, or statistical purposes if the right impairs or makes impossible achieving the objectives of that processing, or the establishment, exercise, or defense of legal claims (Intersoft n.d.i).

What this really means is that all invocations of the Right to be Forgotten requiring organizations to examine the request in detail, and make a decision using informed resources so that the requests can be effectively processed. It is important to understand that Collectors and Processors cannot erase everything, nor should they, and that a cross-disciplinary approach is needed to effectively comply, not just blind erasure.

The Right to Data Portability requires organizations to provide data that a Subject has provided to a Controller or Processor to another provider if feasible, or to the Subject in a usable format (Intersoft n.d.h). The Right to Object to Profiling indicates that subjects have the right not to be subject to a decision solely based on automated processing (Intersoft n.d.k). This has significant ramifications for Artificial Intelligence and Machine Learning as this requires affirmative consent for their data to be used in the AI/ML decision making process, and that there needs to be a nonautomated process alternative in case they do not consent to it.

GDPR is a comprehensive framework that changes how companies operate. It empowers EU citizens to control their data. It has not been fully determined how much non-EU companies have to comply outside of the European Union, such as when EU patients are treated at facilities in the United States, but it is something that will be addressed fully in the future. However, it is a major change requiring companies to plan out how they store, process, and otherwise manage data.

PCI-DSS and How They Apply to Healthcare Organizations

In healthcare, the main way to pay bills is through credit or debit cards. The industry standard that organizations are required to implement to protect credit card data and comply with requirements are the Payment Card Industry – Data Security Standards. These requirements have been developed and maintained by the Payment Card Industry Security Standards Council (PCI-SSC), which include many of the merchant banks in the United States (TechTarget n.d.). There are 12 major requirements and 5 best practices to review to make sure that organizations are in full compliance (PCI SSC 2018).

PCI-DSS requires that organizations install and maintain firewalls configured to protect cardholder data. This does not mean that they just have a firewall. The devices that process credit cards have to be segmented away from the rest of the network. The vendor-supplied default passwords and security parameters need to be changed (TechTarget n.d.).

Stored cardholder data needs to be protected. This doesn't mean just putting it on a file share away from everything else. It needs to be segmented off from the rest of the network, encrypted, and protected with two-factor authentication, approved strong encryption, minimum access, and audit logging. Cardholder data in transit needs to be encrypted across the Internet and public networks. Anti–virus software needs to be used and kept updated as much as possible. Tracking and monitoring of all accesses, changes, and alerts from systems, especially anti–virus, is required (PCI SSC 2018).

Organizations need to develop and maintain secure applications. This doesn't mean to just develop an application or deploy a system once. The expectation is that they will be maintained regularly and kept in a secure state and assessed for risk regularly. As part of this, access will be restricted by need-to-know and documented. Access also needs to be reviewed and removed for those who no longer have a business need promptly. To do so, unique IDs need to be assigned to each person with access. Physical access to cardholder data must be restricted. All of the above needs to be tested and documented at least once a year, and with a corresponding policy which addresses information security (PCI SSC 2018; TechTarget n.d.).

The best practices are to monitor all security controls to ensure they are working effectively. If a control fails, make sure it is detected and responded to. Restore the control, identify the failure, identify and address any security issues that came about due to it, mitigate the issue, and then resume monitoring. Review changes to the environment, determine their impact, identify new requirements, and then update the scope and implement new or changed controls. Review structural changes to the organization and determine if those change PCI scope, and if they do, adjust the scope and requirements, and perform periodic reviews of the environment and organization to make sure that everyone is following the right processes (PCI SSC 2018). PCI-DSS, while not explicitly a healthcare regulation, governs a technology that is in use for financial payments every day. It is also congruent with HIPAA and GDPR in continually assessing and addressing risk.

21st Century CURES Act

The 21st Century CURES Act, Public Law 114–255, effective December 13, 2016, is designed to accelerate medical product development and bring new innovations and advances to patients that need them (114th Congress (Congress) n.d.; FDA n.d.a). As part of this, the Act promotes and supports secure interoperability between providers without special effort on the part of the user, and allow for complete access, exchange, and use of all electronically accessible health information. It establishes the governance needed to do so in Section 4003. Section 4004 defines Information Blocking as the implementation of healthcare information technology in such a way as to inhibit information sharing between providers. This can include development of technology as well as its implementation. Even implementing an Electronic Medical Records system with a high degree of customization can potentially put an organization in violation. Technology developers can be fined up to $1 million dollars per violation, and providers can lose federal incentives and potentially also have civil penalties levied (114th Congress (Congress) n.d.).

What this really means is that US healthcare is being incentivized to share information and develop information portability capabilities similar to what is required by the GDPR for EU citizens. Security and Privacy need to be built into the process, just like GDPR. The draft Trusted Exchange Framework that has been developed by the Office of the National Coordinator also requires two-factor authentication as part of the authentication process (ONC n.d.), further building security into interoperability.

The convergence of HIPAA, HITECH, GDPR, PCI-DSS, and the 21st Century CURES Act lead us to one set of conclusions. Organizations need to have a good unified plan and program to address cyber security threats and requirements. If your organization has a good core security program, they can not only address cyber threats, but also become a more resilient organization capable of meeting regulatory compliance requirements. An example of an attempt to come up with a unified plan is the Health Information Trust Common Security Framework (HITRUST CSF), version 9.2, which combines HIPAA, HITECH, GDPR, PCI-DSS, and Singapore's Personal Data Protection Act (PDPA) into one framework (HITRUST n.d.).

Key Programs and Processes Required for Effectively Managing Cybersecurity Response

Information Security Program

Healthcare organizations need to have several programs in place to effectively manage their cybersecurity response. The first, and most important, is to have an Information Security Program in place with an assigned Security Officer, whose overall responsibilities of this program are to identify, assess, and coordinate addressing of risks to administrative, technical, and physical controls in an organization (HHS n.d.a). While in the past, the Information Security Officer and their team, if they have one, would have day to day responsibility for implementation of solutions, the pervasiveness of the projects required to address cybersecurity needs require the entire organization to own components and tasks of it.

This is because many of the security components that once were separately managed, such as encryption, anti-malware protection, firewalls, and intrusion detection systems are now included in existing components such as switches, routers, desktop management, and desktop and embedded operating systems.

When there are dedicated security components, the level of integration required for tools such as Network Access Control, which allows network access based upon the security posture of a device, requires close communication between the security and network components (Cisco n.d.).

The Security Officer is also responsible for overseeing the administrative processes, such as overseeing review of system users and their access levels, policy and procedure development, assist in enforcement and investigations, provide information security training and security communication, performing HIPAA, PCI-DSS, security, and technical risk assessments and audits as part of the compliance process, and developing the risk management plan. They are also responsible for change management processes and ensuring that they do not reduce security. The Security Rule defines the requirements for Disaster Recovery, Downtime Procedures, and Alternate Mode operations (HHS n.d.j). As such, the Security Officer is also responsible for oversight of those. They are not responsible for the operation of the plans, however they are responsible for documenting them as part of risk assessments and participating in their development, as well as testing of the plans.

With the complexity of environments, they are also responsible for managing security during the asset intake and disposition processes. Devices and systems need to be evaluated for risk before they even come into the environment and need to be deployed securely. The security officer needs to make sure that devices such as medical equipment or systems such as Electronic Medical Record or virtual infrastructures are configured securely before they attach to the network. When the devices or computers are disposed of, PHI needs to be erased from the devices securely. For Cloud-based infrastructures, the Security Officer is responsible for ensuring that the configurations are also secure.

They are also responsible for oversight of the technical processes, including working with the technical teams so that devices, networks, and their configurations are reasonably and appropriately secure to the standards set in the HIPAA Security Rule, HITECH Act, and 21st Century CURES Act. This includes the oversight of encryption and key management processes, oversight of network security, and oversight of infrastructure management to ensure secure configurations.

In addition, they are also now responsible for working with the physical security teams to address physical security requirements to protect assets. This includes physical placement of devices to reduce the security and privacy risks from exposure to PHI, physical protection to reduce the risk of theft, and performing physical security risk assessments (HHS n.d.d). A corollary item some healthcare providers do is to collaborate on Hazard Vulnerability Analysis reports with the Physical Security Team (CHA n.d.).

Since Downtime Procedures, Disaster Recovery, and Alternate Mode Operations are part of the HIPAA Security Rule, a part of the responsibility of the Security Officer is to coordinate with the parties who own those on integrating

Information Security as part of these programs, as cybersecurity attacks causing system downtimes are now very common (HHS 2018d).

Privacy Program

The Privacy Officer is responsible for the implementation of a Privacy Program whose scope is managing the appropriate intake, usage, consents, and disclosures of protected information across an organization to minimize risk, ensure confidentiality, and maintain compliance with international, federal, state, and local laws and requirements for data intake, processing, consenting, and disposition/retention (AHIMA n.d.). This position is usually staffed by a lawyer because of the needed legal knowledge and training to be able to interpret laws, requirements, their applicability, and ethics standards and apply them to develop policies, procedures, and processes that meet organizational requirements.

The Privacy Program requires good organizational governance, including senior management, security, corporate compliance, and risk to be able to address and respond to identified issues. They need to establish solid policies, procedures, and processes to meet the requirements of managing the appropriate usage, consents, processing, intake, and disposition of data in an organization.

They also need to conduct privacy walkthroughs and risk assessments of organizations to make sure that all parties with access to protected information manage and use it in concordance with appropriate rules and regulations, and protect the integrity, security, and privacy of it across the organization. They own the development and maintenance of consent and authorization processes to make sure they are in accordance with applicable laws, regulations, and standards. They are also responsible for investigations and risk assessments of potential data and privacy breaches. They track and investigate potential breaches, and report breaches to the appropriate state and Federal authorities (HHS n.d.d).

They are responsible for the privacy education program, and ongoing education where required as part of it. This includes establishing channels by which organization members or concerned parties can report issues.

With the changing needs of cybersecurity, there is a heightened need to work with Information Security to establish Privacy Monitoring of systems containing Protected Health Information. This involves auditing all accesses to Electronic Medical Record (EMR) systems and reviewing them to see if people are accessing records outside of a defined Business, Payment, or Operations need, or if they are accessing records of friends, family, VIPs, or neighbors. This Privacy Monitoring also can include reviewing where the records were accessed from, such as from certain IP addresses, to see if an account has been compromised. It may also include proactive scanning of audit logs to see what records that an asset which has been compromised with malware accessed to identify potentially breached patients. It may include scanning audit logs to see what records a stolen or missing asset may have accessed and potentially stored (Cyngergistek n.d.). Finally, as part of an investigation, it may include all accesses from a certain employee. This is a constantly evolving process that requires good governance, excellent partnerships with Information Security and the technology teams, and constant monitoring for compliance.

Data Management, Classification, and Monitoring Program

The Data Management, Classification, and Monitoring program ties into the HIPAA Security Rule requirements for Data Management, Information Security, and Privacy. It also is required for PCI-DSS processing. The reason for this is to establish lifecycle management and intake, storage, processing, protection at rest, protection in transit, access requirements, retention, and disposition/disposal of data by an organization of its data. The reason for this program is to provide a single reference that organizations can use to determine the appropriate protection and management requirements for it.

As part of this, organizations need to identify where they store data, how it is stored, and where it flows throughout its lifecycle. This is required for HIPAA, PCI-DSS, and GDPR compliance. Privacy Monitoring is a way of sampling identified data flows for inappropriate access (Cyngergistek n.d.).

Data Loss Prevention (DLP) is a process and technology used to identify data that is not being adequately protected by monitoring data flows to identify data that is not flowing to appropriate destinations or is not being protected adequately (Zhang 2019). Examples of DLP include scanning workstations to see if data is being transferred to flash drives, scanning web traffic to see if unencrypted credit card data is being sent to websites, or scanning to see if Protected Health Information is being sent unprotected across the network or to the Internet. This can also include scanning for information being sent to suspect places or to "phishing" sites.

Cloud Access Security Brokers (CASB), a subset of DLP, are technologies used to monitor for data loss with cloud providers. As a lot of Cloud providers encrypt traffic to and from their environments, it requires organizations to interface with the Cloud providers, and potentially use integration via Application Programming Interfaces to identify potential DLP violations at the cloud providers (Netskope n.d.).

The Data Management, Classification, and Monitoring program either needs its own dedicated resource or needs to be managed by the Privacy and Security teams with input from Risk and Corporate Compliance. There needs to be governance either through data management or through existing Privacy and Security committee(s). This needs the attention and action of senior leadership to be effective.

Asset Management Program

Along with Data Management, there needs to be excellent Asset Management. HIPAA, PCI-DSS, and GDPR all require the identification of assets that will be storing or processing protected information (HHS n.d.d). An Asset Management program tracks the intake, location, information stored and processed, protection, demographic information, and ultimate disposition of an asset through its lifecycle in a company. It also tracks the software, operating system, configurations, and security patches loaded on a device. It can be done either automatically or manually. The use of asset management combined with technology such as Network Access Control, which is a gatekeeper to allow known devices onto the network, to identify assets that have not been identified or catalogued. This program needs to be managed and maintained by a dedicated Asset Management team, and requires synchronization with Supply Chain, Medical Device Management, Server Infrastructure, and Desktop Support along with Privacy and Security.

The cybersecurity benefit of Asset Management is that the cataloguing of assets will help organizations better discover and address vulnerabilities by being able to target and address them. It also provides information on the security posture of the environment at any given time, and what data resides where. This is useful for when machines have been targeted by malware, when you need to identify assets that need to be protected, or when you need to otherwise assess risk.

Risk Assessments

Information Security Risk Assessments, both technical and nontechnical, are absolutely critical for organizations. They are required by HIPAA, HITECH, GDPR, and PCI-DSS at least annually. They should also be done for incoming assets and systems. Ideally, the risk assessment will have a quantitative score from 0 to 5, 0 being no risk, and 5 being critical, to measure the likelihood of each risk occurring, its overall risk, and its impacts on income, patient satisfaction, employee engagement, operations, and reputation. The risk assessment should use a standard format, such as the System Readiness Assessment tool from the Center for Medicare and Medicaid Services (CMS), which gives providers a basic risk management framework they can use to conduct their own risk assessments (HHS n.d.j). They need to be done at least yearly and during major changes such as an EMR implementation to identify known risks so that they can be followed up on. These are excellent at identifying technical, administrative, and physical risks and give you not only a background as to what they are, but what you should address based on the impact.

Risk Management Plan

The Risk Management Plan, which follows up on the Risk Assessment, takes the identified top risks and assigns resources to address them, provides a documented plan to address them within a fixed time period. The risk, identified resources, project plan, completion date, and follow-up dates need to be identified. This helps identify and put timelines on cyber risks and allows organizations to plan remediations to risks and budget money and resources for them.

Ten Emerging Threats More Prevalent in the Healthcare Environment

There are currently ten major types of cyberattacks and threats that cause most of the events you read about on the Internet or in newspapers. We will look at each of these attacks and discuss how they can threaten the security of the network and the data stored on it.

Ransomware

Ransomware, the most common type of attack now, works by utilizing any number of security vulnerabilities in common programs such as document viewers, web browsers, or operating systems to execute itself and encrypt files or databases. To get your files back, you need to pay a "ransom" in cryptocurrency, usually Bitcoin or Ethereum, to get a key to unlock your files. Ransomware works because many organizations do not back up their data, do not have effective downtime procedures, or do not have tested alternate mode operations, and because of that will pay the cost to get their business back. A number of healthcare organizations such as Hollywood Presbyterian Hospital and municipal governments have fallen victim to Ransomware. However, files containing PHI that have been encrypted by Ransomware are considered breached unless you can affirmatively prove that there has been a low probability of a breach (HHS 2016).

Phishing

Phishing works by tricking the user into responding to or clicking on a link in a fake message, usually official looking or sounding, from a malicious third party, and using that to retrieve credentials for the purpose of either impersonating that user to send more phishing emails, using those credentials to log into sites and take control of bank accounts, cryptocurrency wallets, or personal information, or to impersonate someone for more targeted attacks (Fruhlinger 2018). An example is the attack that happened to New York Oncology Hematology, where 15 employees fell for one of these schemes and caused a data breach of 128,000 employees and patients (Davis 2018). Children's Mercy Hospital in Kansas City, MO had to notify 63,049 patients and family members after employees fell for a phishing scheme and caused data breaches (Paavola 2018).

Business Email Compromise

Business Email Compromise is a targeted attack that uses either a compromised account from a successful Phishing attack, a forged email address, or a throwaway free account to impersonate an executive for the purpose of convincing someone with access to make financial transactions to maliciously send money to bank accounts in their control. They normally use a false sense of urgency to attempt to convince their targets to make these transactions without following standard processes or controls that would otherwise thwart this attack (FBI n.d.). Southern Oregon University had a $2 million loss due to this type of attack in 2017 (Dellinger 2017). The Federal Bureau of Investigation, in a Public Service Announcement from July 12, 2018, indicated that they had recorded 78,617 domestic and international incidents of these with an exposed dollar loss of $12,536,948,299 across multiple industries (FBI 2018).

System Currency

System Currency and Keeping Software Updated is the main reason many of these breaches occur in the first place. The healthcare world is well-known for its use of legacy software. WannaCry, Petya/NotPetya, SamSam, and other variants of malware spread by taking advantage of known vulnerabilities in Microsoft software that had been patched. Not keeping software patched has caused a very large number of data breaches and is easily preventable.

Healthcare Data Breaches

Intentional and Unintentional data breaches, such as copying data to an unprotected flash drive and then losing it, having an unprotected laptop stolen, or having a train car full of trash spread papers of unprotected PHI through a neighborhood are a persistent threat. Employees snooping on medical records that they do not have a business need for purposes of Payment, Treatment, or Operations is also a major threat. A major example of a data breach involving a flash drive is with the MD Anderson Cancer Center data breach, where two unencrypted USB Flash drives containing records on 33,500 patients were lost, and they had to pay $4.3 million in penalties (HHS n.d.g). A contemporary example of medical records snooping happened to reality TV star Kim Kardashian, whose medical records were accessed by six people at Cedars-Sinai in Los Angeles after she gave birth to her daughter there (Gorman and Seweel 2013). Those six people were later terminated after an investigation. An example of a data breach involving remote access is with the one that happened at Community Health Systems in 2014, where a security vulnerability in the OpenSSL library used to encrypt data known as Heartbleed was used to access the medical records of 4.5 million patients (Ragan n.d.).

Coding/Programming Errors

Coding errors in software are often exploited by malicious third parties to extract data from databases or systems. SQL Injection attacks, which work by overriding legitimate database commands either programmatically or through exploiting the lack of input validation in web programming, are the most common manifestation of this (OWASP.org. n.d.). However, buffer overflow, heap overflow, and stack overflow attacks, which take advantage of poor input validation techniques, poor bounds checking (often in the C programming language), and insecure coding, are also prevalent. SQL Injection attacks were the cause of a data breach that affected the company Medical Informatics Engineering, and its subsidiary, NoMoreClipboard, who lost 3.9 million personal records in 2015. Attorneys general from 12 states are suing this company, based in Indiana, because they failed to secure their computer systems (Bradbury 2018).

User Management

User management and the failure to remove unneeded or unused accounts is also a consistent threat. A number of cases prosecuted by the US Justice Department's Computer Crimes and Intellectual Property Section have involved terminated employees that have retained access to systems at their previous employers and either utilized that data for gain elsewhere or maliciously attacked their old companies (US DOJ n.d.). Memorial Health System paid a $5.5 million penalty to the US government after a data breach that affected 115,143 individuals caused by the use of a former employee's credentials to access protected health information (HHS n.d.k). It is important to constantly validate who has access to systems, to terminate system and remote access immediately upon termination, and to review access immediately after a job role change.

Wireless Security

Wireless security is also a constant concern. Many wireless devices either still use no encryption, the Wired Equivalent Privacy (WEP) Protocol, which was broken in 2007, or older encryption schemes such as Wi-Fi Protected Access, which does not encrypt data with strong encryption (Schneier 2007). The use of no or weak encryption on wireless allows malicious actors to maliciously eavesdrop on traffic and hijack account credentials. While this example was not in healthcare, TJX, the parent company of TK Maxx, was breached because of lax wireless security, and 45.6 million credit cards were taken from their network because of their use of WEP (Espiner 2007).

Network Hijacking

A newer type of attack is Network Hijacking. Well-funded nation-states such as China and Russia are able to either hack existing network providers or manipulate networks under their control to redirect traffic from the Internet to computers they control for the purpose of impersonating legitimate sites and taking advantage of weak security controls in software for hijacking information or injecting their own. The Amazon Route 53 DNS attack utilized a technique called Border Gateway Protocol (BGP) hijacking to reroute traffic to Amazon Domain Name Server (DNS) servers through servers in Russia. This allowed malicious hackers to steal $23 million in cryptocurrency (Goodin 2018). There is also a malware that targets home routers and redirects traffic to hostile websites which hijack passwords, personal data, and bank account information such as VPNFilter (Symantec 2018).

Medical Device Security

Medical Device Security involves attacks directly aimed at compromising medical devices. A common issue with them is that their software and firmware are never updated for a number of reasons (FDA n.d.b). Many of these devices run either Windows or embedded Linux. Many healthcare providers also do not segment off their networks to protect medical devices. They will run them on their wireless network or will put them on open networks that everyone can see. This allows malicious actors to easily compromise them and use them to either attack other devices, steal information, or further compromise the integrity and behavior of hospital networks. The Wannacry malware attack, attributed to North Korea, which crippled numerous organizations in the United States and United Kingdom, directly attacked medical devices in the United States, including a Bayer Medrad device. This was the first time that a malware attack had directly attacked medical devices (Brewster 2017).

How Can We Address These Threats for Healthcare Organizations and Professionals?

We can take a structured approach and utilize fifteen techniques to increase our defenses to greatly improve security and bring us in compliance with HIPAA,

PCI-DSS, HITECH, and GDPR. These techniques will make your organization more resilient and increase your chances of surviving a cybersecurity attack. These are the both the implementation of the mandatory regulatory requirements of the HIPAA Security Rule, and recommendations from PCI-DSS. The application of these basic controls and best practices will not only bring your health care organization into regulatory compliance, but also significantly reduce risk. The goals of these controls, based on HIPAA and PCI-DSS, are to provide organizations with the basic controls needed to maintain effective cybersecurity and demonstrable evidence of compliance.

Risk Assessments

Conducting risk assessments at least annually will not only bring your organization into regulatory compliance with the HIPAA Security Rule, it will also allow you to identify key areas for improvement. Utilizing a quantitatively scored risk assessment that addresses the items of highest impact and likelihood first will allow you to prioritize what risks to address and how quickly you need to address them. If possible, use the CMS System Readiness Assessment tool in combination with numeric scoring to identify your top risks. PCI-DSS also requires at least an annual risk assessment.

Training/Communication Plan

Your workforce needs to know what to do. The HIPAA Security Rule requires that you have comprehensive training on an ongoing basis that addresses Information Security (HHS n.d.d). This training should be updated at least yearly. There also needs to be at least monthly communication on emergent issues, and if there is an emergency, immediate communication. A good training program will discuss what actions team members need to take, why they need to take them, and what they can do to further assist. Constant communication on how team members can report vulnerabilities or issues to Privacy or Security is also highly recommended.

Asset Management

Having a good asset management program allows you to quickly identify vulnerable assets and develop remediation plans. If you do not know what you own, how can you protect it? You need to be able to identify and locate your assets so you can address risks as they arise and ensure that all assets are reasonably and appropriately protected against high risks. It is also required by the Security Rule to protect physical and digital assets. Physical assets include physical computers, hardware, networks, backup tapes, smartphones, flash drives, and paper files and data. Digital assets include software, medical data, computer files, and images, including pictures and radiology images.

Vulnerability Management

When vulnerabilities occur, they need to be patched or protected against as quickly as possible. A good vulnerability management program involves coordination with the IT department to address identified vulnerabilities through the application of remediation processes to the applicable assets. These processes can include configuration changes, software patches, or network changes. Penetration test, which is where someone attempts to break into a network for the purpose of identifying security holes, is also a form of vulnerability management. Vulnerability management, which is the identification of risks and the plan to remediate them, is written up in the Security Rule as "Protect against any reasonably anticipated threats or hazards to the security or integrity of such information" (HHS n.d.d).

Downtime Procedures/Alternate Mode Operations

When a system downtime or cyberattack happens, your organization needs to be able to continue business without the computers being fully operational, or operational at all. Ransomware is effective because it is disruptive and can cause organizations that do not have downtime procedures, which are steps that an organization can take to remain operational without key critical systems such as EMRs working. These procedures need to be updated and tested with team members at least yearly. The HIPAA Security Rule also requires this.

Data Backup

Ransomware is also successful because many organizations do not back up their data, do not test to see how long it takes to restore it, or do not test the backups at all. In the case of Hancock Regional Hospital in Indiana, ransomware actually affected the backups (Long n.d.). Often the ransomed data is the only copy the organization has. Having good tested data backups and knowing how long it takes to restore backups of critical systems will be of great assistance in successfully recovering from an attack. Also, organizations, in light of the Hancock incident, need to make sure that ransomware does not affect them. The HIPAA Security Rule, finally, also requires them.

Data Loss Prevention/Cloud Access Security Broker

The use of DLP/CASB to identify either the intentional or unintentional exfiltration of protected data outside the assets or cloud services authorized to access it will assist in several ways. First, it will help identify potential exfiltration of data by malware from infected PCs. It can also find misconfigured websites or cloud services not protecting data. It can also find potential insider threats of people sending out data for either personal gain or criminal purposes. The

Security Rule requires this because organizations are required to "Protect against any reasonably anticipated uses or disclosures of such information that are not permitted or required under subpart E of this part" (HHS n.d.d). The use of these technologies helps guard against these potential unauthorized disclosures.

Continual System and Security Monitoring

Having logs and audit data is not enough. You need to look at this data and examine it for potential anomalies. The use of a Security Incident and Event Management (SIEM) system to scan the copious amounts of log files and audit data is required to be able to meet the requirements of HIPAA, PCI-DSS, and GDPR. This is because it is nearly impossible to accurately scan this data and make associations and correlations to find anomalies or issues. Systems also need to be configured to send their log data to the SIEM to be able to give it enough data for analysis to find said anomalies. HIPAA, in particular, requires the review of information systems activity in the Security Rule to determine potential violations of patient privacy and security (HHS n.d.d).

User Access Reviews

As we previously discussed, a number of data breaches have been caused by access to accounts by people that were not currently employed at the companies they pilfered data from or hacked. It is required under HIPAA, PCI-DSS, and GDPR to review access to systems, promptly make sure that people who do not need access have it removed, and review access and entitlements of team members who have changed roles to make sure that they only access what they need to.

Privacy Monitoring

Under both HIPAA and GDPR, it is required to monitor access to systems to ensure that people access Personal Data or PHI for business and job-related purposes. Privacy Monitoring is the overall process by which organizations can detect potential privacy violations and address them. In healthcare, looking at the records of patients you are not involved in treatment, operations, or arranging payment for is a breach. This includes your own family, friends, and neighbors. With GDPR, viewing Personal Data outside of the documented business processes and workflows is also considered a breach.

Encryption and Key Management

The HITECH Act and HIPAA mandate the use of strong encryption to protect PHI in transit and at rest. PCI-DSS requires it to protect cardholder data in transit and at rest. Strong encryption is no longer an addressable issue, it is a standard part of doing business. Encrypting Personal Data, credit card data, and PHI, and protecting the encryption keys from potential compromise is a requirement to handle it. Having good processes to manage encryption keys and passwords will help protect the organization.

Distributed Ledger Technologies

The use of Distributed Ledger Technologies to exchange data between mutually exclusive members of consortiums can greatly help verify and validate who made transactions and when and provide an authoritative record of them. The Security Rule requires that organizations protect the confidentiality, integrity, and availability of data. This is another means by which they can do so using cryptographic hashing, identity management, and encryption.

Digital Distributed Ledger Technologies provide verification and validation services between mutually exclusive and independent participants where there is no central authority, by utilizing a synchronized shared ledger (distributed ledger) replicated and distributed across participants, mutually agreeable record and transaction formats that can be verified and validated by source systems, Strong cryptographic verification and cross-validation of record entries to ensure immutability and prevent tampering or double-entries, processes for ensuring consensus and consistency of the ledger view by all participants utilizing strong cryptography, appropriate consensus mechanisms, and cryptographic hashing algorithms to link entries together, and technologies that allow participants to utilize Application Programming Interfaces or utilities to enable reading or writing to the distributed ledger and enable records and transactions to be transferred between the Distributed Ledger and authoritative data sources and/or systems.

Note: Participants could be further defined, such as size of node, light or heavy processing, light or heavy devices or systems, and so on (Ray n.d.).

Network Segmentation and Design/Zero Trust Networks

Building on the principles of Minimum Necessary from the HIPAA Privacy Rule, this involves only giving minimum network access to assets that explicitly need it and explicitly forbidding any unnecessary communications. This involves the use of firewalls, switching, and Access Control Lists to explicitly permit only required traffic and deny the rest. The use of this technique greatly reduces the attack surface of the network and reduces the potential risk of medical device compromise. It buys time so that limited resources can effectively patch these devices.

Effective Contract Management

Security management of Cloud providers and third parties is not enforceable without good binding legal contracts. The Business Associate Agreement should not only specify the security parameters required, it needs to also specify the methods and processes by which third parties can comply, and what requirements they need to follow. This is so that systems and devices that are going to be used on behalf of your organization are able to be monitored for compliance and security. The HIPAA Security rule also requires that organizations appropriately safeguard the information.

Risk Management Plan

Finally, nothing is effective without a constantly managed and updated Risk Management Plan. Building on the information from the risk assessments, vulnerability management program, security monitoring, and privacy monitoring systems, this is the comprehensive plan that will guide your organization's security program. This plan has to have assigned tasks, resources, plans, and definitive and realistic dates that need to be met. If there is no plan, there is no way of measuring progress across the organization, and it will be significantly more difficult to manage the Information Security program. It is also required by the HIPAA Security Rule.

Conclusion

This chapter's aim was to assist the reader in understanding the regulatory environment behind data protection and cybersecurity in healthcare, including HIPAA, HITECH, PCI-DSS, GDPR, and the 21st Century CURES Act. We discussed how HIPAA started out as providing a basic set of Privacy and Security controls and definitions, specifically of BAs and Covered Entities for organizations, and how they emphasized the need to continually review and monitor processes. We then discussed how the HITECH Act cleared up ambiguity in the HIPAA Security Rule, specifically with the definitions of BAs, specifics of when to notify affected parties of breaches, increased administrative penalties, and acceptable encryption standards for protecting Protected Health Information. The GDPR was then looked at in detail, with the emphasis on the Article 35 Data Protection Impact Analysis, the rights of Natural Persons, and the need for organizations to be able to articulate their data processes and flows. As healthcare relies significantly on payments from patients and families, the Payment Card Industry – Data Security Standards (PCI-DSS) standards, and steps organizations need to take to be compliant, specifically with protection and segmentation of cardholder data, and the required continual review and assessment steps. The 21st Century CURES Act, with its emphasis on secure interoperability using documented processes and standards that require privacy and security to be designed in, and increased financial penalties for organizations that do not comply, was explained, particularly the requirement for organizations to not engage in Data Blocking.

Healthcare organizations need to have six key programs in place to protect themselves against cybersecurity attacks and remain in regulatory compliance. An Information Security program with an assigned Information Security officer to oversee information security efforts is where they need to start. A Privacy Program to effectively monitor how organizations administratively protect patient data is also required by the HIPAA Privacy Rule. Knowing where data resides, where it flows, and monitoring it for appropriate use and disclosure is the responsibility of the Data Management, Classification, and Monitoring program. Knowing what assets a healthcare organization has and what data is stored and processed by them is the mission of the Asset Management program. Risk Assessments are the method and process by which healthcare organizations can effectively quantify and measure risk and determine what the greatest threats are. Risk Management plans take the input of the risk assessment and are used to develop processes and steps by which organizations can effectively mitigate risk.

The ten most dangerous cybersecurity threats to healthcare organizations were then explored. These include Ransomware, Phishing, Business Email Compromise, System Currency (keeping systems current so security risks can be addressed with patches and fixes), Data Breaches, Coding/Programming Errors (especially SQL Injection and Buffer Overflow attacks), User Management, specifically not removing users who no longer need access, Wireless Security, Network Hijacking, and Medical Device Security. The case of MD Anderson Cancer Center, who had a data breach and subsequent $4.3 million-dollar administrative penalty caused by the loss of two unencrypted flash drives, was extrapolated.

Finally, we looked at the fifteen steps an organization can take to help greatly improve their resiliency and reduce their risk of cyberattack, based on the HIPAA Security Rule and PCI-DSS. These are basic steps we covered earlier in the chapter. They include risk assessments, effective training and communication plans, asset management, vulnerability management, tested downtime procedures and alternate mode operations, data backup, DLP/CASB, Continual System and Security Monitoring, User Access Reviews, Privacy Monitoring, Encryption and Key Management, the potential use of Distributed Ledger Technologies, Zero Trust/Segmented Networks, Effective Contract Management, and finally, a Risk Management Plan to tie it all together. The cases of Hancock Regional Hospital and MD Anderson show that there is a need for continual security monitoring.

The most important aspect to take away from this chapter is that security is continuous, and that it takes continual effort and following up to assess and address risk. The work is never done, and threats continue to evolve. However, we can work to make our organizations significantly more resilient from these attacks.

References

American Health Information Management Association (AHIMA). n.d. "Sample (chief) privacy officer job description". Accessed October 31, 2018. http://bok.ahima.org/doc?oid=107672#.W9h4FmJKjxs.

Bosshart, Angie. n.d. "Data breach notification". Accessed February 23, 2019. www.chs.net/media-notice/.

Bradbury, Danny. 2018. "Unencrypted medical data leads to 12-state litigation". Accessed December 7, 2018. https://nakedsecurity.sophos.com/2018/12/07/unencrypted-medical-data-leads-to-12-state-litigation/.

Brewster, Thomas. 2017. "Medical devices hit by ransomware for the first time in US hospitals". Accessed May 17, 2017. www.forbes.com/sites/thomasbrewster/2017/05/17/wannacry-ransomware-hit-real-medical-devices/#275f2e14425c.

California Hospital Association (CHA). n.d. "Hazards vulnerability analysis". Accessed October 31, 2018. www.calhospitalprepare.org/hazard-vulnerability-analysis.

Cisco Systems, Inc. n.d. "What is network access control?" Accessed October 31, 2018. www.cisco.com/c/en/us/products/security/what-is-network-access-control-nac.html.

114th Congress (Congress). n.d. "Public Law 114–255". Accessed October 31, 2018. www.congress.gov/114/plaws/publ255/PLAW-114publ255.pdf.

Cyngergistek, Inc. n.d. "Guide to proactive access monitoring and auditing under the HIPAA security rule". Accessed October 31, 2018. https://cynergistek.com/blog/guide-proactive-access-monitoring-auditing-hipaa-security-rule/.

Davis, Jessica. 2018. "Phishing attack impacts health data of 128K employees, patients". Accessed November 19, 2018. https://healthitsecurity.com/news/phishing-attack-impacts-health-data-of-128k-employees-patients.

Dellinger, AJ 2017. "Fraudulent email: business email compromise attack costs southern Oregon University $2M". Accessed June 3, 2017. www.ibtimes.com/fradulent-email-business-email-compromise-attack-costs-southern-oregon-university-2m-2551724.

Department of Health and Human Services (HHS). 2016. "FACT SHEET: ransomware and HIPAA". Accessed July 11, 2016. www.hhs.gov/sites/default/files/Ransomware FactSheet.pdf.

Espiner, Tom. 2007. "Wi-Fi hack caused TK maxx security breach". Accessed May 8, 2007. www.zdnet.com/article/wi-fi-hack-caused-tk-maxx-security-breach/.

Federal Bureau of Investigation (FBI). 2018. "Business email compromise the 12 billion dollar scam". Accessed July 12, 2018. www.ic3.gov/media/2018/180712.aspx.

Federal Bureau of Investigation (FBI). n.d. "Business E-mail compromise – FBI". Accessed October 31, 2018. www.fbi.gov/news/stories/business-e-mail-compromise-on-the-rise.

Flahive, Paul. n.d. "MD Anderson cancer center fined $4.3 million in data breach". Accessed June 19, 2018. www.tpr.org/post/md-anderson-cancer-center-fined-43-m-data-breach.

Fruhlinger, Josh. 2018. "What is phishing? How this cyber attack works and how to prevent it". Accessed October 31, 2018. www.csoonline.com/article/2117843/phishing/what-is-phishing-how-this-cyber-attack-works-and-how-to-prevent-it.html.

GDPREU.org. n.d. "Who must comply". Accessed October 31, 2018. www.gdpreu.org/the-regulation/who-must-comply/.

Goodin, Dan. 2018. "Suspicious event hijacks Amazon traffic for 2 hours, steals cryptocurrency". Accessed October 31, 2018. https://arstechnica.com/information-technology/2018/04/suspicious-event-hijacks-amazon-traffic-for-2-hours-steals-cryptocurrency/.

Gorman, Anna, and Seweel, Abby. 2013. "Six people fired from Cedars-Sinai over patient privacy breaches". Accessed February 4, 2019. http://articles.latimes.com/2013/jul/12/local/la-me-hospital-security-breach-20130713.

Heimes, Rita. n.d. "Top 10 operational impacts of the GDPR: Part 5 – Profiling". Accessed October 31, 2018. https://iapp.org/news/a/top-10-operational-impacts-of-the-gdpr-part-5-profiling/.

HIPAA Journal. n.d. "HIPAA history". Accessed October 30, 2018. www.hipaajournal.com/hipaa-history/.

HITRUST. n.d. "HITRUST CSF". Accessed February 4, 2019. https://hitrustalliance.net/hitrust-csf/.

Intersoft Consulting. n.d.b. "Art. 1 GDPR subject-matter and objectives". Accessed October 31, 2018. https://gdpr-info.eu/art-1-gdpr/.

Intersoft Consulting. n.d.c. "Art. 2 GDPR material scope". Accessed October 31, 2018. https://gdpr-info.eu/art-2-gdpr/.

Intersoft Consulting. n.d.a. "Art. 4 GDPR definitions". Accessed October 31, 2018. https://gdpr-info.eu/art-4-gdpr/.

Intersoft Consulting. n.d.j. "Art. 17 right to erasure ('right to be forgotten')". Accessed October 31, 2018. https://gdpr-info.eu/art-17-gdpr/.

Intersoft Consulting. n.d.i. "Art. 20 right to data portability". Accessed October 31, 2018. https://gdpr-info.eu/art-20-gdpr/.

Intersoft Consulting. n.d.k. "Art. 22 automated individual decision-making, including profiling". Accessed October 31, 2018. https://gdpr-info.eu/art-22-gdpr/.

Intersoft Consulting. n.d.g. "Art. 25 GDPR data protection by design and by default". Accessed October 31, 2018. https://gdpr-info.eu/art-25-gdpr/.

Intersoft Consulting. n.d.f. "Art. 35 GDPR data protection impact assessment". Accessed October 31, 2018. https://gdpr-info.eu/art-35-gdpr/.

Intersoft Consulting. n.d.d. "Art. 37 GDPR designation of the data protection officer". Accessed October 31, 2018. https://gdpr-info.eu/art-37-gdpr/.

Intersoft Consulting. n.d.e. "Art. 39 GDPR tasks of the data protection officer". Accessed October 31, 2018. https://gdpr-info.eu/art-39-gdpr/.

Intersoft Consulting. n.d.h. "Art. 58 GDPR powers". Accessed October 31, 2018. https://gdpr-info.eu/art-58-gdpr/.

Irwin, Luke. n.d. "Gaining explicit consent under the GDPR". Accessed October 31, 2018. www.itgovernance.eu/blog/en/gaining-explicit-consent-under-the-gdpr-2.

Levinson, Daniel R. n.d. "Department of Health and Human Services – Office of Inspector General – Hospitals largely reported addressing requirements for EHR contingency plans". Accessed January 10, 2019. https://oig.hhs.gov/oei/reports/oei-01-14-00570.pdf.

Long, Steve. n.d. "The cyber attack – From the POV of the CEO". Accessed February 6, 2019. www.hancockregionalhospital.org/2018/01/cyber-attack-pov-ceo/.

McGraw, Deven. n.d. "Summary of health privacy provisions in the 2009 economic stimulus legislation". Accessed October 31, 2018. www.digitalbusinesslawgroup. com/CDT_20090324_ARRA_HIPAA_Privacy_Summary.pdf.

Netskope, Inc. n.d. "What is a cloud access security broker?". Accessed October 31, 2018. www.netskope.com/company/about-casb.

Office of the National Coordinator for Health Information Technology (ONC). n.d. "Trusted exchange framework and common agreement". Accessed October 31, 2018. www.healthit.gov/topic/interoperability/trusted-exchange-framework-and-common-agreement.

OWASP.org. n.d. "SQL injection". Accessed October 31, 2018. www.owasp.org/index. php/SQL_Injection.

Paavola, Alia. 2018. "Phishing attack hits children's Mercy Hospital, 63K people affected". Accessed July 5, 2018. www.beckershospitalreview.com/cybersecurity/phishing-attack-hits-children-s-mercy-hospital-63k-people-affected.html.

PCI Security Standards Council (PCI SSC). 2018. "Payment Card Industry (PCI) data security standard – Requirements and security assessment procedures – Version 3.2.1". Accessed May 16, 2018. www.pcisecuritystandards.org/documents/ PCI_DSS_v3-2-1.pdf.

Ragan, Steve. n.d. "Heartbleed to blame for community health systems breach". Accessed February 4, 2019. www.csoonline.com/article/2466726/data-protection/data-protec tion-heartbleed-to-blame-for-community-health-systems-breach.html.

Ray, Shann. n.d. "The difference between blockchains & distributed ledger technology". Accessed October 31, 2018. https://towardsdatascience.com/the-difference-between-blockchains-distributed-ledger-technology-42715a0fa92.

Schneier, Bruce. 2007. "Breaking WEP in under a minute". Accessed April 4, 2007. www. schneier.com/blog/archives/2007/04/breaking_wep_in.html.

Symantec Corporation. 2018. "VPNFilter: New router malware with destructive capabilities". Accessed May 23, 2018. www.symantec.com/blogs/threat-intelligence/ vpnfilter-iot-malware.

TechTarget. n.d. "PCI DSS 12 requirements". Accessed October 31, 2018. https://searchse curity.techtarget.com/definition/PCI-DSS-12-requirements.

Teichert, Erica. 2018. "Anthem to pay $16M in record data breach settlement". Accessed October16, 2018. www.modernhealthcare.com/article/20181016/NEWS/181019927.

The Internet Society. 2018. "Cyber incident & breach trends report". Accessed January 25, 2018. www.otalliance.org/system/files/files/initiative/documents/ota_cyber_incident_ trends_report_jan2018.pdf.

U.S. Food & Drug Administration (FDA). n.d.a. "21st century cures act". Accessed October 31, 2018. www.fda.gov/regulatoryinformation/lawsenforcedbyfda/significan tamendmentstothefdcact/21stcenturycuresact/default.htm.

U.S. Food & Drug Administration (FDA). n.d.b. "Cybersecurity". Accessed October 31, 2018. www.fda.gov/medicaldevices/productsandmedicalprocedures/ucm373213.htm.

U.S. Department of Health and Human Services (HHS). n.d.a. "45 CFR parts 160 and 164 – Breach notification for unsecured protected health information; Interim final rule". Accessed October 31, 2018. www.gpo.gov/fdsys/pkg/FR-2009-08-24/pdf/E9-20169.pdf.

U.S. Department of Health and Human Services (HHS). n.d.e. "Business associate contracts". Accessed October 29, 2018. www.hhs.gov/hipaa/for-professionals/cov ered-entities/sample-business-associate-agreement-provisions/index.html.

U.S. Department of Health and Human Services (HHS). n.d.b. "HIPAA act enforcement interim final rule". Accessed October 31, 2018. www.hhs.gov/hipaa/for-professionals/ special-topics/hitech-act-enforcement-interim-final-rule/index.html.

U.S. Department of Health and Human Services (HHS). n.d.f. "HIPAA administrative simplification". Accessed October 31, 2018. www.hhs.gov/sites/default/files/ocr/priv acy/hipaa/administrative/combined/hipaa-simplification-201303.pdf.

U.S. Department of Health and Human Services (HHS). n.d.c "HIPAA security series – 6 basics of risk analysis and risk management". Accessed January 10, 2019. www.hhs.gov/ sites/default/files/ocr/privacy/hipaa/administrative/securityrule/riskassessment.pdf.

U.S. Department of Health and Human Services (HHS). n.d.d "HITECH act enforcement interim final rule". Accessed January 10, 2019. www.hhs.gov/hipaa/for-professionals/ special-topics/hitech-act-enforcement-interim-final-rule/index.html.

U.S. Department of Health and Human Services (HHS). n.d.g. "Judge rules in favor of OCR and requires a Texas cancer center to pay $4.3 million in penalties for HIPAA violations". Accessed February 4, 2019. www.hhs.gov/about/news/2018/06/18/judge-rules-in-favor-of-ocr-and-requires-texas-cancer-center-to-pay-4.3-million-in-penal ties-for-hipaa-violations.html.

U.S. Department of Health and Human Services (HHS). n.d.h. "Minimum necessary requirements". Accessed October 31, 2018. www.hhs.gov/hipaa/for-professionals/ privacy/guidance/minimum-necessary-requirement/index.html.

U.S. Department of Health and Human Services (HHS). n.d.i. "Security risk analysis tip sheet: Protect patient health information". Accessed October 31, 2018. www.cms.gov/ Regulations-and-Guidance/Legislation/EHRIncentivePrograms/Downloads/ 2016_SecurityRiskAnalysis.pdf.

U.S. Department of Health and Human Services (HHS). n.d.j. "Security risk assessment". Accessed February 4, 2019. www.healthit.gov/topic/privacy-security-and-hipaa/secur ity-risk-assessment.

U.S. Department of Health and Human Services (HHS). n.d.k. "Summary of the HIPAA security rule". Accessed October 31, 2018. www.hhs.gov/hipaa/for-professionals/ security/laws-regulations/index.html.

U.S. Department of Health and Human Services (HHS) Press Office. n.d.k "$5.5 million HIPAA settlement shines light on the importance of audit controls". Accessed February 16, 2017. www.hhs.gov/about/news/2017/02/16/hipaa-settlement-shines-light-on-the-importance-of-audit-controls.html.

U.S. Department of Justice (US DOJ). n.d. "CCIPS press releases – 2018". Accessed October 31, 2018. www.justice.gov/criminal-ccips/ccips-press-releases-2018.

Winton, Richard. 2016. "Hollywood hospital pays $17,000 in bitcoin to attackers; FBI investigating". Accessed February 17, 2016. www.latimes.com/business/technology/ la-me-ln-hollywood-hospital-bitcoin-20160217-story.html.

Zhang, Ellen. 2019. "What is Data Loss Prevention (DLP)? A definition of data loss prevention". Accessed January 3, 2019. https://digitalguardian.com/blog/what-data-loss-prevention-dlp-definition-data-loss-prevention.

Chapter 10

Mobile Cybersecurity: A Socio-Technical Perspective

Hsia-Ching Chang
Department of Information Science, University of North Texas
Cybersecurity Policy Fellow, New America

Introduction: From the Internet to Mobile Internet

Mobile Internet, one of the disruptive technologies, has transformed the way people communicate, live, and work since the early 2010s (Manyika et al. 2013). As for mobile Internet penetration, the number of mobile-only users has surpassed that of desktop-only users since 2015 in the United States, and the trend has remained stable since 2017 (VPN Mentor 2018). A recent study by the Pew Research Center (2018) indicates that 95% of Americans own mobile devices among which 77% of them are smartphones and 50% are tablets. The wider popularity of smartphones is not surprising because unlike tablets, smartphones offer various ways of communication; they have changed how people do things and interact with others. Interestingly, based on eMarketer's report, users spend more time interacting with smartphones and apps than with tablets and mobile web; apps account for more than 90% of time spent on smartphones (Wurmser 2018). The increasing number of smartphone users and their consumption in apps manifest the importance of mobile cybersecurity. The growing smartphone penetration also demonstrates that we are facing the challenges of preventing increased risks as users continue to install new apps in their mobile devices.

Verizon's Mobile Security Index 2018 reports that mobile security threats are increasing; 79% of surveyed organizations stated that they considered their employees to constitute a greater threat than hacktivists, criminals, and partners. The risks could result from downloading vulnerable mobile apps or visiting malicious websites. This exhibits social sub-systems (i.e., people) and technical sub-systems (i.e., mobile technologies) co-existing in an organization; therefore, the socio-technical perspective emerges as we tackle issues of information systems and technologies (Bostrom and Heinen 1977). A socio-technical perspective refers to a balance between people and technology when accounting for technology development. Goode (2018) suggests the significance of embedding a socio-technical perspective in cybersecurity training and education. In this sense, considering the role of humans and social aspects of cybersecurity is equally important to technical aspects. Therefore, this chapter discusses mobile security from a socio-technical perspective to address humans and socio-technical systems instead of just technology. In the context of cybersecurity, human mistakes are inevitable and application vulnerabilities continue to emerge, which reinforces the importance of socio-technical cybersecurity.

Organizations usually have different layers of protection in place to ensure cybersecurity. For instance, they apply encryption on data as the first step, install anti–virus software in every device, and deploy firewalls to block unauthorized access, deterring cyber criminals from performing attacks. It would be easy for cyber criminals to target individuals (employees, partners, and customers) affiliated with the organizations they attack. In the mobile environment, it would be even easier for cyber criminals to perform attacks and steal sensitive data, intellectual property, or top-secret information because employees and partners might use their mobile devices to access organization data that are classified as critical digital asset, and their mobile devices usually do not have strong security protection.

The chapter is structured as follows. It begins by outlining the layers of the multi-stakeholder mobile ecosystem and the cyber threats within the ecosystem. Following that, the security and privacy risks of mobile apps for mobile devices are laid out. It then touches on mobile users' security behavior and discusses the mobile security best practices. Finally, it concludes with the proverb "prevention is better than cure" and reiterates the importance of layered security to defend mobile security.

Cyber Threats in the Mobile Ecosystem

Mobile devices like tablets and smartphones have a wide range of features that could increase the attack surface, such as sensors, audio, video interfaces, connectivity, data storage, and so on. Mobile security is not only limited to mobile devices, but also the operating system, network communication, and applications. In its study on mobile device security, the Department of Homeland Security (2017) identifies the key components when assessing the security of mobile devices: mobile device technology stack (e.g., mobile operating systems, firmware, hardware, and mobile applications); communication networks (e.g., cellular, Wi-Fi, Bluetooth) and network services; and device physical access. Figure 10.1 shows the key elements in a mobile ecosystem at a glance.

As smartphones become one of the most popular personal mobile devices, various personal data as well as information are generated and saved in the smartphone. From the overview of the mobile ecosystem, the data and information in the smartphone can be leaked through multiple channels, such as the installation of mobile applications, mobile network connection, malware targeting hardware/firmware, and system vulnerability. Figure 10.2 illustrates several concepts that pertain to mobile security risks and their relationships. Vulnerabilities refer to weaknesses where a threat could occur by exploiting the vulnerabilities in a device or service, leading to potential risks.

Serving as an industry-recognized vulnerability database, common vulnerabilities and exposures (CVE), maintained by the MITRE Corporation, is a shared repository housing entries of publicly known cybersecurity vulnerabilities freely available for businesses or organizations that provide cybersecurity-related products and services around the world. Compared to threats and risks, vulnerabilities involve more technical details that only IT security professionals would understand. Therefore, when communicating with average users, we usually focus on threats and risks to address cybersecurity issues. With the daily increase in the newly identified vulnerabilities, cybersecurity threats involve an ever-changing landscape. In the mobile ecosystem, hackers have tried to capitalize on vulnerabilities at different levels, including hardware, firmware, operating system, applications, networks, and so forth. Counter-measures need to be taken in order to mitigate the risks.

The Department of Homeland Security divides the threats to the mobile ecosystem into five categories: (1) operating system/firmware threats, (2) application-based threats, (3) network-based threats, (4) physical device/access-based threats, and (5) threats to the mobile enterprise systems. Mobile enterprise systems refer to enterprise mobile services and infrastructure that comprise mobile device management and enterprise app stores. An organization that provides mobile device management can have an addition layer of security protection against potential threats. Although mobile enterprise systems are one of the key components in the mobile ecosystem, they are excluded as beyond the scope of this chapter. This chapter will focus on mobile security issues from a consumer's perspective rather than an organization's perspective.

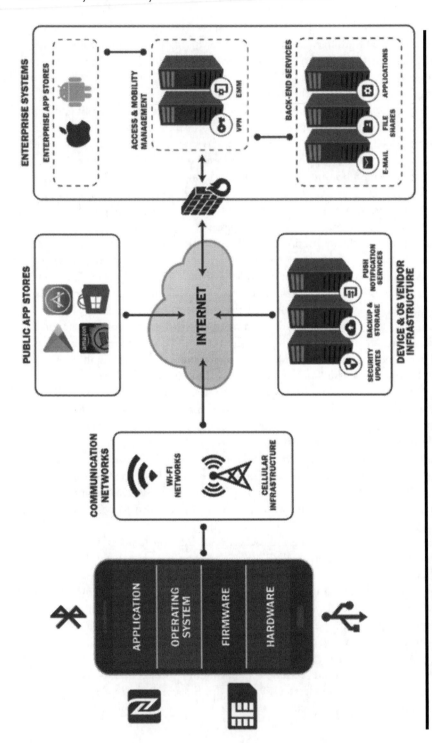

Figure 10.1 Mobile ecosystem (Department of Homeland Security 2017).
Reprinted courtesy of U.S. Department of Homeland Security. Not copyrightable in the United States.

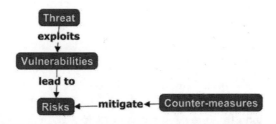

Figure 10.2   The relationships among threat, vulnerability, and risk.

Threats in Mobile Technology Stack (Operating System/Firmware/Hardware)

As shown in Figure 10.3, mobile technology stack represents a mobile device's multiple technology layers consisting of the hardware, firmware, mobile operating system, mobile applications, and embedded components (e.g., sensors and SIM card). Vulnerabilities in any of these components may be targeted by threats. For instance, the memory corruption defects in firmware could be manipulated by the attackers to gain administrative access (Veracode 2013); therefore, the user must regularly update firmware to ensure the device remains up to date. Like desktops or laptops, the mobile devices have operating systems, such as iPhone OS (iOS), Android, Blackberry, and Windows. Modifying the mobile device configuration (e.g., jailbreaking or rooting a phone) is one of the major mobile threats,

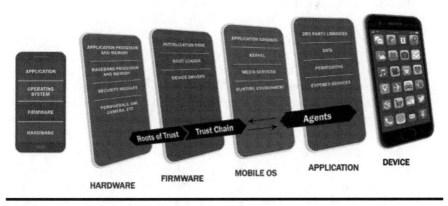

Figure 10.3   Mobile device technology stack (Department of Homeland Security 2017).

Reprinted courtesy of U.S. Department of Homeland Security. Not copyrightable in the United States.

tampering, as it exploits the defects in the phone's operating system (Department of Homeland Security 2017).

Physical Device Threats

The causes of physical device threats include device loss or theft and malicious charging stations. The GSMA, an association that represents the worldwide mobile communications industry, provides security advice to mobile phone users on mobile phone theft (GSMA 2019b). Public charging stations available in the airports, subways, or hotels have become targets for criminals to launch charging attacks. Juice jacking, one type of charging attacks, can access personal data through a USB charger with data transfer capabilities without the user's permission, or install malicious software on the device when the device is unlocked, while juice filming attacks, a more advanced charging attack, can do the same even when the phone is locked (Meng et al. 2015). Most users are unaware of the charging risks and trust public charging stations. Bekker (2018) provides several tips to protect mobile devices from juice jacking, including verifying the power source or using portable power sources instead, plugging one's own cord into a wall outlet, or bringing power-only USB cords to help mitigate data transfer risks.

Network-Based Threats

The network communication layer accommodates different types of mobile connectivity network, such as Bluetooth, cellular, Wi-Fi, near-field communication and services provided by mobile network operators (e.g., AT&T, Verizon, and T-Mobile). The data stored in the mobile devices can be intercepted in any network environment, from short distance within a wireless personal area network (WPAN) to medium distance within a wireless local area network (WLAN) to long distance within a wireless wide area network (WWAN).

WPAN is designed to enable personal devices (e.g., smartphones, laptops, tablets, and wearables) to exchange data and communicate with each other in proximity. Bluetooth is one of the representative technologies and has become a connection feature in smartphones; however, as Bluetooth is easy to hack, it is therefore suggested that mobile users should not leave Bluetooth on unless they are using it.

The data communication using WLAN, also called Wireless Fidelity (Wi-Fi), is commonly seen in offices, homes, and universities. The IEEE 802.11 working group has developed and extended the standards for WLAN, allowing longer-distance and a faster transmitting rate for data communication. Wi-Fi Protected Access 2 (WPA2), a Wi-Fi security standard, aims to use the Advanced Encryption Standard (AES) protocol (password) to prevent unauthorized access to users' wireless data. However, Mathy Vanhoef, a network security researcher, identified the WPA2 standard flaw and was able to hack Wi-Fi networks using a technique called key reinstallation attack (KRACK) (Vanhoef 2017). Wi-Fi Alliance is developing the next-generation Wi-Fi security, WPA3, to simplify and enhance Wi-Fi security (Wiggers 2018).

WWAN goes beyond the borders of cities and countries when it comes to mobile communications. Mobile communication technology has gone through multiple generations, from the first-generation (1G) to the fourth-generation (4G). The third-generation (3G) of mobile communication standard has transformed the way people communicate using mobile phones. Adopting Internet protocol (IP) technology, 3G enables mobile devices to communicate not only through voice, but also through text, images, videos, and multimedia information. Additionally, voice over Internet protocol (VOIP), also called Internet phone, offers an alternative way of calling. With the Internet, mobile users can make phone calls using Wi-Fi calling apps, such as Skype, WhatsApp, Facebook Messenger, and Google Talk. According to the Skype website, the information (e.g., voice, messages, images, and files) exchanged between one Skype account and another Skype account via the Internet is encrypted. However, for calls made from Skype app through the Internet to mobile or landline phones, the communication over the telephone network is not encrypted. In June 2018, Skype added a new feature called "Private Conversations," which provides end-to-end encrypted audio calls, text messages, images, videos, and files to prevent eavesdropping attacks (Skype 2018). Users need to update their Skype app to the latest version on their PCs or mobile devices in order to utilize this feature to make their communication data more secure.

A mobile device, the primary endpoint in cellular networks, interacts with cellular base stations to send and receive information. 4G technology like Long-Term Evolution (LTE) has dominated the current cellular infrastructure; however, it is not completely secure as there are various cybersecurity threats to LTE hardware, software, and networks (Cichonski, Franklin and Bartock 2017). A mobile phone must send authentication data to the base station before making a call; therefore, a potential risk with mobile phones is the attack by fake base stations in LTE Networks (Al Mazroa and Arozullah 2015). In addition, it is difficult for mobile users to detect fake base stations. As the attackers receive the user's authentication data through fake base stations, they gain unauthorized access to the cellular network. Once this scenario happens, short message service (SMS) One-Time-Password (OTP) sent by online banks is not as secure as we believe. The reason is that the OTP sent through SMS is not encrypted and can be intercepted by a hacker with a fake base station or performing Man-in-the-Middle attacks on mobile apps exploiting OS vulnerabilities (Grassi, Garcia and Fenton 2017). The security of SMS OTP greatly depends on network security. Requesting OTP through email or app might be more secure because most major email service providers and OTP authentication apps offer encryptions. According to

NIST's recent special report, Digital Identity Guidelines (SP 800–63), SMS OTP also faces the risk of social engineering.

Application-Based Threats

As shown in the mobile ecosystem (Figure 10.1), the mobile Internet infrastructure is mainly supported by device vendors (e.g., Apple, Samsung, and Motorola) and OS vendors (e.g., Apple and Google) to provide security updates, App stores, data backup and cloud storage, and push notification services. A mobile application (or app) refers to "an application that runs on a mobile device and is context aware" (Arzenšek and Heričko 2014, 1); the nature of context-awareness results in privacy and security challenges in mobile application testing. The application layer is where the user directly interacts with the mobile information systems. Except for the pre-installed apps, users have the freedom to download and install the apps they prefer or need. A variety of apps allow users to do different things that involve managing sensitive personal information. Social media and networking apps enable users to share their status and real-time information at their fingertips. Users can use bank apps to pay bills or transfer money. They can also adopt health-related apps to monitor their sleep patterns or fitness information. Despite the great benefits that various apps offer, users must be aware of major application-based threats and know how to protect their personal data and information generated by using those apps from being manipulated by unauthorized parities. However, taking precautions against mobile threats does not seem to be a common practice across most organizations. For instance, two-factor authentication is a strong authentication method that adds an extra layer of identifying users as they login. OTP, a form of two-factor authentication, adds the "second" factor to the authentication besides the first factor, the account password. According to Verizon's Mobile Index 2018, only 38% of organizations use two-factor authentication on their mobile devices. More than a quarter of respondents whose organizations do not provide mobile security training rated their users as having low awareness of mobile threats.

Different privacy and security issues have been identified in distinct categories of mobile applications. Zlatolas et al. (2017) conducted a systematic literature review on academic publications that propose models illustrating privacy and security issues on mobile applications. They found that privacy variables appear about three times more often than security variables in academic publications from 2009 to 2016. This is not surprising because mobile applications collect user data, and unauthorized secondary use could jeopardize a user's security and privacy. Mobile cloud computing (MCC) apps move computation and data storage from mobile devices to cloud platforms; however, MCC applications face ongoing security challenges, such as authentication, authorization, data/code integrity (Wang, Chen and Wang 2015).

Mobile App Privacy and Security Risks

According to Verizon's mobile security index that surveyed more than 600 mobility professionals responsible for mobile device procurement and management in their organizations, they were most concerned about malware, ransomware, and device theft/loss as the top three threats that were more likely to lead to security incidents in 2018, while they were most concerned about malware, ransomware, and insecure or poor coding practice in 2019 (Verizon 2018, 2019). Device theft/loss is one of the physical device threats, whereas malware, ransomware, and poor coding practice are app relevant threats. Although the general impression about malware and ransomware is that such threats are more frequently seen on Android devices, iOS devices are not immune.

The Open Web Application Security Project (OWASP) has a work-in-progress mobile security project to categorize mobile-specific vulnerabilities, and offers Mobile Apps Checklist for developers and security professionals. Based on an analysis of 2 million mobile applications and 500,000 mobile devices with the US National Vulnerability Database, US Computer Emergency Readiness Team (US-CERT), and OWASP vulnerabilities, Pradeo Lab's Mobile Threat Report (2018) demonstrates that approximately 60% of mobile applications have severe flaws that can lead to data leakage (49%), Denial of Service (DoS) attacks (7%), Man-in-the-Middle attacks (5%), and encryption weaknesses (5%). The report also reveals that the most leaked data that may seriously infringe on privacy is geo-location coordinates, followed by contact lists, user profile information, user files (e.g., photo, video, and documents) and SMS. It is not surprising that data leakage/breach is the major issue among attacks because we are sharing every detail about the usage of the app and mobile device as well as personal data generated by using the app and device. Additionally, if a user owns a wearable (e.g., smart watch, fitness tracker) and it connects to the user's smartphone through an app, the vendors can compound the information stored in the wearable with other information in the smartphone in order to further profile the user. Therefore, due to the connectivity of the mobile devices, the configuration and security, as well as the privacy settings of wearables are equally important in regard to the mobile phone, tablet, and laptop.

The OWASP top 10 checklist can also inform mobile device users about which mobile app issues put users at risks. Table 10.1 presents the most recent top 10 list released in 2016 and summarizes the attack vectors, security weaknesses, as well as technical and business impacts. According to Table 10.1, the exploitability of the attack vectors is rated EASY across nine out of the top ten risks: improper platform usage, insecure data storage, insecure communication, insecure authentication, insufficient cryptography, insecure authorization, code tampering, reverse engineering, and extraneous functionality. The prevalence of security weakness is identified as COMMON across the top ten risks, while the detectability of

Table 10.1   OWASP Mobile Apps Top 10 Risks

	Attack Vectors	Security Weakness		Technical Impacts	Business Impacts
	Exploitability	Prevalence	Detectability		
1. Improper Platform Usage	EASY	COMMON	AVERAGE	SEVERE Three forms of Cross-Site Scripting (XSS) usually target users' Web browsers: Reflected XSS, DOM XSS, and stored XSS	Stealing credentials, sessions, or delivering malware to the victim
2. Insecure Data Storage	EASY	COMMON	AVERAGE	SEVERE Extraction of the app's sensitive information via mobile malware, modified apps or forensic tools	Identity theft; Privacy violation; Fraud; Reputation damage; External policy violation (PCI); or Material loss
3. Insecure Communication	EASY	COMMON	AVERAGE	SEVERE If the adversary intercepts an admin account, the entire site could be exposed. Poor SSL setup can also facilitate phishing and Man-in-the-Middle attack	The privacy violation of a user's confidentiality; Identity theft; Fraud, or Reputational damage
4. Insecure Authentication	EASY	COMMON	AVERAGE	SEVERE Unable to identify the user performing an action request. Unable to log or audit user activity	Reputational damage; Information theft; Unauthorized access to data
5. Insufficient Cryptography	EASY	COMMON	AVERAGE	SEVERE Unauthorized retrieval of sensitive information from the mobile device	Privacy violations; Information theft; Code theft; Intellectual property theft; Reputational damage

(Continued)

Attack Vectors	Security Weakness			Technical Impacts	Business Impacts
	Exploitability	Prevalence	Detectability		
6. Insecure Authorization	EASY	COMMON	AVERAGE	SEVERE Over-privileged execution of remote or local administration functionality may result in destruction of systems or access to sensitive information	Able to execute over-privileged functionality; Reputational damage; Fraud; Information theft
7. Poor Code Quality	DIFFICULT	COMMON	DIFFICULT	MODERATE Foreign code execution or denial of service on remote server endpoints	Information Theft; Reputational Damage; Intellectual Property Theft; Degradations in performance
8. Code Tampering	EASY	COMMON	AVERAGE	SEVERE Unauthorized new features; Identity theft; Fraud	Revenue loss due to piracy; Reputational damage
9. Reverse Engineering	EASY	COMMON	EASY	MODERATE Reveal information about back end servers; Reveal cryptographic constants and ciphers; Steal intellectual property; Perform attacks against back end systems; Gain intelligence needed to perform subsequent code modification	Intellectual property theft; Reputational damage; Identity theft; Compromise of backend systems
10. Extraneous Functionality	EASY	COMMON	AVERAGE	SEVERE Exposure of how backend systems work; Unauthorized high-privileged actions executed	Unauthorized access to sensitive functionality; Reputational damage; Intellectual property theft

Source: Information compiled from www.owasp.org/index.php/Mobile_Top_10_2016-Top_10

security weakness in one risk is ranked EASY and eight risks are rated AVER-AGE. These imply that most mobile apps are very likely to be exposed to the OWASP mobile top 10 risks. The level of technical impacts demonstrates SEVERE in eight out of ten risks. Perhaps the terminology used in technical impacts appears to be a bit difficult for non-security professionals to understand. However, certain patterns emerge among the resultant business impacts across different top risks, such as different types of theft (identity theft, information theft, code theft, and intellectual property theft), reputational damage, fraud, unauthorized access to data or sensitive functionality, and privacy violations. Insecure mobile apps can jeopardize users' data/information assets and privacy; mobile app users should not fully trust the apps; they need to be more aware of the business impacts and cautious about protecting their own mobile security.

The top 10 issues can be grouped into five major categories. The first is exploiting the wrong use of Web platform feature or exploiting extraneous functionality, like a backdoor, by injecting malicious code (#1 and #10). The second is insecure data storage and insufficient cryptography (#2 and #5), which could cause a data breach. Data storage and cryptography should go hand in hand because solid cryptography principles and techniques can secure data storage. Encryption is one of the key components in cryptography, as well as an effective data security technique. The third is insecure authentication and authorization (#4 and #6), which could lead to unauthorized access and use. Authentication and authorization are two essential elements in identity and access management. While authentication verifies user identity at system login, authorization determines the user role or permission to access the system. The fourth is insecure communication and reverse engineering (#3 and #9); these pertain to authentication compromise. The fifth is poor code quality and code tampering (#7 and #8); they could influence the executing and running of code. Though several categories relevant to platform, communication protocol, reverse engineering, and coding might not be controlled by the mobile users, mobile app developers and mobile users can select the apps and the services that provide secure data storage/authentication/authorization and use strong cryptography.

Mobile Users' Security Behavior

Mobile devices are exposed to greater risks than are personal computers (McGill and Thompson 2017). According to Symantec Corporation (2019), 55% of Internet traffic involves mobile devices, which increases the likelihood of mobile users being targeted by cyber criminals. The advances of mobile technology and applications have been the focal points of the research community. However, research on mobile users' security behavior is still in its nascent state (Verkijika 2018). Protecting mobile devices against cybersecurity threats requires considering both human and technological factors (Alsaleh, Alomar and Alarifi 2017). A socio-technical perspective of cybersecurity accounts for cybersecurity issues from both people and technology aspects, not just technology alone. Drawing on such a perspective, we can balance the understanding of both technical and social challenges in relation to mobile cybersecurity and find feasible ways to address them.

User Interactions with Mobile Apps

The life cycle of mobile apps involves five stages: install, update, uninstall, open, and close (Böhmer et al. 2011). Unless the apps were pre-installed in the mobile device, a user has to initiate the installation. What are the major factors that drive a user's installation of an app? A recent study reports that trust and security of the mobile applications significantly influence users' intention to install mobile applications, while risk and privacy have rather insignificant effects on users' intention of installations (Chin, Harris and Brookshire 2018). It is worrisome that users seem to over-trust the public app store; therefore, perceived risk and privacy concerns play unimportant roles in users' intent to install mobile apps. A survey comparing the Android and iPhone users' security and privacy awareness shows that Android users appear more conscious of the security and privacy risks than iPhone users while adopting a new app (Reinfelder, Benenson and Gassmann 2014). The reason might be that Apple App store has more rigorous review processes than Google Play store; thereby, their users tend to trust the apps reviewed by the Apple App store and worry less about the potential privacy and security risks. In addition, most users do not have the good security habit of updating their mobile device as soon as a new operating system (OS) version is available. Consequently, OS vulnerabilities become the most common mobile device threat (Pradeo Lab 2018).

Interestingly, regardless of the number of apps installed, mobile users only pay attention to several apps they installed for major tasks, and the main user-mobile phone interactions occur in social apps (Jesdabodi and Maalej 2015). As social apps have access to users' personal sensitive information and data stored in the mobile devices and the third party cloud platforms, users need to be cautious about data breach issues since unintentional data leakage through mobile applications has been identified as the number-one threat to organizations; location coordinates as well as contact lists are the most leaked private data (Pradeo Lab 2018). Malicious apps are often camouflaged with social or gaming apps taking different forms to gain access to users' data or information. Four primary types of malicious mobile apps identified by the application security company include: (1) Spyware: tracking user location and mobile activities, (2) Trojans: appearing legitimate but designed to disguise its dangerous intent, such as taking control of the device or stealing personal data/information, (3) Phishing links: tricking users into fake websites similar to the original ones to steal user credentials, and (4) Hidden processes: running in the background of the mobile device as certain actions are initiated, for example, launching a bank app (Veracode n.d.). Spyware and Trojans are common forms of malicious software (i.e., malware) and could be the tools used for technical cybercrimes (Martens, De Wolf and De Marez 2019). It is worth noting that mobile scams through phishing links involving social interactions can be disseminated by cyber criminals via social media or SMS (Symantec Corporation 2019).

Mobile Users' Security Awareness and Behavior

A hierarchical taxonomy of mobile users' security awareness, developed by Bitton et al. (2018), provides a socio-technical overview of users' mobile security behavior; it combines technical and psychological dimensions to measure the security awareness of smartphone users. The four technological focus areas and ten sub-focus areas of the taxonomy encompass: (1) Applications (application installation/handling), (2) Browsing and Communication (browser, virtual communication, and accounts), (3) Communication Channels (networks and physical channels), and (4) Device (operating system, data privacy, and security systems). The technical focus areas in this taxonomy align with the elements illustrated in the mobile ecosystem, except for Browsing and Communication, which involve user interaction with the interfaces of browsing and communicating. Bitton et al. (2018) note that the abovementioned technical focus areas could change over time due to the ongoing development of new mobile technologies and threats. The other part of the taxonomy centers on a user's psychological dimensions and sub-dimensions, ranging from attitude (perceived probability of a threat, perceived severity of a threat, and perceived usefulness of recommended behavior) to knowledge (declarative knowledge and procedural knowledge) to security behavior (preventive behavior and confronting behavior). This taxonomy indicates that mobile users' interactions with those technical components are driven by their attitude, knowledge, and security behavior.

From a psychological perspective, Kraus, Wechsung and Möller (2017) suggest that security is not the only or even major motivation that steers users' security and privacy actions on smartphones. They recommend considering other psychological needs (e.g., meaning, stimulation, autonomy, and competence) when designing security and privacy features for smartphones. In this sense, while organizing user training or awareness programs for promoting mobile security, designing a motivational intervention accounting for psychological needs can be more effective than a regular intervention without a psychological component. In a recent study on extending the protection motivation theory in the context of smartphone security, anticipated regret, a common emotion in decision making, is considered as a mediator between users' threat perceptions and security intentions/behaviors (Verkijika 2018). The study results demonstrate that users' perceived vulnerability and perceived severity have a positive effect on anticipated regret, while self-efficacy and anticipated regret directly impact smartphone security intentions. Thus, anticipated regret and security intentions positively influence smartphone security behaviors.

There are two general types of mobile security behaviors: preventive behavior refers to the user's actions to reduce risk, whereas confronting behavior relates to the user's actions while facing a security risk (Bitton et al. 2018). Although installing mobile security tools could be a good preventive behavior to protect users from malware and viruses, not every tool can guarantee perfect protection (AV-Test 2018). Several organizations, like Virus Bulletin, a member of the Anti-Malware Testing Standards Organization (AMTSO), and AV-Test Institute, evaluate anti-virus software for computers and mobile devices, and make the reports freely available to the public. Mobile users are encouraged to remain informed of current anti-virus and anti-malware software for protecting mobile devices.

Smartphone Security Best Practices: Protecting Security at Every Layer

Benjamin Franklin's famous quote, "An ounce of prevention is worth a pound of cure," was intended to raise the awareness of fire prevention (UShistory.org. 1995). However, his foresight axiom is very relevant to cybersecurity and can be applied to cybersecurity prevention and intervention. When it comes to mobile cybersecurity prevention within a multi-stakeholder ecosystem, the stakeholders need to share their responsibilities and protect mobile security at every layer. Only by working together can we make achieving this goal possible. We can start as mobile phone consumers to do what we need to do to strengthen the weakest (human) link.

Layered security is a concept combining multiple types of security measures and posture, which helps to protect against different attack vectors (see Table 10.1). The more security measures in place, the more difficult for attackers to hack into the mobile system.

Compared to other mobile devices, smartphones have become indispensable in most people's daily life. With smartphones, mobile users can perform multitasks efficiently, including writing emails, surfing on the Internet, listening to podcasts/radio/music, using social media/networking apps to catch up with family and friends, playing mobile games, and so on. The more versatile the smartphones become, the more likely they will gradually replace certain roles that personal computers and laptops play in our work and personal life. However, the capability of smartphones supporting both voice and data communication makes users face the daunting risk of disclosing personal data or information that is mostly unprotected. The largest barrier of smartphone security is that users are often prone to focus on the convenience and overlook the mobile security and privacy concerns, compared to computer security. The reason might be that mobile users still view their smartphone as a conventional mobile phone and remain unaware of the multiple layers of a smartphone going beyond the physical layer to app and network layers. For example, not only does the app layer contain the applications that a user installed, but it also has access to users' data and information saved in the device and online cloud storage services. Designed by Federal Communications Commission (FCC), the smartphone security checker (see Table 10.2) provides helpful

Table 10.2   Ten Steps to Smartphone Security for Two Major Mobile Operating Systems

Steps to Smartphone Security	Android	Apple iOS	Involved Mobile Security Layer
1	Set PINs and passwords	Set PINs and passwords	Physical Device
2	Do not modify your smartphone's security settings	Do not modify your smartphone's security settings	Physical Device
3	Backup and secure your data	Backup and secure your data	Data, Application
4	Only install apps from trusted sources	Only install apps from trusted sources	Application
5	Understand app permissions before accepting them	Understand app permissions before accepting them	Application
6	**Install security apps that enable remote location and wiping**	**Use the free, built-in security features**	Application
7	Accept updates and patches to your smartphone's software	Accept updates and patches to your smartphone's software	Application, Physical Device
8	Be smart on open Wi-Fi networks	Be smart on open Wi-Fi networks	Network
9	Wipe data on your old phone before you donate, resell or recycle it	Wipe data on your old phone before you donate, resell, or recycle it	Data
10	Report a stolen smartphone	Report a stolen smartphone	Physical Device

Source: Adapted from Federal Communications Commission (2015)

Note: The only different step to smartphone security between Android and Apple iOS is highlighted in bold.

guidance for achieving the best practices. The ten steps suggested by FCC can be mapped and divided into four layers of security: physical device, app, network, and data security. Multiple security layers involve the growing diversity of mobile devices, applications, and services in communication networks. It would be helpful to decompose the layers when discussing the best practices of mobile security.

Physical Device Protection

As for physical device security, setting up a lock screen PIN or strong password is the first step to protect our smartphones from being accessed by others. Beyond the PIN and password, newer mobile devices support alternative biometric features to guard their devices with fingerprint, face recognition, or iris scanning. A longer password is a better choice because the four-digit PIN can be easily compromised. The rule of thumb of creating a password is to avoid using sequential number (e.g., 123456) and short or common names/words that can be identified in a dictionary. If a user has difficulty creating a strong password, password manager tools have been designed to assist in generating complex passwords. Diceware, an open source tool invented by Arnold Reinhold, can help construct a strong passphrase or password. Diceware was developed by utilizing entropy theory to achieve a higher level of randomness with encryption and security. To use Diceware, one needs dice and the Diceware wordlist which is available from either the Reinhold's Diceware Passphrase page (Reinhold 2018) or from the Electronic Frontier Foundation's Dice-Generated Passphrases page (EFF 2019). A passphrase generated by Diceware can also be used to encrypt information for mobile devices. For protecting device security, the second step is to avoid modifying the default security settings unless it is necessary, because the default security settings provide features to run automatic security updates on hardware, software, and OS. Users need to pay attention to how the security settings of their smartphones are set up. It is important to ensure that the latest security patch updates and OS version have been installed. In case the smartphone is lost, users need to set up Find Device function to remotely locate the smartphone. Another protection tip relevant to physical device (see Table 10.2) is reporting to the mobile network operator and local police department when the smartphone is stolen or lost. In so doing, the mobile network operator will suspend the user's service immediately to avoid unauthorized usage.

It is worth noting that setting up a password to lock the screen can only prevent others from directly accessing the mobile device. If the mobile device is not encrypted, the person who stole or held the device can access the data in the mobile device using a USB cable physically connected to a computer or software tools to access the data in the device.

App-Level Security Protection

As shown in Table 10.2, five out of the ten steps pertain to application security. It is known that most Android devices lack built-in security features. Therefore, the steps suggested by FCC for both Apple iOS and Android are almost identical except for step six, where Apple iPhones have built-in security features that users can leverage and rely on, while Android devices might need the installation of mobile security apps to secure the device and data privacy. Samsung, one of the dominating Android device vendors, has developed a built-in security tool, Knox, for its latest devices to address data security and privacy issues (Samsung Knox 2019). Google also has Google Play Protect built-in the Android devices to regularly scan the apps and detect harmful apps; users can check their app security status by opening the Google Play store app. Mobile users need to be vigilant about the availability of built-in defense features that support different mobile security layers and take precautions when interacting with their mobile device, in order to protect their data/information privacy and security on their mobile device.

One of the tips suggested is only installing apps from trusted sources. Unlike Google Play store for Android apps, Apple store has detailed guidelines to approve apps, and has earned a good reputation for protecting users' privacy. A well-known example was Apple removing Facebook's Onavo Protect app from the App Store in 2018 due to its violation of Apple's privacy guideline about collecting user data on the usage of other apps. Facebook Research app reused the code from Onavo Protect app to recruit paid users aged between 13 and 35 in the United States and India, using a virtual private network (VPN) and granting root network access to monitor and collect their mobile usage data. VPN is meant to protect the data transmitted to and from the mobile device with encryption; users are encouraged to use it to protect the security of data exchange when connecting to free public Wi-Fi, and to understand app permissions before accepting them. Paid users of Facebook Research app were asked to grant permission to access the "root" level of their phones, which allows Facebook to retrieve all the data in their phones, including encrypted data and private messages (Constine 2019). Due to the controversial use of the security app for spying on users, Onavo Protect app was eventually removed by Facebook from Google Play store in early 2019. Considering an analogy of a door, if users give permission to change the configuration or access their personal data, it is like either opening or unlocking the door to invite the strangers to enter their house, which is an unwise decision that could become a major threat.

Similar to physical device security, the apps require users to accept updates and patches in order to keep their systems safely up to date. Based on a recent survey, approximately 80% of US adults believe that updating apps is crucial, while 33%

do not regularly update or check if their apps need to be updated (Google and Harris Poll 2019). This implies that information professionals can help provide more information access, training, or education to transfer the best practices internally within our served communities.

Data Protection

Two tips listed in Table 10.2 pertain to data protection; they suggest not only backing up and securing personal data, but also erasing data on old devices before reselling or recycling them. It would be safe to backup with two copies of the same files in the mobile devices on different devices, such as a PC, laptop, or external hard disk. To secure sensitive personal data, users might consider encrypting the device as well as backing up data. While handling old mobile devices, users can set them back to their original factory settings to wipe personal data saved on the device.

If mobile users lack awareness of data protection using their smartphones, they will not take precautions to prevent their personal data from being transferred or disclosed to social networks or third parties. Consequently, mobile users who do not have a cybersecurity habit will not be motivated to install anti–virus and anti-malware software tools in their mobile devices, not to mention ensuring that the applications and the mobile OS are up to date. Nevertheless, users constantly install a variety of apps with diverse features promoted by the businesses and organizations, which may put them at great risk of data breach. For instance, retailers have developed mobile apps to engage online customers and extend their in-store shopping experience to e-commerce and mobile commerce (i.e., m-commerce). Users can make online purchases and set up pick up orders with grocery store apps, check account balance and make payments with banking apps, and use ride sharing apps to arrange transportation. In general, these apps will not work without the Internet because the apps with the capability of context-aware mobile computing need to gather data about the user's location in real-time and analyze the data to adapt interactions accordingly. On the one hand, the collection of user data makes it easier to offer personalized services and targeted advertisement. On the other hand, users might not know how much data or information have been collected and sent back to the retailers, banks, and ride sharing service providers, or sold to advertisers. Based on the app usage data and the data/information that users gave away, those companies and app service providers can piece the data together into a whole to profile users. The problem is that consumers cannot seem to opt out for privacy reasons, and they are not asked for consent for data collection and sharing between third parties. As smartphones gradually turn into a repository of biometrics and associated data, data security and privacy should be a big part of individual's efforts to protect their data. Considering security and privacy together is often essential for protecting either.

Network-Level Security Protection

Mobile devices are equipped with the wireless connectivity via Bluetooth, cellular data, or Wi-Fi network to access the Internet. In terms of network security, the tip suggests being smart on open Wi-Fi networks, which could be done in different ways. Open or public Wi-Fi networks available in cafés, airports, or hotels are usually free and handy, but unfiltered and unsecured. Not all free public Wi-Fi requires a password, making data communication vulnerable to hacker attacks. Thus, before connecting to any free Wi-Fi access points, mobile users need to cautiously ensure they connect to the recognized wireless network with the name given by the café, airport or hotel instead of an unknown or bogus one. The mobile users also have to think twice before using open Wi-Fi network to access sensitive or confidential information as it reveals one's identity (e.g., banking, credit card information, and health records); data or information can be easily intercepted and captured in public W-Fi network by cybercriminals. As a result, mobile device users' personal information and digital identity might be stolen.

To securely access the systems or resources at work, mobile device users should use a VPN when connecting to an open Wi-Fi network to prevent a data breach. Additionally, it is safer to avoid keeping copies of sensitive/confidential data on personal mobile devices. Furthermore, while accessing social networking or online shopping websites, mobile users could consider using their data plan supported by their mobile phone network rather than choosing open Wi-Fi networks. The anti–virus and anti-malware software installed in mobile devices could help detect and protect against malicious attacks; however, users have to ensure that software is updated in a timely fashion.

More and more websites employ encryption to protect user information. If the web address starts with "https" instead of "http," it is encrypted. As a mnemonic, Federal Trade Commission (FTC) suggests considering the "s" in "https" as "secure." Therefore, paying attention to the secure "https" can ensure that the data transfer is protected by encryption. Combining crowdsourcing and machine learning, Web of Trust (WOT) (www.mywot.com) is a website reputation service facilitating safer browsing and searching with web reputation icons and risk notifications to prevent users from scams, malware, phishing, rogue websites, and malicious links. WOT is not only an add-on in an Internet browser, but also a mobile app on both iOS and Android to prevent users from inadvertently clicking on a problematic link through mobile Internet. The user of WOT can contribute to the community by rating a website, which identifies the threats that anti–virus or anti-malware tools might not catch.

Using a smartphone as a hub, smart mobile Internet of Things (M-IoT) connects a variety of applications, such as smart wearables, smart home, smart campus, smart city, smart health, and so on. Smart M-IoT devices could put the user's network at high risk owing to the context-aware connectivity and artificial intelligence (AI), extending communication from human-mobile interaction to invisible machine-to-machine communications. The growing prevalence of smart devices increases the attack surface targets; however, security, privacy and trust remain the major concerns in connected M-IoT networks (Sharma et al. 2019). Internet Society (2018) offers top tips for the consumers of connected IoT devices to protect their security and privacy, such as purchasing the connected devices that support updates for devices and apps, enabling encryption, reviewing the privacy settings on the devices and their apps, using strong passwords, avoiding reusing identical passwords, disconnecting the devices when not in use, and making home Wi-Fi networks more secure with firewalls.

Conclusion: Prevention Is Better than Cure

As we continue to become more connected with a variety of ubiquitous mobile devices, it is inevitable that the attack surface spreads wider and will eventually become limitless, which presents considerable challenges regarding mobile cybersecurity. Different groups of stakeholders, including device makers, mobile network operators, software companies, equipment providers, and Internet companies in the mobile ecosystem, actively participate in shaping and altering the future of mobile Internet, mobile apps, and mobile security. In light of the worrisome statistics on users' low awareness of mobile security, information professionals should not only consider the impact of mobile devices on information behavior and mobile device use to access information, but also help communicate and disseminate the major threats and risks in today's dynamic mobile landscape.

Mobile security needs to be considered and implemented at multiple layers, including hardware security, OS security, network security, app security, and data/information security. To enhance mobile security, ideally, those multiple layers should be combined into the security design of future mobile devices. Although there is still long way to go, mobile industries have united through GSMA to initiate a global open standard, named Mobile Connect, which aims to offer a secure universal identity solution and enable mobile users to authenticate themselves with an industry-wide standard API supporting multiple factors authentication (GSMA 2019a).

Mobile devices are equipped with sensors to scan and collect a user's biometrics (e.g., voice, iris, face, and fingerprints), geo-coordinates, personal information, and mobile usage data. On the one hand, users need to provide certain information to connect to a mobile network or an app. The service providers and vendors in the mobile ecosystem collect all kinds of user information as well as usage data to know more about their users in order to provide better user experience and improve apps or services. On the other hand, users have no clue as to whether those companies that collected the user data sell their data to unauthorized third parties for revenue. With all the features, functions, and conveniences that users have to give up in exchange for their data security and privacy, studies have shown that users are usually willing to compromise their data privacy for convenience or preferred features and functions. However, users need to be more critical about the privacy and security settings of their mobile devices. After all, the awareness of privacy and security risks will drive a user's decisions concerning the level of privacy and security settings, or permission of data disclosure.

If mobile devices lack built-in security features like iPhones do, mobile users need to install mobile antivirus and anti-malware apps on their mobile devices to help detect and defend against malicious attacks. Furthermore, mobile users need to ensure that the mobile OS and apps are kept updated. They also need to be cautious about the network connections of their mobile devices and the possibility of leaking sensitive information by checking if the device is running unnecessary functions or requesting unnecessary data.

Ongoing security training in the organizations is very important to raise public awareness of mobile security and privacy risks. If it is unavailable, mobile users need to stay aware of the latest security threats and risks. As information professionals, it is essentially important for us to help mobile users seek or access risk information in the mobile landscape.

References

Al Mazroa, Alanoud, and Mohammed Arozullah. "Detection and Remediation of Attack by Fake Base Stations in LTE Networks." *International Journal of Soft Computing and Engineering* 5 (2015): 12–15.

Alsaleh, Mansour, Noura Alomar, and Abdulrahman Alarifi. "Smartphone Users: Understanding How Security Mechanisms Are Perceived and New Persuasive Methods." *PLoS One* 12, no. 3 (2017): e0173284.

Arzenšek, Bostjan, and Marjan Heričko. "Criteria for selecting mobile application testing tools." In *CEUR Workshop Proceedings*, 1–8, 2014.

AV-Test. "Best Antivirus Software for Android." 2018. Accessed March 18, 2019. www.av-test.org/en/antivirus/mobile-devices/.

Bekker, Eugene. "Fake Charging Stations Can Hack Your Smartphone." *Fighting Identity Crimes*, 2018. Accessed March 18, 2019. www.fightingidentitycrimes.com/fake-charging-stations-hack-smartphone/.

Bitton, Ron, Andrey Finkelshtein, Lior Sidi, Rami Puzis, Lior Rokach, and Asaf Shabtai. "Taxonomy of Mobile Users' Security Awareness." *Computers & Security* 73 (2018): 266–293.

Böhmer, Matthias, Brent Hecht, Johannes Schöning, Antonio Krüger, and Gernot Bauer. "Falling Asleep with Angry Birds, Facebook and Kindle: A Large Scale Study on Mobile Application Usage." In *Proceedings of the 13th International Conference on Human Computer Interaction with Mobile Devices and Services*, 47–56. ACM, 2011.

Bostrom, Robert, and Stephen Heinen. "MIS Problems and Failures: A Socio-Technical Perspective. Part I: The Causes." *MIS Quarterly* 1, no. 3 (1977): 17–32.

Chin, Amita Goyal, Mark A. Harris, and Robert Brookshire. "A Bidirectional Perspective of Trust and Risk in Determining Factors that Influence Mobile App Installation." *International Journal of Information Management* 39 (2018): 49–59.

Cichonski, Jeffrey, Joshua Franklin, and Michael Bartock. "Guide to LTE Security." *NIST*, 2017. Accessed March 18, 2019. https://csrc.nist.gov/publications/detail/sp/800-187/final.

Constine, Josh. "Facebook Will Shut down Its Spyware VPN App Onavo." *TechCrunch*, 2019. Accessed March 18, 2019. https://techcrunch.com/2019/02/21/facebook-removes-onavo/.

Department of Homeland Security. "Study on Mobile Device Security." 2017. Accessed March 18, 2019. www.dhs.gov/sites/default/files/publications/DHS%20Study%20on%20Mobile%20Device%20Security%20-%20April%202017-FINAL.pdf.

Electronic Frontier Foundation. "EFF Dice-Generated Passphrases." 2019. Accessed March 18, 2019. www.eff.org/dice.

Federal Communications Commission. "FCC Smartphone Security Checker." 2015. Accessed March 18, 2019. www.fcc.gov/smartphone-security.

Goode, Jodi. "Comparing Training Methodologies on Employee's Cybersecurity Counter-measures Awareness and Skills in Traditional Vs. Socio-Technical Programs." PhD diss., Nova Southeastern University, 2018.

Google & Harris Poll. "Online Security Survey Infographic." 2019. Accessed March 18, 2019. http://services.google.com/fh/files/blogs/google_security_infographic.pdf.

Grassi, Paul, Michael Garcia, and James Fenton. "Digital Identity Guidelines." *NIST*, 2017. Accessed March 18, 2019. https://csrc.nist.gov/publications/detail/sp/800-63/3/final.

GSMA. "Introducing Mobile Connect – The New Standard in Digital Authentication." 2019a. Accessed March 18, 2019. www.gsma.com/identity/mobile-connect.

GSMA. "Security Advice for Mobile Phone Users – Mobile Phone Theft." 2019b. Accessed March 18, 2019. www.gsma.com/aboutus/workinggroups/working-groups/fraud-security-group/security-advice-for-mobile-phone-users/mobile-phone-theft.

Internet Society. "Top Tips for Consumers: Internet of Things Security and Privacy." 2018. Accessed June 24, 2018. www.internetsociety.org/resources/doc/2018/top-tips-for-consumers-internet-of-things-security-and-privacy/.

Jesdabodi, Chakajkla, and Walid Maalej. "Understanding Usage States on Mobile Devices." In *Proceedings of the 2015 ACM International Joint Conference on Pervasive and Ubiquitous Computing*, 1221–1225. ACM, 2015.

Kraus, Lydia, Ina Wechsung, and Sebastian Möller. "Psychological Needs as Motivators for Security and Privacy Actions on Smartphones." *Journal of Information Security and Applications* 34 (2017): 34–45.

Manyika, James, Michael Chui, Jacques Bughin, Richard Dobbs, Peter Bisson, and Alex Marrs. *Disruptive Technologies: Advances That Will Transform Life, Business, and the Global Economy*. Report. Washington, DC: McKinsey Global Institute, 2013.

Martens, Marijn, Ralf De Wolf, and Lieven De Marez. "Investigating and Comparing the Predictors of the Intention towards Taking Security Measures against Malware, Scams and Cybercrime in General." *Computers in Human Behavior* 92 (2019): 139–150.

McGill, Tanya, and Nik Thompson. "Old Risks, New Challenges: Exploring Differences in Security between Home Computer and Mobile Device Use." *Behaviour & Information Technology* 36, no. 11 (2017): 1111–1124.

Meng, Weizhi, Wang Hao Lee, R. Murali, and S. P. T. Krishnan. "Charging Me and I Know Your Secrets!: Towards Juice Filming Attacks on Smartphones." In *Proceedings of the 1st ACM Workshop on Cyber-Physical System Security*, 89–98. ACM, 2015.

Pew Research Center. "Mobile Fact Sheet." 2018. Accessed March 18, 2019. https://www.pewinternet.org/fact-sheet/mobile/.

Pradeo Lab. "Mobile Threat Report." 2018. Accessed March 18, 2019. https://www.pradeo.com/media/Mobile_Security_Report_S22018.pdf.

Reinfelder, Lena, Zinaida Benenson, and Freya Gassmann. "Differences between Android and iPhone Users in Their Security and Privacy Awareness." In *International Conference on Trust, Privacy and Security in Digital Business*, 156–167. Cham: Springer, 2014.

Reinhold, Arnold G. "The Diceware Passphrase Home Page." 2018. Accessed March 18, 2019. http://world.std.com/~reinhold/diceware.html.

Samsung Knox. "Devices Built on Knox." 2019. Accessed March 18, 2019. www.samsungknox.com/en/knox-platform/supported-devices.

Sharma, Vishal, Ilsun You, Karl Andersson, Francesco Palmieri, and Mubashir Husain Rehmani. "Security, Privacy and Trust for Smart Mobile-Internet of Things (M-IoT): A Survey." *arXiv Preprint arXiv:1903.05362*, 2019.

Skype. "What are Skype Private Conversations?" 2018. Accessed March 18, 2019. https://support.skype.com/en/faq/FA34824/what-are-skype-private-conversations.

Symantec Corporation. "Mobile Scams: How to Identify Them and Protect Yourself." 2019. Accessed March 18, 2019. https://us.norton.com/internetsecurity-mobile-mobile-scams-how-to-identify-them-and-protect-yourself.html.

UShistory.org. "Philadelphia: In Case of Fire, Independence Hall Association." 1995. Accessed March 18, 2019. https://www.ushistory.org/franklin/philadelphia/fire.htm.

Vanhoef, Mathy. "Key Reinstallation Attacks (KRACK): Breaking WPA2 by Forcing Nonce Reuse." 2017. Accessed April 24, 2019. https://www.krackattacks.com/.

Veracode. "The Rise of Malicious Mobile Applications." Veracode App Sec Knowledge Base, n.d. Accessed March 18, 2019. www.veracode.com/security/rise-malicious-mobile-applications.

Veracode. "Understanding the Risks of Mobile Apps." Veracode Whitepaper, 2013. Accessed March 18, 2019. https://www.ndm.net/sast/pdf/vcode_wp_mobile_elect.pdf.

Verizon. "Mobile Security Index 2018." 2018. Accessed March 18, 2019. https://enterprise.verizon.com/resources/reports/2018/mobile_security_index_2018.pdf.

Verizon. "Mobile Security Index 2019." 2019. Accessed March 18, 2019. https://enterprise.verizon.com/resources/reports/mobile-security-index/.

Verkijika, Silas Formunyuy. "Understanding Smartphone Security Behaviors: An Extension of the Protection Motivation Theory with Anticipated Regret." *Computers & Security* 77 (2018): 860–870.

VPN Mentor. "Internet Trends 2019. Stats & Facts in the U.S. and Worldwide." 2018. Accessed March 18, 2019. www.vpnmentor.com/blog/vital-internet-trends/.

Wang, Yating, Ray Chen, and Ding-Chau Wang. "A Survey of Mobile Cloud Computing Applications: Perspectives and Challenges." *Wireless Personal Communications* 80, no. 4 (2015): 1607–1623.

Wiggers, Kyle. "Wi-Fi Alliance Introduces WPA3 and Wi-Fi Easy Connect." 2018. Accessed March 18, 2019. https://venturebeat.com/2018/06/25/wi-fi-alliance-intro duces-wpa3-and-wi-fi-easy-connect/.

Wurmser, Yoram. "Mobile Time Spent 2018 – Will Smartphones Remain Ascendant?" *eMarketer*, 2018. Accessed March 18, 2019. www.emarketer.com/content/mobile-time-spent-2018.

Zlatolas, Lili Nemec, Tatjana Welzer-Druzovec, Marjan Heričko, and Marko Hölbl. "Models of privacy and security issues on mobile applications." In *Mobile Platforms, Design, and Apps for Social Commerce*, edited by Jean-Éric Pelet, 84–105. Hershey: IGI Global, 2017.

Chapter 11

Psychophysiological and Behavioral Measures Used to Detect Malicious Activities

Dr. Yassir Hashem

Department of Computer Science and Engineering, University of North Texas
Denton, Texas

Introduction

Insider threat is one of the greatest concerns for information security that could cause more significant financial losses and damages than any other attack. The insider threat has received significant attention and has been studied extensively. However, implementing an efficient detection system is a very challenging task.

Insiders usually have authorized access to the organization's computer systems, information, and assets and are capable of infiltrating an organization's sensitive information, stealing, or damaging data and sabotaging facilities, equipment, and IT systems (Cummings et al., 2012). The computer emergency response team (CERT) defines a malicious insider as "a current or former employee who has or had authorized access to an organization's information systems and has intentionally used that access to influence the confidentiality, integrity, or availability of the organization's information systems" (Silowash et al., 2012). Little is known about the insider threat, and the threat of insider activities continues to be of paramount concern in both the public and private sectors (Glasser & Lindauer, 2013; Ponemon Institute LLC, 2016). However, many surveys have been done that demonstrate the severity of insider threats. For example, the cybercrime report by PwC states that the most serious fraud cases were committed by insiders (PWC, P. 2015). A report produced by the security management company, AlgoSec, found that a significant proportion of security professionals view insider threats as their greatest organizational risk (AlgoSec, 2014). The results of the 2014 US state of cybercrime survey conducted by the CERT Division of the Software Engineering Institute at Carnegie Mellon University shows that 32% of the cybercrime events have been carried out by insiders (CERT. (2014, 7)). In 2015, the federal cybersecurity survey of 200 federal IT managers shows that 76% of the participants are concerned about leaks from insider threats (SolarWinds, 2015).

The research community has been striving to provide solutions and approaches that are capable of mitigating the insider threat problem. Many research studies have been investigating and analyzing these types of threats and raising many research questions such as what is the scope for involving human and psychological factors, and what are the proper approaches to detect and effectively protect the information system. Most of these studies and proposed solutions are based on technological and behavioral theories and are intended to detect attacks before they impact a system and pose irreparable damage (Bowen, Salem, Keromytis, & Stolfo, 2010).

Many approaches have been proposed including anomaly detection, segregation of duties, and security awareness approaches (Kaghazgaran & Takabi, 2015; Park & Stolfo, 2012; Salem & Stolfo, 2009; Thompson, 2004). However, it is recognized that solutions to insider threats are mainly user-centric, and several psychological and psychosocial models have been proposed (Greitzer & Frincke, 2010; Theoharidou, Kokolakis, Karyda, & Kiountouzis, 2005). Most of these approaches monitor the insider's voluntary activities on using the network or the organization's computer systems and resources; however, a skilled insider can always forge these activities and deceive the detection system.

Psychological Aspects of Cybersecurity

With the growing interest in psychological aspects of cybersecurity, researchers are concerned with identifying predictors of these behaviors. However, most of these models rely on humans to recognize the signs and record the behaviors to detect insider threats. Further, these models tend to rely on detecting insider's voluntary behaviors that could potentially fail to recognize individuals who are capable of feigning normal behavior. Other behavioral approaches aim to assess cybersecurity violations through changes in user's activities and require the user to interact with computer peripherals (e.g., mouse; keyboard).

On the other hand, the psychophysiological measures such as electroencephalography (EEG), electrocardiogram (ECG), and eye movement dynamics can provide rich knowledge about user behaviors and can provide an excellent source to start a new type of insider threat detection system. These measures are involuntarily generated, continuously available, and can be measured automatically, which is not available by other traditional behavior measures. The human brain processes legitimate and malicious activities differently. By capturing the neural activities and user's brain state, we can develop methods to automatically separate malicious activities from benign behavior. The neurophysiological methods can provide knowledge of a user's brain state by accessing and recording neural activities.

The ECG is another psychophysiological measure; it is the electrical activities of the heart that have been used in the medical domain for diagnosing many cardiac diseases (Polat & Güneş, 2007). It also provides valuable information about the state of the user such as the emotional stress, depression, fear, and anxiety that affects a user's thinking and behaviors (Agrafioti, Hatzinakos, & Anderson, 2012; Cai, Liu, & Hao, 2009; Guo, Huang, Chien, & Shieh, 2015; Xu, Liu, Hao, Wen, & Huang, 2010).

Eye movement and pupil behaviors are other measures that capture spontaneous responses that are unrefined by the conscious mind and generate a distinctive feature space of voluntary, involuntary, and reflexive eye movement. They can help to identify the changes in the user's behavior. As the conscious and unconscious mental processes generate different patterns of eye movement and when an insider engages in malicious activities, his/her brain activity will translate to changes in eye movement behavior compared to when he/she performs non-malicious activities.

As the above psychophysiological measures may not always be available, finding a correlation between these measures and the user's computer-based behaviors would provide another source to detect the malicious activities. The computer-based behaviors, when a user interacts with computer peripherals (e.g., mouse; keyboard), can provide dynamic traces of an insider's mind and can reveal concealed cognitive states that cannot be achieved using the traditional behaviors (a user's behaviors when using the network or the organization's computer systems and resources). The mouse and keystroke dynamics have been used widely for analyzing user's behavior such as deception and fraud detection, mental state, and emotion prediction (Kaklauskas, Krutinis, & Seniut, 2009; Lali, Naghizadeh, Nasrollahi, Moradi, & Mirian, 2014; Lim, Ayesh, & Stacey, 2015; Salmeron-Majadas, Santos, & Boticario, 2014).

In this chapter, we propose a multi-modal framework based on the user's psychophysiological measures and computer-based behaviors to distinguish between a user's behavior during regular activities versus malicious activities. We utilize several psychophysiological measures such as EEG, ECG, and eye movement and pupil behaviors along with the computer-based behaviors such as the mouse movement, mouse clicks, and keystroke dynamics to build our framework for detecting malicious insiders. We conduct human subject experiments to capture the psychophysiological measures and the computer-based behaviors for a group of participants while performing several computer-based activities in different scenarios. We analyze the behavioral measures, extract useful features, and evaluate their capability in detecting insider threats. We investigate each measure separately, and then we use data fusion techniques to build two modules and a multi-modal framework.

The focus of this study is on the malicious insider where an employee or other insider tries to abuse the system intentionally. The adversary in this scenario has an access to the system and a good knowledge of the internal organizational processes and structures.

Psychophysiological Measures

Insider threats have been extensively studied and many approaches based on behavioral, psychological, physiological, and technical perspectives have been proposed to tackle the threat and provide solutions (Greitzer & Frincke, 2010; Kaghazgaran & Takabi, 2015; Park & Stolfo, 2012; Salem & Stolfo, 2009; Theoharidou, Kokolakis, Karyda, & Kiountouzis, 2005; Thompson, 2004).

The psychophysiological measures such as the ECG and EEG have mainly been used for emotion detection and prediction (Chanel, Ansari-Asl, & Pun, 2007; Kim, Bang, & Kim, 2004; Vaish & Kumari, 2014; Wang, Nie, & Lu, 2014). More recently, some studies started to investigate the possibility of using users' psychophysiological measures such as the ECG, EEG, voice, and skin conductivity for data leakage prevention frameworks. For example, Almehmadi et al. investigated the use of physiological signals as a measurement to detect insider threats (Almehmadi & El-Khatib, 2014). Similar to this work, Lee et al. presented an internal information leakage prevention system based on insider biometrics signals such as heart rate variability, core body temperature, and skin temperature to find unusual changes (Lee et al., 2014). These two papers are closely related to this work, where both used psychophysiological signals to develop a data leakage prevention and insider threat monitoring framework. However, our study used different techniques, provided more extensive research, and utilized more psychophysiological measures. Also, our experiments included a larger number of participants and more complex activity tasks. For example, in Lee et al.'s (2014) study, authors simply measured the heart rate variability for one subject while watching a horror movie using a simple software [BioGraph Infiniti Software (Thought Technology Ltd, 2017)] to detect the change in the biometric signals. In the other study (Almehmadi & El-Khatib, 2014), authors collected ECG, galvanic skin response, and skin temperature measures for 15 subjects while performing two tasks (normal and malicious) and measured the abnormal deviation rate for these measures by simply calculating the difference in the average rate of the sensors raw data. There was no feature extraction nor any extensive analysis done over the recorded data in both studies.

Eye-tracking and gaze-tracking technologies have recently been attracting more interest in the computer security domain as relatively low-cost eye-tracking devices have become widely available in the market. They have mainly been used for authentication and identification (Cantoni, Galdi, Nappi, Porta, & Riccio, 2015; De Luca, Denzel, & Hussmann, 2009; Liang, Tan, & Chi, 2012). Closely related to this work are two recent studies that utilize the eye movements for insider threat mitigation.

The first study by Eberz et al. used eye-tracking to propose a biometric based on distinctive eye movement patterns with the goal of mitigating the insider threat in so-called lunchtime attack scenarios where a person temporarily gains physical access to a workstation that he/she is not supposed to use (e.g., using a coworker's workstation while he/she is at lunch) (Eberz, Rasmussen, Lenders, & Martinovic, 2015). Although they target a particular case of insider threat, the approach they proposed was essentially a biometric-based authentication mechanism that detect if someone else other than the authenticated user has physical access to a device. This work, on the other hand, has an entirely different threat model than this work, which focuses on changes in user's eye movement patterns to distinguish between malicious activities and non-malicious activities and detect the threat before it occurs. The second study proposed by Neupane et al. who conducted a three-dimensional study of phishing detection to detect the unintentional insiders based on EEG and eye gaze patterns (Neupane, Rahman, Saxena, & Hirshfield, 2015). Unlike this work, which focuses on intentional malicious insider threats, this study is related to unintentional insiders (apathetic insiders) and the eye movements are used only to identify whether users look at the security indicators.

In addition, the computer-based behaviors are soft biometrics that have not been explored for the insider threat detection and mitigation purposes. However, they have mainly been used for authentication and identification (Bergadano, Gunetti, & Picardi, 2002; Monrose & Rubin, 2000; Pusara & Brodley, 2004; Valacich, Jenkins, Nunamaker, Hariri, & Howie, 2013). Further, the mouse and keystroke dynamics were explored for emotional and mental state prediction as well as deception and fraud detection (Lali, Naghizadeh, Nasrollahi, Moradi, & Mirian, 2014; Lim, Ayesh, & Stacey, 2015; Salmeron-Majadas, Santos, & Boticario, 2014). Closely related to this work, Valacich et al. proposed a polygraph technique that utilizes an online survey platform to differentiate between innocent users' and guilty insiders' reaction to a concealed information test using specific mouse movement features (Valacich, Jenkins, Nunamaker, Hariri, & Howie, 2013). This study targeted a very specific case of insider threat using a controlled experiment where the participants reacted to visual stimuli displayed on the screen (online survey). However, in this work, we did not use any stimuli in the experiments; instead, we focused on developing tasks that are close to real-world scenarios.

Experimental Design

This section provides an overview of the design goals and then describes the experimental design developed to meet those goals. We record the EEG signals, ECG signals, eye movements, mouse movements, and keystroke dynamics for participants while they perform several activities, both benign and malicious. In general, there are two types of experiments that can be conducted for our purpose. The first is the events control experiment where the participant reacts to visual stimuli displayed on the screen, and the device records his/her reaction. The second type is the free experiment where there is no specific stimuli or event. In order to closely imitate real-world insider threat scenarios we aim to address in this study, we have chosen the second type of experiments. We do not use any stimuli in the experiments and instead we focus on developing tasks that are close to realistic situations.

The study involves human subject experiments to collect the psychophysiological and the computer-based behavioral measures. We have conducted two experiments on different groups of subjects. Each experiment emulates a real-life scenario that is very similar to a typical work environment. In the first experiment (the pilot study), we use a small group of participants (10 subjects) to establish our initial analysis and investigate the capability of using the psychophysiological measures to detect malicious insiders. Based on the results, we extend our analysis and conduct the second experiment (the main study) which includes a larger number of participants (25 subjects), more complex activity tasks, and more advanced recording devices. Also, we include the eye movement dynamics, as well as the mouse movement and keystroke dynamics in our data acquisition phase.

Ethical Considerations

The experiments were conducted with the approval of the Institutional Review Board (IRB) from the University of North Texas, and the participants were compensated $30 for one hour of their time. Their participation was voluntary, and they were given the option to withdraw from the experiment at any time. We followed the standard best practices to protect the participants' data collected during the experiments.

Pilot Study

This experiment was our first step to analyze the psychophysiological measures and investigate the effectiveness of using these measures for insider threat detection purposes. More specifically, we captured the ECG and EEG signals for a group of participants (10 subjects) while they performed several computer-based activity tasks for both malicious and benign scenarios. The experiments were conducted in our lab room which was set up to keep the same environmental conditions for all tasks and all participants. Figure 11.1a shows the experimental setup for our pilot study. We used the following devices in our experiments:

- To record the participant's EEG signals, we used a noninvasive Brain-Computer Interface Emotive EPOC headset (Emotiv, 2014) that contains 14 active electrodes (channels) recording the signals from different parts of the brain with a sampling rate of 128 Hz.
- To record the participant's ECG signals, we used the OpenBCI 32-bit Board (OpenBci, 2014) with 256 sampling rates, a high-quality bio-sensing platform that provides a high-resolution recording of electromyography (EMG), ECG, and EEG signals.

The experiments were done for each participant separately and at different times during the day. The experiment was divided into 10-minute tasks for a total of three tasks: one regular activity task and two malicious activity tasks. Once the participants were seated in a comfortable chair in front of a computer used for the experiments, the investigators then proceeded to mount the emotive device on the participants' heads. Then, they were given a brief verbal explanation of the tasks. Five minutes were given to participants to relax and to feel comfortable with the recording device and the test environment.

(a)   (b)

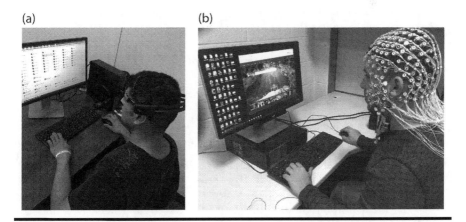

Figure 11.1   (a) Pilot experiment setup. (b) Main experiment setup.

Experiment Tasks

In the first scenario, the participants performed regular office job activities such as browsing the Internet, using computer applications, or using the email account. This scenario observes the brain and heart reactions to the regular daily activities done by most of the employees in any organization. In the second and third scenarios, the participants were asked to perform malicious activities by trying to access to information they were not authorized to access. We used two realistic scenarios in which an employee uses remote access or network to access unauthorized information. We recorded the ECG signal and the EEG signals in reaction to these malicious activities during each experiment.

Main Study

In this experiment, we extend our previous pilot experiment by conducting a human study that includes a larger number of participants and more complex activity tasks, and includes the eye movement dynamics, as well as the mouse movement and keystroke dynamics. In addition, we use more advanced EEG recording device that includes 256 channels to cover the entire skull and record the ECG signals from each part of the brain. We aim to deeply investigate the use of the EEG for the purpose of insider threat detection and mitigation. Also, we want to study each channel and identify which part of the brain can reflect the best knowledge and can provide better detection accuracy. We recruited a total of 30 participants, we used data of 25 participants in our approach evaluation; the record of 5 participants was incomplete, and we decided to remove it from our evaluation process. Out of 25 participants, 15 were male and 10 were female. The entire group of participants were between the age of 18 and 34 years and were graduate or undergraduate students at the University of North Texas. Figure 11.1b shows the experimental setup for our main study. We used the following guidelines in our recruiting process:

(1) Include participants with different levels of programming and cybersecurity knowledge. As insider skills range from script kiddie (an unskilled user who uses codes or programs developed by others) to highly skilled insider, our participants vary in the levels of programming and cybersecurity knowledge from novice to intermediate and advanced.
(2) Include participants with prescription eyeglasses to emulate real-world scenarios for a typical work environment where some employees may wear eye glasses (10 of our participants had eyeglasses and we made sure they wore their glasses during the experiments).
(3) Have a fair distribution of participants with respect to gender, race, and age.

We used the following devices and applications to record our experiment data:

- To record the EEG signals, we used a medical grade device, Geodesic EEG System with the Geodesic Sensor Net, which includes a total of 256 sensors (channels) that capture the EEG signals from all around the skull. Also, the device provides an excellent recording quality with a sampling rate of 1000 Hz (Electrical Geodesics Inc., 2017).
- To record the ECG signals, we used the Open BCI 32-bit Board (OpenBci, 2014), with 256 sampling rates.
- To record the participant's eye movements during the experiment, we used a Tobii Pro X2-60 device, a screen-based eye tracker capturing gaze data at 60 Hz (Tobii, 2017).
- For mouse and keyboard data collection, we used a Mini Mouse Macro Pro software. The software records the X and Y coordinates of the mouse position on the screen, the mouse and keyboard event (drag, click, move, key press), and the event time- stamp in milliseconds (ms) (Turns soft, 2016) .

Experiment Procedure

The experiment was divided into six different tasks for a time period of 10 minutes per task. There were four benign activity tasks and two malicious activity tasks. In the following, we describe each of the six tasks in detail:

Task 1 users perform benign daily activities: This task emulates the benign daily activities performed by most employees in any organization such as browsing the Internet, using computer applications, or using an email account. In this task, the participant received through email an Excel sheet containing names of students who participated in a previous survey and their associated information. The participant needed to use the browser to login to the database system and find out the students' names and update their missing data using the Excel sheet.

Task 2 users perform benign daily activities under stress: In this task, we repeat the previous task with some changes to introduce stress to the participants in order to emulate the scenario where employees work under pressure or emotional stress. The reason behind this experiment is to differentiate between the regular work pressure, fatigue, or stress and the malicious intent of the users. To do this, we conducted the experiment at the end of the working day, so the participants came to the lab after attending classes, exams, or labs during the day causing them to have a greater mental workload compared to Task 1. The participants were also given twice the number of the records compared to the first task and asked to make sure to finish all the records during the 10-minute experiment. In addition, we removed some of the students' names from the database to add more pressure as the participants were not able to find those names. The participants were told that there would be a special prize for the participant who finishes his/her report faster than the others, adding more stress and time pressure to the experiment.

Task 3 users perform high mental workload activities: This task emulates the professional job activities when the employees perform some activity involving high mental interaction. We tried to show that our approach covered all the possible activities. The participants were asked to complete a short coding project (designing a calculator), which required more mental workload compared to the benign daily activities in Task 1. The participants were allowed to choose any programming language they felt comfortable with (C++, Java, or Python). They were also allowed to browse the Internet for help if needed. However, the participants were told that copying the code from the Internet was not allowed and they had to write their own code. The participants were encouraged to finish the code project. However, the task did not require them to complete the project and there was no pressure or stress introduced to the experiment.

Task 4 users perform high mental workload activities under stress: In this task, we repeated the previous task with some changes to introduce stress to the participants in order to emulate situations where employees work under pressure or emotional stress. To do this, we conducted the experiment at the end of the working day, so the participants came to the lab after attending classes, exams, or labs during the day, causing them to have a greater mental workload compared to Task 3. The participants were asked to repeat the same project but with another programming language. Participants were still able to browse the Internet for help. However, the participants had to complete and test the code within 10 minutes. The participants were told that there would be a special prize for the participant who completes his/her code faster than the others.

Task 5 users perform remote access attack: This task emulates the malicious activities that could be performed by an insider. We used the remote access scenario where an insider accesses another computer in the network which he/she is not authorized to access. The insider can steal the credential information using shoulder surfing and use it to login to the victim's device. So in this task, the participants were asked to remotely access the teaching assistant's computer and steal data related to the exams, quizzes, projects, and so on. Participants were instructed to perform this task without the teaching assistant noticing and to incentivize them, they were told that they would receive an extra reward if they could complete these tasks without leaving any trace. For this task, we added a timer on the computer desktop showing the time remaining to complete the task.

Task 6 users perform SQL injections attack: In this task, we emulate the scenario of the script kiddie insider (an unskilled system user who uses codes or programs developed by others to attack computer systems and networks participants). More specifically, we used the scenario when an insider uses SQL injections to access to information he/she has no permission to access. In this task, the participants were asked to use SQL injection to bypass the authentication for the database system, then update, copy, and delete the student's information in the database. The participants were told that successfully completing this task without leaving any trace would result in an extra reward.

Analysis of Signal Psychophysiological Measures and Results

In this section, we present our analysis and results for the signal part of our recorded psychophysiological measures (i.e., EEG, ECG) for both the pilot study and the main study. One of the essential steps in our data analysis is to preprocess and filter the acquired EEG and ECG signals. This step requires specific signal processing and filtering procedures. The EEG and the ECG signals, as we explained before, are low-frequency signals with very low signal intensity measured in microvolts (μV). Therefore, these signals will naturally include some noises that may corrupt their purity. There are two types of noises we should take into consideration. First, the EMG artifacts that happen due to involuntary muscle movements around the device electrodes. Second, the electrooculogram (EOG) artifacts that are generated as a result of eye blinking when the EEG signals are recorded. Our first step is to remove these artifacts and remove any other noises that may influence the results. We use a low-pass filter to remove the higher frequency noises, high-pass filter to cut off the lower frequency, and select a frequency range from 0.1 to 30 Hz. Next, we extract features from the recorded signals. EEG and ECG signals can be presented in time and frequency domains. A combination of frequency information features and time domain information features can be extracted from both domains and improve the classification performance (Mensh, Werfel, & Seung, 2004). There are many feature extraction techniques that can be applied to these signals and they can obtain features from either the time or frequency domains. In this work, we use a wavelet packet decomposition technique (Khushaba, Kodagoda, Lal, & Dissanayake, 2011; Li, Shen, & Beadle, 2004) to extract the features. The wavelet transform algorithms have gained substantial consideration in analyzing non-stationary signals (signal's statistical characteristics change with time) (Li, Shen, & Beadle, 2004). Wavelets are localized in both the time and frequency domains, and these characteristics of wavelets make it a useful tool for the purpose of feature extraction. In more detail, the wavelet packet decomposes the originally recorded signals into a specific number of sub-bands using the wavelet function and generates a sub-band tree. Each level of the tree is composed by passing the previous approximation coefficients over high-pass and low-pass filters. Then, we take the energy of the wavelet coefficients using the normalized filter bank energy for each sub-band.

Pilot Study Experimental Results

The results for the pilot study are divided into two parts: the first one presents the results of using the EEG signal recorded during the experiments, and the second one presents the results of combining the EEG and ECG signals for the same subjects.

After the experiments were performed and the EEG signal was collected, we organized the recorded signals into three groups: regular activity group, malicious I activity group, and malicious II activity group. Each group consists of 10 signal samples represented by each participant in the experiments. These signals were then analyzed and pre-processed by removing the high-frequency noise and the EOG artifacts. Next, we applied our feature extraction algorithm for each five-second time frame and labeled the result features vector with its represented group type. We used the wavelet packet decomposition method to decompose the EEG frequencies into different frequency subsets and extract features from each subset. We decomposed the filtered EEG signals with the frequency of 30 Hz to smaller sub-bands and got the energy for each band. To do this, we used three levels of decomposition that gave us seven different bands. For the ECG signals, we used the same wavelet packet decomposition but with six levels of decomposition. We also extracted other features such as microvolts (μV) mean value, maximum μV, minimum μV, number of peaks, and the distance between the high and low μV from each five-second time frame. Then, we applied the Support Vector Machine (SVM) classifier to our labeled feature vectors using k-fold cross validation ($k = 5$) to classify the normal and malicious activities.

The SVM classifier is one of the most practical and powerful classifiers that have been widely used in machine learning and previous related work (Cortes & Vapnik, 1995). Table 11.1 shows the results in detail for the EEG dataset containing the first malicious scenario. The results indicate that our approach can detect malicious activities with 84% accuracy. In addition, the results show that our framework recognizes regular activities with 84% accuracy. The results also show around 0.84 precision and about 0.85 recall in detecting malicious activities. Similar to the first test, we applied the SVM on the EEG dataset containing the second malicious activity scenario. The results show 90% accuracy in detecting both regular and malicious activities for the EEG dataset containing the second malicious activity scenario as shown in Table 11.2. In the third test, we applied the SVM on both

Table 11.1 Results of EEG dataset containing the first malicious scenario

Class Type	Classification Accuracy	Precision	Recall	F-measure	Error Rate
Normal class	84.45%	0.8383	0.8231	0.8513	0.1555
Malicious class	83.90%	0.8430	0.8557	0.8320	0.1610

Table 11.2   Results of EEG dataset containing the second malicious scenario

Class Type	Classification Accuracy	Precision	Recall	F-measure	Error Rate
Normal class	89.95%	0.8874	0.8929	0.8969	0.1005
Malicious class	89.60%	0.9090	0.9045	0.8985	0.1040

malicious scenarios to evaluate the detection accuracy for normal and malicious activities in general. The results show up to 86% accuracy for detecting both regular and malicious activities as shown in Table 11.3. In general, our results show a detection accuracy above 84% for all the scenarios. In order to show how the results can improve when we add features of another psychophysiological measure (such as the ECG) to our original EEG feature set, we chose five subjects with the lowest detection accuracy from the EEG results. We then combined their EEG signals with their ECG signals recorded during the same experiment to improve the detection accuracy. We ran our classifier for each case and compared the results. Note that all the ECG signals have been pre-processed, filtered, and passed the same process as the EEG signals and applied our feature extraction algorithm for each five-second time frame and labeled the result feature vector with its represented group type. We applied the SVM to the EEG features dataset containing the regular activities and both malicious activities for five subjects using k-fold cross validation ($k = 5$), then we added the ECG features and ran the classification process again.

Table 11.4 shows the results for the two cases: the first one using the EEG signal features only, and the second one using both EEG and ECG signal features for the same subjects and the same time frames. The results show up to a 5% increase in

Table 11.3   Results of EEG dataset containing both malicious scenarios

Class Type	Classification Accuracy	Precision	Recall	F-measure	Error Rate
Normal class	85.01%	0.8865	0.7734	0.8876	0.1499
Malicious class	85.55%	0.7797	0.8877	0.7856	0.1445

Table 11.4   Results of combined EEG and ECG signals

Experiment	Class Type	Accuracy	Precision	Recall	F-measure	Error Rate
EEG	Normal class	81.91%	0.8856	0.8228	0.8531	0.1809
	Malicious class	81.16%	0.7066	0.8194	0.7588	0.1884
EEG + ECG	Normal class	86.20%	0.9266	0.8720	0.8984	0.1380
	Malicious class	86.91%	0.7490	0.8483	0.7956	0.1309

the classification accuracy, precision, and recall for both normal and malicious activity scenarios using both psychophysiological measures. The accuracy improved from 81.91% to 86.20% in detecting the normal activities and from 81.16% to 86.91% in detecting the malicious activities. The results also show much better *F*-measure and Error rate values compare to the values when using the EEG data individually.

Main Study Experimental Results

After completing the experiments for all participants, we used a 10-second time frame (epoch) to extract our feature vector. We applied our feature extraction algorithm using three levels of decomposition for the EEG signal and six levels of decomposition for the ECG signals. We organized the recorded signals (i.e., EEG, ECG) in two ways:

- Setup 1: we label the feature vectors extracted from each participant by the task activity. The feature vectors extracted from the benign activities' tasks (tasks 1, 2, 3, 4) have a label "0" (negative), and the feature vectors obtained from the malicious activity tasks (tasks 5, 6) have a label "1" (positive). This setup is used to evaluate our approach's capability in distinguishing the malicious activities from the benign activities.
- Setup 2: we only use four tasks in this setup by labeling the feature vectors extracted from the benign activities under stress (task 2, 4) as "0" (negative), and the feature vectors extracted from the malicious activities (tasks 5, 6) as "1" (positive). We only consider the malicious tasks and the benign tasks performed under stress to measure our approach's performance in differentiating between the malicious activities and the benign activities performed under stress.

We perform training and testing on each participant's data. We split each participant's data into 70% tuning and training set, and 30% testing set by dividing the ten minutes of task time into seven minutes for training and three minutes for testing. In order to ensure a good separation between the training and testing sets, we ignore the 30-second time window between the training and testing time (the last 15 seconds of the training minutes and the first 15 seconds of the testing minutes). We utilize three different classification algorithms, namely: the SVM (Cortes & Vapnik, 1995), the *K*-nearest-neighbors (*k*-NN) (Altman, 1992), and the Random forests (Breiman, 2001) to classify the examples.

Malicious Activities Detection

We use the data on setup 1, which contains data from all experiment tasks and evaluates the performance of the classifiers in detecting the malicious activities. On the ECG data, we run the three classifiers over the ECG datasets that were extracted from the ECG signals. Table 11.5 shows the average results for our

Table 11.5   ECG data average results for detecting malicious activities using six experiment tasks

Classifier	Number of Participants	Accuracy	TP Rate	FP Rate	F-Measure
SVM	25	83.71	0.84	0.26	0.81
Random Forests	25	91.90	0.92	0.10	0.92
k-NN	25	91.98	0.92	0.10	0.92

participant's group and shows an average detection accuracy of up to 92% using the Random Forest and the *k*-NN classifiers. The SVM classifier shows the lowest performance with an average of 83% detection accuracy. These results conclude that we can reach a detection accuracy of up to 92% by using only the ECG signals. For the EEG signals, we use the data recorded by the all 256 channels in our dataset then we run the classifiers and calculate the average results. Table 11.6 presents the results of our three classifiers. As shown, all classifiers show a good average accuracy. However, the SVM classifiers outperformed the other three classifiers and achieved up to 99.77% average classification accuracy and a 0.99 average true positive rate while the Random Forest and the *k*-NN classifiers show an average classification accuracy of up to of 97%.

Channels Location and Brain Regions

The brain is a complex and magnificent organ in the human body that contains around 100 billion neurons to promote and control our perception and interaction with the world. Thoughts, feelings, behaviors, and plans are controlled by our brain. In neuroscience, the largest part of the brain is the cerebrum, which is associated with higher brain exercises such as thoughts and actions (Goldberg, 2002; Posner & Petersen, 1990; Roland, Roland, & Roland, 1993). The cerebrum consists of four regions (lobes) as shown in Figure 11.2: the frontal lobe, parietal lobe, occipital lobe, and temporal lobe (Apperly, Samson, Chiavarino, &

Table 11.6   EEG data average results for detecting malicious activities using six experiment tasks

Classifier	Number of Participants	Accuracy	TP Rate	FP Rate	F-Measure
SVM	25	99.77	0.99	0.02	0.10
Random Forests	25	97.78	0.98	0.03	0.98
k-NN	25	97.93	0.98	0.03	0.98

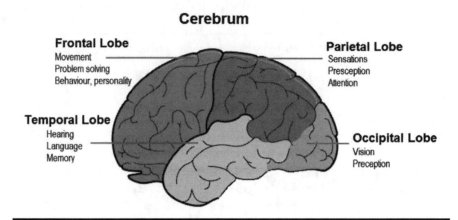

Figure 11.2   Human brain regions.

Humphreys, 2004; Bajada et al., 2016; Chayer & Freedman, 2001; Fogassi et al., 2005; Stuss & Knight, 2002). Each lobe controls specific functions, for example, the frontal lobe constitutes two-thirds of the human brain and controls functions associated with problem solving, behavior, some emotion as well as muscle movement and physical reactions. The parietal lobe functions are associated with orientation, recognition, perception, and appreciation of form through touch (stereognosis). The occipital lobe is responsible for vision and reading functions, and the temporal lobe correlates with visual memories, speech and language, and some perception and recognition functions. As insider threats involve different behaviors and activities associated with the insider's brain, we deeply investigate our results and identify if there is a correlation between the detection accuracy we achieve and the brain regions. We first investigate each channel's location to find out which area of the brain the best channels belong to. Then, we calculate the average accuracy for each region using the results from each channel in that area.

To do that, we label our 256 channels to five brain regions based on the sensor's location on the skull: Frontal, parietal, temporal, and occipital lobe regions, we add to that the central region which is shared between the frontal and parietal lobes and we call it the "Frontal-Parietal Lobe" region.

We run our classifiers using the data of each channel separately for the entire participant's group and sort the average detection accuracy from each channel. Table 11.7 presents the top-ten channels, which show the best average detection accuracy. The first column shows the channels number based on the Geodesic Sensor Net used in our recording. As shown, we can reach up to 82% average accuracy using only one channel (single sensor). However, all the ten best channels are located in the frontal and parietal lobes, as both lobes are associated

Table 11.7   The top-ten channels and their associated brain regions

Channel No.	Accuracy	Brain Location
132	82.39	Frontal-Parietal Lobe
89	80.75	Parietal Lobe
185	80.51	Frontal-Parietal Lobe
81	80.48	Frontal-Parietal Lobe
79	79.67	Frontal-Parietal Lobe
17	79.62	Frontal-Parietal Lobe
198	79.38	Frontal Lobe
45	79.38	Frontal-Parietal Lobe
53	79.28	Frontal-Parietal Lobe
186	79.15	Frontal-Parietal Lobe

with functionality that can be correlated to insider practices and behavior, such as the fluctuation of behavior, emotional state, and attention.

Furthermore, we investigate each channel and each region's performance in detecting the malicious insider. However, the accuracy when using a single channel is much lower than using the entire group of channels (256 channels); using 256 channels in real-life scenarios seems unrealistic, as this number of channels only available with medical devices and most of the consumer-grade brain computer interface devices have only a small number of channels (1 to 16 channels).

Therefore, it is better to find the right trade-off between the detection accuracy and the applicability of the approach. To find a smaller group of channels that can provide acceptable detection accuracy, we use the best performing channels and generate a combination of a small number of channels. We start by using one channel (best channel) then combine the channel with other channels and measure the performance. We use four groups of channels: two channels, three channels, four channels, and five channels. Table 11.8 presents the average results using the different groups of channels start with one channel to five channels group. As we can see, the average accuracy improves by increasing the number of channels for all the classifiers.

By only using a group of five channels, we can reach an average detection accuracy of 95.64% using the Random Forest classifier. These results indicate that we can reduce the number of channels to five channels and still achieve an excellent detection accuracy.

Table 11.8 Average results for detecting malicious activities using small groups of channels (EEG data)

Classifier	Number of Channels	Accuracy	TP Rate	FP Rate	F-Measure
SVM	1	79.58	0.796	0.370	0.760
Random Forests	1	82.39	0.824	0.236	0.823
k-NN	1	80.53	0.805	0.240	0.808
SVM	2	85.88	0.858	0.194	0.848
Random Forests	2	89.31	0.893	0.136	0.893
k-NN	2	87.57	0.876	0.162	0.876
SVM	3	87.77	0.878	0.174	0.871
Random Forests	3	92.12	0.921	0.100	0.921
k-NN	3	90.82	0.908	0.124	0.908
SVM	4	89.99	0.900	0.138	0.899
Random Forests	4	93.51	0.935	0.086	0.935
k-NN	4	91.91	0.919	0.108	0.920
SVM	5	92.49	0.925	0.0899	0.925
Random Forests	5	95.64	0.956	0.060	0.956
k-NN	5	94.85	0.948	0.0721	0.949

Distinguishing Malicious Activities from Benign Activities Performed Under Stress

One may argue that changes in neural activity might be due to emotional stress. To address this issue, we examined the capability of our approach in distinguishing between benign activities that are performed by users under emotional stress or high mental workload and the malicious activities of an insider. To do that, we use the second setup which contains four tasks: tasks 2 and 4 which are benign activities performed under stressful conditions, and tasks 5 and 6 which are the insider threat tasks. We evaluate the performance of the classifiers using the data with this setup. We choose to use the feature vectors extracted from five channels as using the entire group of channels will always show relatively better results. We want to make it slightly difficult for the classifiers by using less number of channels. Table 11.9 presents the results for each classifier. As shown, the three classifiers are able to distinguish between the malicious and the benign stressful activities with about 90% detection accuracy on average. However, the Random Forest classifiers show the best result with 91.92% detection accuracy on average. Using the ECG data, our approach reaches up to a 91% average detection accuracy in distinguishing between benign activities and malicious activities. Table 11.10 shows the results in more detail. As shown, the

Table 11.9   Average results for detecting malicious activities from benign activities performed under stress (EEG data)

Classifier	Number of Channels	Accuracy	TP Rate	FP Rate	F-Measure
SVM	5	89.53	0.90	0.10	0.89
Random Forests	5	91.92	0.92	0.08	0.92
k-NN	5	90.71	0.91	0.10	0.91

Table11.10   Average results for detecting malicious activities from benign activities performed under stress (ECG data)

Classifier	Number of Participants	Accuracy	TP Rate	FP Rate	F-Measure
SVM	25	84.12	0.84	0.17	0.82
Random Forests	25	91.38	0.91	0.09	0.91
k-NN	25	91.09	0.91	0.10	0.91

Random Forest and the *k*-NN classifiers still show an average detection accuracy close to 91%. These results clearly show that our study does not look for stress or emotional symptoms but can differentiate between malicious and benign activities even when they involve emotion or stress.

Analysis of Non-signal Psychophysiological Measure and Results

In this section, we analyzed the eye tracking and the mouse and keystroke dynamics data from our main experiment. The Tobii Pro eye-tracking device captures the eye gaze locations on the screen (x,y) coordinates, and produces raw data at a sampling rate of 60 times per second including the saccade, fixation locations (x,y), and pupil diameter. We analyze the raw samples to extract the features. We sample the eye-tracking raw data into 10-second time frames (epochs), each time frame representing one feature vector. We extract all possible features from the eye-tracking data: the movement features, which can be categorized as spatial and temporal features, as well as the pupil features.

We obtain a total of 38 movement features and 8 pupil features that include temporal features such as the saccade and fixation duration, the pairwise speed and acceleration, spatial features such as the pairwise distance, the distance from the center of the screen, and the direction of saccades. We also consider the frequency of eye blinking, and the saccades and fixations frequencies. For the pupil's data, we obtain the pupil diameter and apply statistic measures such as the minimal, maximal, mean, and standard deviation of the diameter. The same statistics were applied to the eye movement features. Since we use classification to distinguish between malicious and benign activities, we need to ensure only good-quality features are used in order to achieve a good accuracy. Therefore, using feature selection techniques and choosing the best feature set is an essential process before doing the classification. In our approach, we choose three feature selection algorithms to choose a group of best-quality features: forward greedy search algorithm, backward greedy search algorithm, and information gain feature evaluation algorithm. We combine the three feature selection techniques to evaluate the features and select the best, as a result, we reduce our feature set from 46 to 17 features which are selected as best features.

Malicious Activity Detection on Eye Tracking Data

After completing the experiments for all participants, we organized the data into two groups as we explained in our main study experimental results section. We use the datasets from the first setup to evaluate the performance of our approach in distinguishing the malicious activities from the benign activities in general. The datasets from the second setup, on the other hand, we use to evaluate the ability of our approach in differentiating between the malicious activities and the benign activities performed under stress. The goal is to show that our approach is not only able to differentiate between benign behavior and the malicious behavior, but also can differentiate between benign and malicious behavior even when the benign behavior is performed under stress. After generating the datasets, we divide each one into a 70% tuning and training dataset and a 30% testing dataset. In particular, we split the ten-minute task time to seven minutes for training and three minutes for testing, and we disregard the last ten seconds of the seven minutes and the first ten seconds of the three minutes to ensure a good separation between the training and testing sets. We use the same classification algorithms in our main study experiment and consider each participant individually and generate a separate profile for each using only their data to train and test the classifiers.

We use the dataset containing data from all experiment tasks (setup 1) to measure the performance of the classifiers. We run the classifiers for each participant two times. First, we use the complete feature set (the original 46 extracted features). Then, we run the classifiers again using the best feature set (the 17 selected features) and compare the results. Table 11.11 shows the results of our three classifiers (SVM, Random Forest, and k-NN). The top part of the table shows the average results over our 25 participants using the complete feature set and the selected feature set.

Table 11.11   Results of detecting malicious activities using the complete feature set (47 features) and the selected feature set (17 features)

Classifier	Subjects	Complete Feature Set				Selected Feature Set			
		Acc.	TP Rate	FP Rate	F-Measure	Acc.	TP Rate	FP Rate	F-Measure
SVM	25	78.10	0.78	0.39	0.76	79.23	0.79	0.36	0.79
Random F	25	81.99	0.82	0.27	0.81	83.18	0.83	0.25	0.83
k-NN	25	75.00	0.75	0.37	0.74	76.29	0.76	0.37	0.76
SVM	10*	75.20	0.75	0.40	0.73	76.20	0.76	0.37	0.75
Random F	10*	75.84	0.76	0.33	0.75	78.54	0.79	0.29	0.78
k-NN	10*	71.31	0.71	0.38	0.70	73.79	0.74	0.38	0.73
SVM	15	80.02	0.80	0.38	0.78	81.25	0.81	0.35	0.79
Random F	15	86.10	0.86	0.24	0.85	86.27	0.86	0.24	0.86
k-NN	15	77.46	0.77	0.36	0.76	77.95	0.78	0.37	0.76

* Participants who wore eyeglasses during the experiments.

The results indicate that the Random Forest classifier outperformed the other two classifiers and was able to achieve up to 81.99% classification accuracy on average when using the complete feature set. Also, it provides a better average true-positive rate, average false-positive rate, and average *F*-measure. When using the selected feature set, the accuracy improved to 83.18%.

In the middle part of the table, we present the average results for 10 participants who were wearing eyeglasses during the experiments. As shown, the average accuracy for these participants is below the average results from the entire participants set in both the complete and selected feature set for all the classifiers. These results are expected as eyeglasses can reflect the light and impact the recording accuracy of the eye-tracking device. In contrast to these results, the bottom part of the table shows a higher average detection accuracy when we use just the 15 participants who had healthy eyes and did not wear eyeglasses during the experiments. The average accuracy is up to 86% using the Random Forest classifier with 0.86 average true-positive rate and 0.24 average false-positive rate, while the average *F*-measure is 0.85.

In general, the results show that Random Forest classifier outperformed the other two classifiers, and the selected feature set improved the results by about 2–3% for all the classifiers.

In order to distinguish malicious activities from activities performed under stress, we measure the performance of our approach using the second setup as explained in our main study experimental results section. In this setup, we only use four tasks for training and testing: tasks 2 and 4, which are the benign tasks performed under stressful conditions and tasks 5 and 6, which are the insider threat attack tasks. The goal is to demonstrate that our approach is capable of distinguishing between the activities that are performed by legitimate users under emotional stress or high mental workload (since they might affect eye movement patterns) and the malicious activities of an insider. Table 11.12 shows the results from each classifier using three different groups of participants. The top part of the table indicates that the best average detection accuracy using the entire group of 25 participants is 78.39% and is achieved using the Random Forest classifier. The middle part of the table shows 73.65% average accuracy using the Random Forest classifier on the group of participants who were wearing eyeglasses. The same classifier shows the highest average accuracy (up to 81.55%) on the group of 15 participants with healthy eyes. The results demonstrate that our approach does not simply look for stress signs and can distinguish between malicious activities and other activities, even the ones involving emotion or stress.

Mouse and Keystroke Dynamics Data Processing and Results

We recorded the X and Y coordinates of the mouse position on the screen, the mouse and keyboard event (drag, click, move, key press), and the event time-stamp in milliseconds (ms). We analyzed the data by feeding the file to our feature extraction algorithm to extract useful features that represent the user behavior and can be used for detecting malicious insiders. We sample the raw data into a 10-second time frame (epoch), each time frame represents one feature vector. We extracted all possible

Table 11.12 Results for detecting malicious activities from activities performed under stress

Classifier	Number of Participants	Accuracy	TP Rate	FP Rate	F-Measure
SVM	25	72.35	0.72	0.29	0.71
Random Forest	25	78.39	0.78	0.23	0.78
k-NN	25	69.38	0.69	0.32	0.69
SVM	10*	67.74	0.68	0.32	0.67
Random Forest	10*	73.65	0.74	0.26	0.74
k-NN	10*	67.46	0.67	0.32	0.67
SVM	15	75.42	0.75	0.27	0.74
Random Forest	15	81.55	0.82	0.21	0.81
k-NN	15	70.66	0.71	0.32	0.70

* Participants who wore eyeglasses during the experiments.

mouse and keystrokes dynamic features that include spatial and temporal mouse movement features, mouse click features, and keystroke features. To calculate the mouse movement speed, we measured the length of the mouse path by adding the total distances between all adjacent path coordinates and dividing by the total time the mouse path took (the summation of the time-stamps in the path). As our experiment does not record the mouse and keystroke actions regarding stimuli, and it is entirely free (participants can move the mouse and press the keyboard keys freely during the experiment), we do not have a predefined start and end point to the mouse movement path. To address that, we choose our start and end points by the value of the movement event's time-stamp. We use 800 ms as our threshold to identify the start and end points. A mouse movement event with a value of 800 ms or above is considered as the stop position, and the next movement event would be the start point for the next path. We also calculate the mouse movement distance or the length of the mouse movement path and subtract that from the direct distance (the Euclidean distance between the first point in the path and the last point) to extract the mouse travel distance feature. In addition, we calculate the mouse left click duration and keystroke duration. We also consider the direction of the mouse movements, the backspace, the numeric keypad usage, and the frequency of direction change.

After completing the experiments for all participants, we analyze the recorded mouse and keystroke data for each participant separately. Then, we sample the data into a 10-second time frame (each time frame represents one sample). We extract the features and in order to investigate which features are more valuable and useful in detecting the malicious intent among the extracted features, we run the statistical analysis over the mouse and keystroke extracted features and investigate each feature separately. We calculate the mean for each feature among the group of our participants for each task. We evaluate our results by dividing our experiment into three different tasks: the benign tasks, the benign under stress tasks, and the malicious tasks. Of all the features we tested, four features show statistically significant differences between the mean value of the tasks, namely the mouse movement speed, the mouse travel distance, the left mouse click duration, and the keystroke duration. As shown in Table 11.13, the malicious tasks were associated with mouse movement speed at a mean of 0.97 (pixels/ms) (SD 0.34). By comparison, the benign tasks and the benign under stress tasks were associated with slower mouse movement speed at a mean of 0.75 (pixels/ms) (SD 0.26) for the benign tasks, and a mean of 0.82 (pixels/ms) (SD 0.42) for the benign under stress tasks. To test the hypothesis that there is a statistically significant difference between mouse movement speed of malicious tasks and the benign tasks, a related t-test was performed. The related t-test shows a statistically significant effect on p-value = 0.0000316. Thus, the malicious tasks were associated with significantly larger mean than the benign tasks. We also run the related t-test between the mean of the mouse movement

Table 11.13   Results of the mouse and keystroke dynamics

Features	Benign		Benign Under Stress		Malicious	
	Mean	SD	Mean	SD	Mean	SD
Mouse movement speed	0.75	0.26	0.82	0.42	0.97	0.34
Mouse distance travel	145.93	44.75	146.57	61.53	181.13	72.10
Mouse left clicks duration	167.63	69.60	226.69	224.20	293.01	152.32
Keystroke duration	218.52	463.96	506.65	937.79	469.32	792.49

speed for the malicious tasks and the benign under stress tasks. The results show that there is a statistically significant difference with p-value = 0.034. These results suggest that users will move the mouse at a relatively higher speed when performing a malicious act than performing a benign act even when they are involved in stressful conditions.

On the other hand, the participants performing the malicious tasks show longer travel distances than the other two tasks. As shown in the table, the malicious tasks were associated with the longest mouse movement distance among all the tasks with a mean of 181.13 pixels (SD 72.10).

To ensure these differences are statistically significant, we performed the related t-test over the malicious and benign tasks, and the p-value was about 0.000832. We repeated the same test over the malicious and benign under stress tasks, and the p-value was about 0.00114. From these results, we can conclude that users performing a malicious act will tend to make longer mouse movement paths than their normal patterns even when they experience stressful conditions.

The mouse left click duration feature also showed interesting results. As shown in the table, the mouse left click duration for the malicious tasks were the longest among all the tasks with a mean of 293.01 ms (SD 152), while the benign tasks show the shortest mouse left click duration with a mean of 167.63 ms (SD 69.60). We performed the related t-test over the two tasks, and the results show there was a statistically significant effect with p-value = 0.000015. The related t-test was also applied to test the difference between the mean of the mouse left click duration on both the malicious tasks and benign under stress tasks, and the p-value was about 0.0451. These results indicate that individuals will click the mouse (left click) at a slower speed when they perform a malicious act than their normal click patterns. However, individuals experiencing stressful conditions will also click the mouse (left click) at a slower speed than normal but this speed is still faster than performing a malicious act.

For the keystroke part, the benign under stress tasks showed the highest keystroke duration with a mean of 506.65 ms (SD 937). However, the results of the malicious tasks were very close to the benign under stress tasks with a mean of 469.32 ms (SD 792) and were much longer than the benign tasks. We performed the related *t*-test over the malicious and benign tasks, and the results show the differences were statistically significant with *p*-value = 0.0385. However, there was no statistically significant difference between the benign under stress, and the malicious tasks – the *p*-value = 0.418571. Thus, users experiencing stressful conditions and users performing malicious acts may have similar keystroke duration and slower than their regular patterns. In addition, looking at this feature participants were varied on their keystroke speed as they came from different backgrounds and different computer skills which can be seen clearly from the high standard deviation value.

In conclusion, participants show different mouse movements and keystroke behavior patterns when they perform malicious acts than their normal behavior pattern. These changes in behavior include high-speed and high-distance mouse movements, and long-lasting left clicks and keystroke duration. Participants performing malicious tasks showed faster speed and longer mouse movements, and long-lasting click and keystroke duration than the benign tasks.

Data Fusion

This section investigated how to combine and integrate multiple sources of behavioral measures to increase the performance and provide more accurate results. To do this, we use data fusion strategies in two different levels of our data analysis process, the feature level and decision level. Data fusion is the procedure of integrating data from numerous domains or sources to deliver more efficient and potentially more accurate learning than those produced by any of the single data sources (Hall & McMullen, 2004).

We fuse the EEG and the ECG data collected from our participants during the same period of time. The goal is to find if there is any correlation between these psychophysiological measures while users perform malicious activities versus normal activities that can increase the detection performance. To do this, we used a canonical correlation analysis (Haghighat, Abdel-Mottaleb, & Alhalabi, 2016; Sun, Zeng, Liu, Heng, & Xia, 2005) over our EEG and ECG feature sets that were extracted from a specific time window and generated a new feature set that maximizes the correlation across the two feature sets. Canonical correlation analysis has been traditionally applied to examine relationships between two sets of variables (Hotelling, 1936). It is described as the process of obtaining two sets of basis vectors, one for A and the other for B, that maximize the correlations between the projections of the variables in these basis vectors (Borga, 2001; Sun,

Zeng, Liu, Heng, & Xia, 2005). In this work, the result feature vector is generated by performing canonical correlation analysis over the synchronized EEG and ECG feature vector.

For the non-signal psychophysiological measures fusion, we use the participants' eye-tracking, mouse, and keystroke dynamics data. We conduct the feature extraction process for both eye-tracking data and mouse and keystroke dynamics data independently for the same participant. However, this process should be synchronized. In other words, the feature vector obtained from the eye-tracking dynamics data and the feature vector derived from the mouse and keyboard dynamics data are extracted from the same time window (epoch). Then, both feature vectors are combined into one feature vector that represents the user's non-signal psychophysiological behavior for that specific epoch.

As we generate different modules that employ different psychophysiological measures and as each of the modules will have its own prediction results, a decision fusion is needed to combine the results and generate the final prediction. To do this, we use a majority voting technique (Zuev & Ivanov, 1999) that consolidates the decisions of multiple classifiers in one final decision. In this work, we have three modules represented by three classifiers, each one of them utilizes one psychophysiological measure. Namely, the EEG signal module, the ECG signal module, and the non-signal behavior module that uses the eye-tracking and mouse and keystroke dynamics data. We take the synchronized decisions from each module and feed them to our majority voting function as the following: Suppose that C_{EEG} represents our EEG signals classifier, C_{ECG} represents our ECG signals classifier, and $C_{\text{Non-signal}}$ represents our eye-tracking, mouse and keystroke dynamics classifier. Then the final prediction is calculated as following:

$$\widehat{P} = V\left(C_{\text{EEG}}, \ C_{\text{ECG}}, \ C_{\text{Non-signal}}\right) \tag{11.1}$$

where $V(\ )$ is our voting function and $\widehat{P}$ is the final prediction. For example, assuming that we combine our three classifiers that classify synchronized training samples as $C_{\text{EEG}} = 0$, $C_{\text{ECG}} = 1$, $C_{\text{Non-signal}} = 1$, the voting function $\widehat{P} = V(0, 1, 1)$ and the final decision will be $= 1$.

Multi-modal Framework and Results

This section discusses the results of integrating our psychophysiological measures together and building a multi-modal framework that can efficiently detect the malicious intent of a user. We categorize the psychophysiological behavior to signal and non-signal measures. We build two detection modules, namely the signal psychophysiological behavioral module which integrates the synchronized EEG and

ECG signals as explained, and the non-signal psychophysiological behavioral module which utilizes the eye-tracking data combined with the computer-based behaviors data (the mouse movements and keystrokes data). Then, we consolidate all measures in one framework by calculating the decisions fusion for each classifier.

For our signal module, we use the synchronized EEG and ECG extracted features from participants' data. After extracting the features for each of the signals separately, we use our canonical correlation analysis method to fuse the pair of synchronized feature vectors and generate a new feature vector that represents the correlated features. Then, we run the classifiers for each user individually and calculate the average results. Table 11.14 presents our signal features fusion results. As shown, the fused feature set shows a detection accuracy of up to 95.16% using the SVM classifier; this result is about 2% higher than using the EEG signal alone and about 6% higher than using the ECG data individually. The Random Forest classifier shows an average detection accuracy of up to 95.93% which is also higher than the detection accuracy achieved by the feature set of each signal separately. We also investigate the detection accuracy improvement on combining the features extracted from our eye-tracking data with the features derived from the mouse movement and keystroke data. For each participant, we consider both eye-tracking and mouse movement and keystroke features in building our non-signal module.

Table 11.15 shows the results of using the non-signal psychophysiological behaviors. The top part of the table shows the results for all participants. As shown, the best detection accuracy is achieved using the Random Forest classifier with up to 94.37% average detection accuracy, 0.94 true-positive rate, and 0.08 false-positive rate. The middle part of the table presents the results for participants with eyeglasses where the best average detection accuracy is about 90.81% using the Random Forest classifier. This is about 4% lower than the average detection accuracy for all participants as eyeglasses may reflect the light and impact the recording accuracy of the eye-tracking device.

Table 11.14   Average results for detecting malicious activities using the EEG and ECG fused features set and the individual signal features set

	EEG feature Set			ECG feature Set			Fused feature Set		
Classifier	Accuracy	TP Rate	FP Rate	Accuracy	TP Rate	FP Rate	Accuracy	TP Rate	FP Rate
SVM	93.27	0.93	0.09	88.97	0.89	0.14	95.16	0.95	0.05
Random F	94.75	0.95	0.05	92.10	0.92	0.1	95.93	0.95	0.07
k-NN	93.16	0.93	0.09	91.19	0.92	0.10	91.57	0.94	0.07

Table 11.15 Average results of detecting malicious activities using the combined eye-tracking data and mouse and keystroke features

Classifier	Number of participants	Accuracy	TP Rate	FP Rate	F-Measure
SVM	25	91.67	0.92	0.11	0.92
Random F	25	94.37	0.94	0.08	0.94
k-NN	25	87.89	0.88	0.15	0.88
SVM	10*	89.76	0.90	0.15	0.89
Random F	10*	90.81	0.91	0.11	0.91
k-NN	10*	85.69	0.86	0.19	0.85
SVM	15	92.95	0.93	0.08	0.93
Random F	15	96.74	0.97	0.05	0.97
k-NN	15	89.35	0.90	0.13	0.89

* Participants who wore eyeglasses during the experiments.

The bottom part of the table presents the results for the participants who did not wear eyeglasses during the experiments. The results show an average detection accuracy of up to 96.74% using the Random Forest classifier and 92.95% using the SVM classifier.

The above results demonstrate that feature fusion for the psychophysiological behavior measures results in improved the detection accuracy compared to using each measure separately. However, our detection approach can be customized to fit an organization's needs based on the availability of the tools to acquire the behavior measures. Finally, we formed our multi-modal framework that utilizes all the synchronized psychophysiological behavior measures using the decisions fusion. We took the output of each of our three classifiers (EEG classifier, ECG classier, and the eye-tracking, mouse movements and keystrokes classifier) and fed that to our hard majority voting function to calculate the final prediction. Figure 11.3 shows our multi-modal framework diagram layout. When a participant performs the task, each classifier will return its decision for that specific time window. Then, the final decision will be calculated for the synchronized sub decisions. We used the data of the participants that have all the synchronized psychophysiological behavior measures available, ran the classifiers, and fed their output to the decision fusion function. Table 11.16 presents the results of our multi-modal framework.

As shown, we achieved 97.55% average detection accuracy and 0.95 true-positive rate using the Random Forest classifier. The results also show a false-positive rate as low as 0.01. The k-NN and the SVM classifiers show very similar average detection accuracy up to 95.86% and a false-positive rate as low as 0.02. These results and the results we presented in our study before demonstrate that the psychophysiological

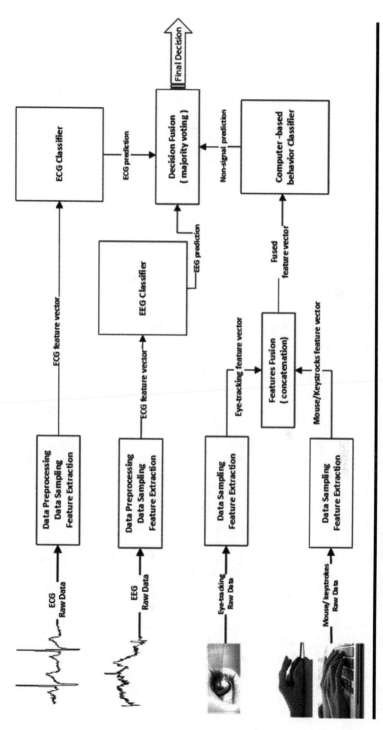

Figure 11.3 The multi-modal framework.

Table 11.16 Average results of detecting malicious activities using decisions fusion

Classifier	Number of Participants	Accuracy	TP Rate	FP Rate	F-Measure
SVM	20	95.86	0.91	0.02	0.93
Random *F*	20	**97.55**	0.95	0.01	0.96
k-NN	20	95.83	0.92	0.02	0.93

behavior measures can be an excellent source of information to help detect and predict malicious user behaviors and his/her intent to conduct an insider threat attack that may cause damage to an organization's data and information system.

Conclusion

The goal of this work is to propose a new insider threat detection approach based on users' psychophysiological and computer-based behavior measures. Our results indicate that both EEG and ECG signals carry out a good resolution on detecting the malicious activities as shown in our signal detection module. While the availability of the hardware and its applicability in real-world work environments introduce some challenges for this module, this module can be used in a highly secured work environment where employees deal with highly sensitive information. Moreover, with recent advances in wearable technologies, this limitation will not exist in the near future. For example, most smartwatches and fitness trackers have sensors that can record ECG signals. Also, EEG technology is growing, and many new devices have been introduced recently such as the smart cap (SmartCap Tech, 2018) and the mindset smart headphones (kickstarter, 2018). Furthermore, we provide the performance of each signal separately, so, based on the availability, the organization can always choose to use any of the signals that fit their needs.

The non-signal detection module provides the best choice as it is easy to setup and does not require any expensive hardware, a video-based eye-tracking device can be used, also, there are many new implementations of eye-tracking systems that do not require complicated hardware and can achieve good accuracy (Gómez-Poveda & Gaudioso, 2016; Mazhar, Shah, Khan, & Tehami, 2015). In addition, the mouse movements and keystrokes tracker does not require any hardware. The reason behind evaluating each single psychophysiological behavior measure individually then combining them together is to provide flexibility in the design as some measures may not be available as explained before. Therefore, our proposed insider threat detection framework can be a combination of the measures or just a single measure. It also can be combined with other approaches which use user's activity behaviors when using the network or the organization's computer systems and resources to improve the detection accuracy.

As we conducted human research experiments to evaluate our approach, we were very cautious in choosing the participants in terms of gender, age, education level, and the cybersecurity knowledge. We did our best to have diverse participants that would represent a real-world workplace environment. However, it would be valuable to include participants with less education and computer skills. Furthermore, our main challenge was the experiment scenarios, mainly the malicious scenarios. Emulating real-world insider threat scenarios is a very complicated task as it is associated with the psychological and behavioral aspects of the users. We tried to simulate real-world workplace environments in our experimental environment and tested the participants in conditions that are similar to real-world insider attack scenarios. We chose two attacks that involved the network and the application level, and mostly used by insiders in practice. It is possible that what we were actually discriminating between the specific tasks the participant was performing rather than whether that task was malicious. To address this issue with statistical significance, more data needs to be collected from more tasks and tested based on tasks not involved in the training.

References

Agrafioti, F., Hatzinakos, D., & Anderson, A. K. (2012). ECG pattern analysis for emotion detection. *IEEE Transactions on Affective Computing, 3,* 102–115.

AlgoSec. (2014, April 16). *AlgoSec survey: state of network security 2014.* Retrieved from www.algosec.com.

Almehmadi, A., & El-Khatib, K. (2014). On the possibility of insider threat detection using physiological signal monitoring. In *Proceedings of the 7th International Conference on Security of Information and Networks* (p. 223).

Altman, N. S. (1992). An introduction to kernel and nearest-neighbor nonparametric regression. *The American Statistician, 46,* 175–185.

Apperly, I. A., Samson, D., Chiavarino, C., & Humphreys, G. W. (2004). Frontal and temporo-parietal lobe contributions to theory of mind: neuropsychological evidence from a false-belief task with reduced language and executive demands. *Journal of Cognitive Neuroscience, 16,* 1773–1784.

Bajada, C. J., Haroon, H. A., Azadbakht, H., Parker, G. J., Ralph, M. A., & Cloutman, L. L. (2017). The tract terminations in the temporal lobe: their location and associated functions. *Cortex, 97,* 277–290.

Bergadano, F., Gunetti, D., & Picardi, C. (2002). User authentication through keystroke dynamics. *ACM Transactions on Information and System Security (TISSEC), 5,* 367–397.

Borga, M. (2001). Canonical correlation: a tutorial. *Online Tutorial, 4,* 5. http://people.imt.liu.se/magnus/cca.

Bowen, B. M., Salem, M. B., Keromytis, A. D., & Stolfo, S. J. (2010). Monitoring technologies for mitigating insider threats. In *Insider Threats in Cyber Security* (pp. 197–217). Springer, Boston, MA.

Breiman, L. (2001). Random forests. *Machine Learning, 45*, 5–32.

CERT. (2014, 7). *Us State of Cybercrime Survey*. Retrieved from https://resources.sei.cmu.edu/asset_files/Presentation/2014_017_001_298322.pdf.

Cai, J., Liu, G., & Hao, M. (2009). The research on emotion recognition from ECG signal. In *Information Technology and Computer Science, 2009. ITCS 2009. International Conference on, 1*, (pp. 497–500).

Cantoni, V., Galdi, C., Nappi, M., Porta, M., & Riccio, D. (2015). GANT: gaze analysis technique for human identification. *Pattern Recognition, 48*, 1027–1038.

Chanel, G., Ansari-Asl, K., & Pun, T. (2007). Valence-arousal evaluation using physiological signals in an emotion recall paradigm. In *Systems, Man and Cybernetics, 2007. ISIC. IEEE International Conference on* (pp. 2662–2667).

Chayer, C., & Freedman, M. (2001). Frontal lobe functions. *Current Neurology and Neuroscience Reports, 1*, 547–552.

Cortes, C., & Vapnik, V. (1995). Support-vector networks. *Machine Learning, 20*, 273–297.

Cummings, A., Lewellen, T., McIntire, D., Moore, A. P., & Trzeciak, R. (2012). Insider threat study: Illicit cyber activity involving fraud in the US financial services sector (No. CMU/SEI-2012-SR-004). Carnegie-Mellon Univ Pittsburgh PA Software Engineering Inst.

De Luca, A., Denzel, M., & Hussmann, H. (2009, July). Look into my eyes! can you guess my password? In *Proceedings of the 5th Symposium on Usable Privacy and Security* (pp. 1–12).

Eberz, S., Rasmussen, K. B., Lenders, V., & Martinovic, I. (2015). Preventing lunchtime attacks: fighting insider threats with eye movement biometrics. In *Proceedings 2015 Network and Distributed System Security Symposium (NDSS)*.

Electrical Geodesics Inc. (2017). Clinical Geodesic EEG System 400. Retrieved from www.egi.com.

Emotiv, Inc. (2019). Emotive system. Retrieved from https://www.emotiv.com/.

Fogassi, L., Ferrari, P. F., Gesierich, B., Rozzi, S., Chersi, F., & Rizzolatti, G. (2005). Parietal lobe: from action organization to intention understanding. *Science, 308*, 662–667.

Glasser, J., & Lindauer, B. (2013, May). Bridging the gap: A pragmatic approach to generating insider threat data. In *2013 IEEE Security and Privacy Workshops* (pp. 98–104). IEEE.

Goldberg, E. (2002). *The Executive Brain: Frontal Lobes and the Civilized Mind.* Oxford University Press, USA.

Gómez-Poveda, J., & Gaudioso, E. (2016). Evaluation of temporal stability of eye tracking algorithms using webcams. *Expert Systems with Applications, 64*, 69–83.

Greitzer, F. L., & Frincke, D. A. (2010). Combining traditional cyber security audit data with psychosocial data: towards predictive modeling for insider threat mitigation. In *Insider Threats in Cyber Security* (pp. 85–113). Springer, Boston, MA.

Guo, H. W., Huang, Y. S., Chien, J. C., & Shieh, J. S. (2015). Short-term analysis of heart rate variability for emotion recognition via a wearable ECG device. In *Intelligent Informatics and Biomedical Sciences (ICIIBMS), 2015 International Conference on* (pp. 262–265).

Haghighat, M., Abdel-Mottaleb, M., & Alhalabi, W. (2016). Fully automatic face normalization and single sample face recognition in unconstrained environments. *Expert Systems with Applications, 47,* 23–34.

Hall, D. L., & McMullen, S. A. (2004). *Mathematical Techniques in Multisensor Data Fusion.* Artech House. Inc., Norwood, MA.

Hotelling, H. (1936). Relations between two sets of variates. *Biometrika, 28,* 321–377.

Kaghazgaran, P., & Takabi, H. (2015). Toward an insider threat detection framework using honey permissions. *Journal of Internet Services and Information Security (JISIS), 5,* 19–36.

Kaklauskas, A., Krutinis, M., & Seniut, M. (2009). Biometric mouse intelligent system for student's emotional and examination process analysis. In *Advanced Learning Technologies, 2009. ICALT 2009. Ninth IEEE International Conference on* (pp. 189–193).

Khushaba, R. N., Kodagoda, S., Lal, S., & Dissanayake, G. (2011). Driver drowsiness classification using fuzzy wavelet-packet-based feature-extraction algorithm. *IEEE Transactions on Biomedical Engineering, 58,* 121–131.

kickstarter. (2019). Mindset: smart headphones that improve your concentration. Retrieved from https://www.kickstarter.com/projects/mindset/headphones.

Kim, K. H., Bang, S. W., & Kim, S. R. (2004). Emotion recognition system using short-term monitoring of physiological signals. *Medical and Biological Engineering and Computing, 42,* 419–427.

Lali, P., Naghizadeh, M., Nasrollahi, H., Moradi, H., & Mirian, M. S. (2014). Your mouse can tell about your emotions. In *Computer and Knowledge Engineering (ICCKE), 2014 4th International eConference on* (pp. 47–51).

Lee, H., Jung, J., Kim, T., Park, M., Eom, J., & Chung, T. (2014, April). An application of data leakage prevention system based on biometrics signals recognition technology. In *The 3rd International Conference on Networking and Technology* (Vol. 63, pp. 1–5).

Li, Z., Shen, M., & Beadle, P. (2004, August). Classification of EEG signals under different brain functional states using RBF neural network. In *International Symposium on Neural Networks* (pp. 356–361). Springer, Berlin, Heidelberg.

Liang, Z., Tan, F., & Chi, Z. (2012). Video-based biometric identification using eye tracking technique. In *Signal Processing, Communication and Computing (ICSPCC), 2012 IEEE International Conference on* (pp. 728–733).

Lim, Y. M., Ayesh, A., & Stacey, M. (2014, August). The effects of typing demand on emotional stress, mouse and keystroke behaviours. In *Science and Information Conference* (pp. 209–225). Springer, Cham.

Mazhar, O., Shah, T. A., Khan, M. A., & Tehami, S. (2015). A real-time webcam based eye ball tracking system using MATLAB. *Design and Technology in Electronic Packaging (SIITME), 2015 IEEE 21st International Symposium for* (pp. 139–142).

Mensh, B. D., Werfel, J., & Seung, H. S. (2004). BCI competition 2003-data set Ia: combining gamma-band power with slow cortical potentials to improve single-trial classification of electroencephalographic signals. *IEEE Transactions on Biomedical Engineering, 51,* 1052–1056.

Monrose, F., & Rubin, A. D. (2000). Keystroke dynamics as a biometric for authentication. *Future Generation Computer Systems, 16,* 351–359.

Neupane, A., Rahman, M. L., Saxena, N., & Hirshfield, L. (2015). A multi-modal neuro-physiological study of phishing detection and malware warnings. In *Proceedings of the 22nd ACM SIGSAC Conference on Computer and Communications Security* (pp. 479–491).

OpenBci, I. (2014). OpenBCI Board.

Park, Y., & Stolfo, S. J. (2012). Software decoys for insider threat. In *Proceedings of the 7th ACM Symposium on Information, Computer and Communications Security* (pp. 93–94).

Polat, K., & Güneş, S. (2007). Detection of ECG Arrhythmia using a differential expert system approach based on principal component analysis and least square support vector machine. *Applied Mathematics and Computation, 186,* 898–906.

Ponemon Institute LLC. (2016). Cost of Cyber Crime 2016: reducing the risk of business innovation. Retrieved from https://saas.hpe.com/en-us/marketing/cyber-crime-risk-to-business-innovation.

Posner, M. I., & Petersen, S. E. (1990). The attention system of the human brain. *Annual Review of Neuroscience, 13,* 25–42.

Pusara, M., & Brodley, C. E. (2004). User re-authentication via mouse movements. In *Proceedings of the 2004 ACM workshop on Visualization and data mining for computer security* (pp. 1–8).

Roland, P. E., Roland, P. E., & Roland, P. E. (1993). *Brain Activation.* Wiley-Liss, New York.

Salem, M. B., & Stolfo, S. J. (2009). Masquerade attack detection using a search-behavior modeling approach. Columbia University, Computer Science Department, Technical Report CUCS-027-09.

Salmeron-Majadas, S., Santos, O. C., & Boticario, J. G. (2014). An evaluation of mouse and keyboard interaction indicators towards non-intrusive and low cost affective modeling in an educational context. *Procedia Computer Science, 35,* 691–700.

Silowash, G. J., Cappelli, D. M., Moore, A. P., Trzeciak, R. F., Shimeall, T., & Flynn, L. (2018). *Common Sense Guide to Mitigating Insider Threats.* 4th Edition. figshare. Journal contribution. https://doi.org/10.1184/R1/6572639.v1.

SmartCap Tech. (2018). Life smart cap. Retrieved from http://www.smartcaptech.com/.

SolarWinds. (2015). SolarWinds survey investigates insider threats to federal cybersecurity. Retrieved from www.solarwinds.com/company/newsroom/press_releases/threats_to_federal_cybersecurity.aspx.

Stuss, D. T., & Knight, R. T. (Eds.). (2013). *Principles of Frontal Lobe Function.* Oxford University Press.

Sun, Q.-S., Zeng, S.-G., Liu, Y., Heng, P.-A., & Xia, D.-S. (2005). A new method of feature fusion and its application in image recognition. *Pattern Recognition, 38,* 2437–2448.

PWC, P. (2015). Managing Cyber Risks in an Interconnected World: Key Findings from the Global State of Information Security Survey 2015.

Theoharidou, M., Kokolakis, S., Karyda, M., & Kiountouzis, E. (2005). The insider threat to information systems and the effectiveness of ISO17799. *Computers & Security, 24,* 472–484.

Thompson, P. (2004, September). Weak models for insider threat detection. In *Sensors, and Command, Control, Communications, and Intelligence (C3I) Technologies for Homeland Security and Homeland Defense III* (Vol. 5403, pp. 40–48). International Society for Optics and Photonics.

Thought Technology Ltd. (2017). BioGraph Infiniti Software. Retrieved from https://bio-medical.com/biograph-infiniti-version-6-0-upgrade.html.

Tobii. (2017). Tobii Pro X2-60 eye tracker. Retrieved from https://www.tobiipro.com/product-listing/tobii-pro-x2-30/.

Turns soft. (2016). Mini Mouse Macro PRO. Retrieved from www.turnssoft.com/mini-mouse-macro-pro.html.

Vaish, A., & Kumari, P. (2014). A comparative study on machine learning algorithms in emotion state recognition using ECG. In *Proceedings of the Second International Conference on Soft Computing for Problem Solving (SocProS 2012), December 28–30, 2012* (pp. 1467–1476).

Valacich, J. S., Jenkins, J. L., Nunamaker, Jr, J., Hariri, F., & Howie, J. (2013). Identifying insider threats through monitoring mouse movements in concealed information tests. In *Hawaii International Conference on System Sciences. Deception Detection Symposium.*

Wang, X.-W., Nie, D., & Lu, B.-L. (2014). Emotional state classification from EEG data using machine learning approach. *Neurocomputing, 129,* 94–106.

Xu, Y., Liu, G., Hao, M., Wen, W., & Huang, X. (2010). Analysis of affective ECG signals toward emotion recognition. *Journal of Electronics (China), 27,* 8–14.

Zuev, Y. A., & Ivanov, S. K. (1999). The voting as a way to increase the decision reliability. *Journal of the Franklin Institute, 336,* 361–378.

Chapter 12

Cybersecurity in the Software Development Life Cycle

Johnson Kinyua, PhD

College of Information Sciences and Technology, The Pennsylvania State University
University Park, PA, USA

12.1 Introduction

Many organizations use software development processes to create software products, and these include developers of major enterprise software packages, firms implementing commercial off-the-shelf software (COTS), and custom in-house development. Developing a large software product is a complex process and to increase the chances of success organizations should use some standard software development methodology. There are indeed a number of variations in approaches to the development of software, which are normally called software development methodologies or software development life cycle (SDLC) models. The range of software development life cycle models includes the waterfall model, agile software development, scrum, rapid application development, joint application development, and the spiral model (Royce 1970, Boehm 1988, Phleeger and Atlee 2010).

Within these life cycle models, there should be variations that should meet an organization's software development needs. Many of the life cycle models are built around the traditional waterfall model, which we discuss in section 12.2. The waterfall model is used extensively, especially in the development of large enterprise software systems. It is important to note that these models vary in end-user involvement, budget, and project implementation time. Traditional software development methodologies do not normally integrate security into the software development process, although this has been changing over the years (Alberts et al. 2003, Mead, Hough and Stehney 2005, Kissel et al. 2008, Noopur 2013, Yuan et al. 2014). Traditionally, life cycle models are focused on creating software to meet the functional and nonfunctional requirements of users. More recently, these life cycle models are being augmented to integrate security at each phase, adding protection to software (Kissel et al. 2008). There are also attempts to integrate security into design and testing techniques, secure coding best practices, and other cybersecurity practices across the curriculum of computer science programs in the context of IoT systems (McManus 2018). Security is no longer a requirement that should be baked into the software after the fact. Indeed security should be considered to be an important functional requirement right from the start.

In order to understand how security can be built into the software product, it is necessary to have an understanding of software development methodologies. In section 12.2 we briefly discuss various software development methodologies. In section 12.3 we provide details on integrating security into software development in the context of SDLC. The conclusion is provided in section 12.4. It is noteworthy that all the software development methodologies discussed in section 12.2 did not traditionally integrate security in any of the phases. It could, however, be argued that the risk analysis step in the Spiral Model (Boehm 1988) could be interpreted in a security context, although the model does not explicitly state so.

With the onset of the Internet, we have seen an exponential growth in the use of the Internet for e-commerce, the Web, email communications, web-enabled enterprise information systems, social media, and the Internet of Things (IoT). Building secure software gradually became a primary concern. The need to build secure software was further exacerbated by the phenomenal growth in the frequency and severity of cyberattacks. Today, the cybercrime ecosystem is a $1.5 trillion dollar industry that costs the US economy alone $57 to $109 billion per year (Advisers 2018, O'Connor 2018). The need to build more secure software is therefore beyond debate. Initial attempts to secure software developed using the traditional insecure methodologies consisted of "bolting on" security to an application after the fact or late in the development life cycle. This is no longer an option and software must be *secure by design*. This is the philosophy adopted in section 12.3.

12.2 Software Development Methodologies

12.2.1 The Waterfall Model

The waterfall model, also called the SDLC model, is the oldest and a popular software development methodology (Royce 1970). It consists of a number of phases where the result of each phase is called a deliverable, which flows into the next phase, as shown in Figure 12.1. The SDLC usually includes the following six phases: requirements analysis, system design, coding, testing, deployment, and operation and maintenance.

After a phase is completed, the process moves to the next phase. In practice, it may not be feasible to work independently on each phase. It is usually required to revisit previous steps based on changes to the requirements or the need for improvements. Hence, in practice the model is iterative in nature because of the need to revisit previous phases.

Prior to the system analysis phase, a system investigation or preliminary investigation is conducted. Some depictions of waterfall models include this as the first phase, although we have not done so. The need to build a software product usually emanates from a business opportunity or problem. The purpose of this investigation is, therefore, to assess or evaluate the IT-related business opportunity or problem. The system investigation is a crucial step because the outcome will affect the entire development process. A key part of system investigation is a feasibility study that evaluates expected costs and benefits and recommends a course of action based on technical, economic, organizational, operational, and time factors. After the system investigation, a decision is made on whether to proceed with the project or not. The process continues to the requirements analysis phase if the decision was to go ahead with the project otherwise further work ceases.

During the requirements analysis phase, the functional and nonfunctional requirements of the software system are gathered, prioritized, and documented.

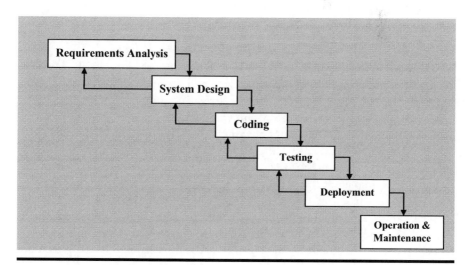

Figure 12.1   The Waterfall Model.

The requirements define the scope of a project and must reflect the desires and needs of users and stakeholders and *not* developers. Unfortunately, requirements can and will change over the life cycle of a project, and this complicates the software development process.

There are two types of requirements: functional and nonfunctional. The functional requirements describe what a software system will do, while nonfunctional requirements describe requirements not directly related to the services that the software will provide to its users. Examples of nonfunctional requirements include ease-of-use, user response time, software memory usage, reliability, and user interactions. Sometimes there is an overlap between functional and nonfunctional requirements. It should be noted that errors in requirements are among the most problematic and expensive in software development projects. This is because their effects cascade throughout the project, causing errors in later development phases. For example, an ambiguous or incomplete requirement might result in wasted design and development work, which in turn might lead to time spent testing software components that are not really needed. The deliverable of this phase is the system requirements document, which is also called the requirements specifications.

The next phase is the system design phase in which the software components of the software product are created based on the requirements gathered in the previous phase. This effectively defines the technical design or architecture of the software, and this architecture describes *how* the system will solve the business problem. The deliverable of the design phase is the architecture that specifies

- system modules
- system inputs, outputs, and user interfaces
- software, hardware, databases, networking, personnel, and procedures
- module integration

The deliverable for this phase is the system design specification, which should be presented to management and users for review and approval; this is required before proceeding to the next phase. The system design specification consists of a logical system design that states what the system will do, using abstract specifications, and a physical system design that states how the system will perform its functions with actual physical specifications. The system design specification will be used by programmers to transform the logical system design into program modules and code.

During the implementation phase, programmers create the computer code for the new system based on the design specifications. Large systems development projects can require millions of lines of computer code, and many programmers organized into teams would be needed. These teams often include some key users to help programmers focus on the business problem at hand. The deliverable for this phase is software modules and associated documentation.

During the testing phase, the software modules are tested to make sure they satisfy the requirements before the software is released into the production environment for use by the customers or users. Testing verifies that the code works correctly under various conditions, and tests are designed to detect errors or bugs in the code and subsequently remove them. Testing is time-consuming and expensive, especially if it is thorough. Inadequate testing, which can occur for a number of reasons, will lead to higher maintenance costs later. Inadequate testing may occur due to time-to-market pressures, reduced testing time due to schedule issues, project management issues that propagate to the testing phase, and so on.

It is necessary to create a comprehensive test plan to conduct systematic testing of the software. The different kinds of tests that are performed on the software include *unit testing, system testing, integration testing,* and *user acceptance testing.* In unit testing, every module is tested alone in order to discover any bugs or errors in its code. In system testing, *all* of the modules that comprise the system are brought together and tested as a whole. With integration testing, various modules are brought together for testing purposes in order to verify that the interaction among them works as expected. With user acceptance testing, a relatively small group of users or other stakeholders are involved directly in the testing process. Users and stakeholders frequently have a very different point of view than developers or even testers. User acceptance testing should include unit testing or system testing. The intended outcome is a decision about whether or not the unit being tested meets user and stakeholder requirements. All tests should be carefully documented and this becomes the deliverable of this phase.

After the testing is completed, the software product is deployed into the production environment for business use by the organization. This is known as the deployment phase. After this point, the software product has entered the operation and maintenance phase. The software will be used by the owning organization to conduct business operations, and it is during this period that maintenance must take place.

The software system needs different kinds of maintenance. From time to time, software errors that were not discovered and removed during the testing phase will occur during normal use of the software. These errors must be removed so that the software operates correctly and this continues throughout the life of the product. The second type of maintenance is updating the software to accommodate changes in business or operations or conditions. For example, a new government regulation might require a software update for the organization to stay compliant. These corrections and updates usually do not really add any new functionality because they simply keep the software functioning as required. The third type of maintenance is that which adds new functionality to the software by adding new features to the existing software without disturbing its operation.

The waterfall model can be very useful in assisting developers in laying out what they need to do. Its simplicity makes it easy to explain to customers who are not familiar with software development because it also makes it explicit which intermediate deliverables are necessary in order to begin the next phase of development. Most other models are really just embellishments of the waterfall model incorporating feedback loops and extra activities. However, it should be noted that with the waterfall model, traditionally building secure software was not the intention; instead, the main consideration was to deliver working software to meet an organization's business goals. This also applies to the other methodologies discussed in the following subsections.

12.2.2 Agile Software Development

Most of the software development processes proposed and used in the 1970s through the 1990s imposed some form of rigor in the way in which software is conceived, documented, developed, and tested. In the late 1990s, some developers who did not like this rigor formulated their own principles and highlighted the role that flexibility could play in producing working software in short periods of time. This gave birth to the Agile software development process codified through an "Agile Manifesto," with four value statements or tenets and twelve principles (Alliance 2001, Amber 2012, Zain 2017):

- Individuals and interactions over processes and tools. Teams organize themselves and communicate through face-to-face interaction rather than through documentation.
- Working software over comprehensive documentation. The primary measure of success is the degree to which the software works properly.
- Customer collaboration over contract negotiation. They focus on customer collaboration rather than contract negotiation, thereby involving the customer in key aspects of the development process.
- Responding to change over following a plan. They concentrate on responding to change rather than on creating a plan and then following it because they believe that it is impossible to anticipate all requirements at the beginning of development.

The principles of agile development are defined as follows (Amber 2012):

1. Early and continuous delivery of working, valuable software
2. Expect and embrace changing requirements
3. Deliver working software frequently (every 2 weeks to 2 months, shorter is better)
4. Developers and business people work together throughout the project
5. Build projects around motivated individuals and support them
6. Most effective project communication is face-to-face

7. Working software is the primary measure of progress
8. Nurture sustainable development to maintain pace indefinitely
9. Promote continuous attention to technical excellence and good design
10. Simplicity is essential – do as little possible while delivering value
11. The best work emerges from self-organizing groups
12. Regularly reflect on how to be more effective and adjust

The Agile software development process is an iterative SDLC model typically used for small- to medium-sized enterprise projects. The essential idea behind the agile software development movement is that the focus of software development should be on creating value for users by providing them with working software in the shortest possible timeframe. There are other tenets, as indicated earlier, for example, that users should be closely involved in the requirements gathering, design, and testing activities associated with a project. Therefore, the Agile process is more focused on the current customer demands, customer satisfaction, and feedback, rather than a stable, long-term project.

Instead of planning tasks or expectations for phases, testing is performed regularly with prototypes during each phase to demonstrate the application's functionality. At this point, the customer's feedback can be considered, so that the end product satisfies the customer. Unlike other development models, this model has the lowest cost associated with changes to the scope or requirements because of how often the customer provides feedback. Therefore, this methodology can result in greater functionality and efficiency, creating tailored software. Although the Agile methodology places less emphasis on documentation, it should be noted documentation is essential. The general approach is that this methodology generates less documentation than the waterfall model, which is heavy on documentation.

Agile software development has a number of different flavors: Scrum, Crystal, Extreme programming (XP), and Agile Software Development. We discuss Scrum and XP briefly in the following sections.

12.2.3 Extreme Programming (XP)

Extreme Programming (XP) is a flavor of Agile software development that improves a software project in five essential ways: communication, simplicity, feedback, respect, and courage (Wells 2009). Extreme programmers keep their design simple and clean; they constantly communicate with their customers and fellow programmers.

XP uses test-driven development and refactoring to improve the quality of the code and produce the most effective design. High-quality code is an essential aspect of software developed using XP. Techniques that enhance quality include test-driven development, refactoring, pairwise programming (explained subsequently) and full test coverage at different levels (unit, system, etc.). XP developers get feedback from customers by testing their software starting on day one and deliver the system to the customers as early as possible and implement changes as suggested.

First, automated tests are created in order to establish the project's objectives. Then, development is performed by programmer pairs so that work can be checked as it is written, providing two sets of eyes on each step. Once all automated tests are done and other testing ideas have been exhausted, the code is written. Then, the same two developers structure and incorporate the design and architecture of the system into the existing code. Once the design and new work are implemented into the existing system, a functioning prototype of the software is created and shown to a group of users, including the customers and at least one developer from the programming pair. If the test proves successful, the process starts over with the next important aspect of the system for development and integration into the existing system. These characteristics are embedded in what is known as the twelve facets of XP (Wells 2009, Phleeger and Atlee 2010).

12.2.4 Scrum

Scrum, which was created at Object Technology in 1994, uses iterative development where each 30-day iteration is termed a "sprint cycle" (Schwaber and Beedle, Agile Software Development with Scrum 2002). The sprint cycles are used to implement the product's backlog of prioritized requirements where multiple self-organizing and autonomous teams implement product increments in parallel. Coordination is done at a brief daily status meeting called a "Scrum."

In the Scrum method, there are three central roles on a project: product owner, scrum master, and developer (Schwaber and Sutherland, The Scrum Guide™ 2017). Each of these will be discussed briefly here. The product owner is responsible for the overall "shape" and feature set of the software product. The product owner identifies and prioritizes features to implement, tracks progress of the overall product, and accepts or rejects work completed by the development team. The scrum master is responsible for managing the development team's work on each sprint. The scrum master identifies impediments to the project process and helps eliminate them. The scrum master ensures the development process as defined for the project is being followed.

Project group members are responsible for designing, implementing, and testing the features that are developed as part of the project. They are also responsible for estimating their work on each product feature as defined for a given sprint, communicating their estimates to the product owner and scrum master, and demonstrating how their software meets requirements as specified in the corresponding user story (discussed subsequently). Once a project is identified and a project group assembled, the scrum method defines a series of activities and iterations toward completion of the project.

User stories are "lightweight," that is, short, textual narratives describing how one or more users or actors will interact with an envisioned or imagined piece of working software to accomplish some realistic goal. Generally, user stories should come from users, or at the very least from a designer's discussion with a prospective user. In cases where this is not possible, the designer should assume the perspective of the user and imagine a realistic goal and the support the software will provide to help accomplish it. The entire set of user stories for a given software project suggests a set of features required to support the users' goals. This set of features makes up the project's *product backlog*. The product backlog is, in essence, all of the development work that needs to be done in order to construct a complete software product.

The core deliverables in a Scrum consist of those shown in Figure 12.2 and are produced in the order suggested in the diagram.

Note that the second of the four steps identified in the figure represents a decomposition of the preceding step. In other words, the Product Backlog decomposes the User Stories into a set of features to design, implement, and test. Each project sprint will decompose Product Backlog items further into the discrete tasks that will be performed as part of the sprint. Planning a sprint involves moving items from the product backlog to the sprint backlog and providing estimates of the effort required to complete each item in the sprint.

12.2.5 The Rapid Application Development (RAD) Model

The rapid application development (RAD) model is another variant of the SDLC. RAD is defined as a methodology created to radically decrease the time needed to design and implement information systems by relying on extensive user involvement, prototyping, integrated CASE tools, and code generators (Hoffer, George, and Valacich 2014). According to Federal Financial Institutions Examination

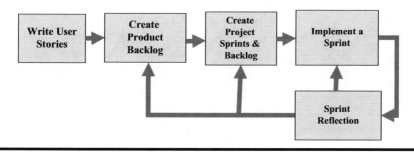

Figure 12.2   The Scrum Deliverables.

Council (FFIEC), this methodology "is a software development technique that emphasizes short development times (30–90 days). The technique is inappropriate for developing complex applications or applications that quickly process significant volumes of transactions, such as batch processing environments." (InfoBase 2013).

RAD has the following steps as shown in Figure 12.3:

Step 1: Outline the requirements for the system with the customer.
Step 2: Develop a prototype.
Step 3: Evaluate the prototype with the customer.
Step 4: Redevelop the prototype and have the customer reevaluate it until all requirements are met.

Specific RAD techniques such as taking advantage of reusable program objects may be used in conjunction with the waterfall model for development of the final system.

12.3 Secure Software Development

12.3.1 Introduction

Secure software development is a proactive process of designing, building, and testing software that incorporates security into each phase. It aims to reduce security risks introduced at each stage of the secure SDLC. Developing secure software poses several challenges. Using software improvement processes such as the Capability Maturity Model Integration (CMMI 1-5), the Personal Software Process (PSP), the Team Software Process (TSP), and other techniques; we have made significant improvements in reducing the number of bugs per thousand lines of code (KLOC) (Boehm 1988, Humphrey 2000a, Humphrey 2000b, Goldenson and Gibson 2003, McGraw 2006). Many IT vendors have been

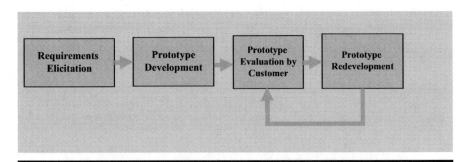

Figure 12.3   The Rapid Application Development (RAD) Model.

using these approaches for several years now, resulting in much "cleaner" software.

In addition, Security Quality Requirements Engineering (SQUARE), Microsoft's Security Development Life Cycle [SDLC], NIST SP 800-64, and application threat modeling are deliberate efforts to address the security aspects of software development (Mead, Hough and Stehney 2005, Howard and Lipner 2006, Howard and LeBlanc 2007, Kissel et al. 2008, Mead 2013, Microsoft 2018). Application threat modeling recognizes that to develop secure software, software developers need to have the mindset of an attacker. Attack patterns are valuable resources that can help software developers to think like an attacker. Attack patterns capture an attacker's perspectives and approaches when the attacker tries to exploit software (Mitre 2018a).

Software architectural risk analysis is the process of identifying and ranking risks applied to the software architecture and design-level artifacts using techniques such as the spoofing, tampering, repudiation, information disclosure, denial of service, and elevation of privileges (STRIDE); and damage potential, reproducibility, exploitability, affected users, and Discoverability (DREAD); these are part of Microsoft's threat modeling approach (Meier et al. 2003). It is the process of hypothesizing potential security threats, evaluating the threats, ranking the threats, and suggesting mitigation strategies. Such techniques can be complemented with attack patterns such as Common Attack Pattern Enumeration and Classification (CAPEC) (Meier et al. 2003, Yuan et al. 2014), to develop more secure software.

Implementation of security in the software development life cycle refers to the processes and activities executed to identify and mitigate threats. According to the National Institute of Standards and Technology (NIST):

> To be most effective, information security must be integrated into the SDLC from system inception. Early integration of security in the SDLC enables agencies to maximize return on investment in their security programs
>
> (Kissel et al. 2008).

NIST also defines steps that should be undertaken, including early identification and mitigation of vulnerabilities, awareness of potential engineering, reuse of security strategies, and facilitation of informed executive decisions. In addition, the Open Web Application Security Project (OWASP) has developed the Software Assurance Maturity Model for SDLC security (OWASP 2018b).

To set the stage for the techniques for integrating security into the software development life cycle, we first discuss the Security Quality Requirements Engineering (SQUARE) (Mead, Hough and Stehney 2005, Mead 2013). Capturing and documenting requirements is the first stage of SDLC, but as was stated in section 12.2, integrating security into SDLC in order to build more secure software requires that we continue with the security process into the design, coding and testing stages of SDLC. We shall discuss SQUARE for developing more secure requirements in section 12.3.2. Securing other phases of SDLC is discussed in other sections.

12.3.2 SQUARE

SQUARE is a nine-step methodology (Table 12.1) created at Carnegie Mellon University to assist organizations in building security in the early stages of development. According to the United States Computer Emergency Readiness Team (US-CERT) (Mead 2013):

> Security Quality Requirements Engineering (SQUARE) provides a means for eliciting, categorizing, and prioritizing security requirements for information technology systems and applications. The focus of this methodology is to build security concepts into the early stages of the development life cycle. The model can also be used for documenting and analyzing the security aspects of fielded systems and for steering future improvements and modifications to those systems.

The nine steps focus on eliciting, categorizing, and prioritizing security requirements for information technology systems and software applications (Mead, Hough and Stehney 2005, Mead 2006, Mead 2013). The goal of SQUARE is to categorize and prioritize security requirements. Steps 1–4 are actually activities that precede security requirements engineering but are necessary to ensure that the process is successful.

Collecting requirements is dependent on representation from the project's stakeholders. To collect requirements using the SQUARE process, it is important to first identify all stakeholders for the project. SQUARE is most effective and accurate when conducted with a team of requirement engineers, security experts, and the stakeholders of the information system or software application project (Mead, Hough and Stehney 2005). The table shows each step explained in detail to satisfy the security goals of the organization. We elaborate on these steps further:

Step 1: Agree on definitions: Project team members and stakeholders are gathered to enable the definition of the most consistent set of topics for security requirements related to the project (Mead 2013). Owing to different background experiences of team members, it is inferred that security definitions related to security of the project may mean different things to different members. Consistent definitions with clearly defined terms should ensure success going forward.

Table 12.1 The Steps in the SQUARE Process (Mead, Hough and Stehney 2005)

Step Number	Step	Input	Techniques	Participants	Output
1	Agree on definitions	Candidate definitions from IEEE and other standards	Structured interviews, focus group	Stakeholders, requirements engineer	Agreed-to definitions
2	Identify assets and security goals	Definitions, candidate goals, business drivers, policies and procedures, examples	Facilitated work session, surveys, interviews	Stakeholders, requirements engineer	Assets and goals
3	Develop artifacts to support security requirements definition	Potential artifacts (e.g., scenarios, misuse cases, templates, forms)	Work session	Requirements engineer	Needed artifacts: scenarios, misuse cases, models, templates, forms
4	Perform risk assessment	Misuse cases, scenarios, security goals	Risk assessment method, analysis of anticipated risk against organizational risk tolerance, including threat analysis	Requirements engineer, risk expert, stakeholders	Risk assessment results
5	Select elicitation techniques	Goals, definitions, candidate techniques, expertise of stakeholders, organizational style, culture, level of security needed, cost/benefit analysis, etc.	Work session	Requirements engineer	Selected elicitation techniques

(Continued)

Table 12.1 (Cont.)

Step Number	Step	Input	Techniques	Participants	Output
6	Elicit security requirements	Artifacts, risk assessment results, selected techniques	Joint Application Development (JAD), interviews, surveys, model-based analysis, checklists, lists of reusable requirements types, document reviews	Stakeholders facilitated by requirements engineer	Initial cut at security requirements
7	Categorize requirements as to level (system, software, etc.) and whether they are requirements or other kinds of constraints	Initial requirements, architecture	Work session using a standard set of categories	Requirements engineer, other specialists as needed	Categorized requirements
8	Prioritize requirements	Categorized requirements and risk assessment results	Prioritization methods such as Analytical Hierarchy Process (AHP), Triage, Win-Win	Stakeholders facilitated by requirements engineer	Prioritized requirements
9	Inspect requirements	Prioritized requirements, candidate formal inspection technique	Inspection method such as Fagan, peer reviews	Inspection team	Initial selected requirements, documentation of decision-making process and rationale

Step 2: Identify assets and security goals: Identifying asset protection and security goals should be aligned with the organization's security policy and environment (Mead 2013). The identification of assets may be different for groups as they have separate goals to achieve. Once all assets have been identified along with the security goals by the various stakeholders, prioritization needs to take place to ensure the most critical assets are protected. Executive decisions can be used to resolve conflicts concerning a goal's priority.

Step 3: Develop artifacts to support security requirements: This step will help identify the area of operations and issues of misuse and scenarios in which various controls or configurations compromise may arise particular to the area of operation. The goal of this step is to identify artifacts that will support the security requirements (Mead 2013).

Step 4: Perform risk assessment: An expert risk assessment is performed based on the artifacts discovered in the previous step. The NIST SP 800-30 publication guide for conducting a risk assessment can be used for this step.

The team then performs a risk assessment by listing vulnerabilities and threats that may affect the system, quantifying their likelihood and determining potential consequences to ensure security requirements developed later are valid and can be mapped to a vulnerability or threat. A semi-quantitative assessment will be performed to provide the qualities of a cost/benefit and qualitative approach. An expert in risk assessment methods should be brought on board during this step to avoid a scenario in which a solution is developed without a real understanding of what problem is being solved by that solution (Mead 2013).

Step 5: Select elicitation techniques: There are different elicitation techniques that can be used for gathering requirements such as interviews, questionnaires, surveys, and focus groups. Cultural backgrounds and past experiences with certain topics can become important in having various stakeholders. Formal elicitation techniques such as structured interviews can help overcome potential communication issues. This may also be solved by simply sitting down with the various working groups to understand the group's specific security requirements need.

Step 6: Elicit security requirements: This is the actual elicitation process that utilizes the technique selected in the previous step. It is important to ensure that the output of this step is a list of requirements and not implementation recommendations.

Step 7: Categorize requirements: A security requirements engineer will need to distinguish essential requirements, desired goals, and architectural constraints that may be present. Some requirements may actually be constraints, and this allows assessment of the risks associated with these constraints. This categorization also helps in the prioritization activity that follows in step 8.

Step 8: Prioritize requirements: This step's dependency relies on what the semi-quantitative risk assessment highlights. Special consideration should be given to the cost/benefit analysis portion to determine which security requirements have a higher payoff relative to the cost. Other factors such as loss of consumer confidence, inability to meet regulatory compliance, and security breach likelihood shall also be considered.

Step 9: Inspect requirements: Requirements inspections (as called Fagan inspections) serve as quality assurance checkpoints for the building of a software product. The project team should have an initial set of prioritized security requirements after the inspections are completed; it should also understand which areas are incomplete so that they can be revisited later. Finally, the project team should understand which areas are dependent on specific architectures and implementations and should plan to revisit those as well (Mead 2013).

12.3.3 Security Analysis and Design

The main goal of the "Security analysis and design" phase is to reduce the number of vulnerabilities in the design prior to coding. It is an iterative review process that analyzes the design and architecture of an application from a security perspective. The review includes an analysis of critical areas of the application responsible for integrating basic security principles; for example, confidentiality, authentication, authorization, and integrity. The main objective is to understand the security needs of a user and the application, which are not necessarily the same. A number of methodologies can be used in reviewing the design and architecture, including threat modeling using techniques such as Microsoft's STRIDE and DREAD (Meier et al. 2003), OWASP (2018a), and OCTAVE (Alberts et al. 2003). A brief discussion of threat modeling is provided in section 12.3.3.1. Studies have shown that it is easier and less costly to eliminate software errors in the earlier phases of SDLC (Marasco 2007). In a study conducted by the IBM System Science Institute in order determine the relative cost of fixing defects within the SDLC, the findings show that the costs escalate further down the SDLC phases (Mukesh 2009), which is almost similar to results presented in the earlier study (Marasco 2007). Based on these studies, the following conclusions can be drawn:

■ Errors in requirements engineering are among the most problematic and expensive.
■ Requirements errors cost US businesses at least $30 billion per year (Marasco 2007) and often result in failed or abandoned projects and damaged careers.
■ It is necessary to find and fix requirements errors early in the life cycle of a project.

These studies do not specifically identify the relative costs of fixing security flaws in SDLC. However, it has been reported that the cost to resolve a security issue by development and application of patch is approximately 60 times the cost of fixing the security bug in an early stage of the SDLC (Hoo, Sudbury, and Jaquith 2001).

12.3.3.1 Threat Modeling

Threat modeling is a proactive process of identifying threats to assets, assessing them, and then designing countermeasures to mitigate them. There are several threat models that have been developed and some of them include Microsoft Threat Model, Trike, OCTAVE, Generic Threat Model, and OMG Threat Model (Alberts et al. 2003, Meier et al. 2003, OWASP 2018a).

A threat is any circumstance or event with the potential to adversely impact organizational operations and assets, or individuals, through an information system through modification of information, unauthorized access, destruction, disclosure, or and/or denial of service. The first step is to identify threats that might damage the software product, and then to identify the vulnerabilities responsible for the threats. The next step is to assess the impact of those threats if they were to be realized. Quantitative ranking techniques such as Microsoft's STRIDE and DREAD (Meier et al. 2003) can be used. OWASP identifies a process for assessing the likelihood of threats in a generic risk model (OWASP 2018a) that uses a qualitative approach for ranking threats. The approach of this model is the calculation: *risk = likelihood × impact*. Likelihood is the possibility of the attack, and impact is a combination of factors ranging from reputational and financial impact to systems impacts. After ranking the threats, countermeasures must be developed for those threats whose risk ranking is unacceptable, especially those that are ranked as high risk and medium risk. However, all risks should be addressed while taking into account the costs and benefits considerations.

12.3.3.2 Secure Design Considerations

As the architecture and design of the application are defined, security considerations should be assessed. The architecture should be assessed from both a security perspective and a performance perspective. It is in this phase that hard-to-correct security problems can be fixed at the time they are easiest to address and less costly. When addressing vulnerabilities, the basic concept behind each vulnerability and attack needs to be understood in order to create a secure design. The design can be constructed while keeping in mind all the security prerequisites of the application. Then the detailed design specifications are created that show developers exactly which security controls must be included and how the components will interact with the overall application. You should also develop the security test plans and misuse cases that will be used during the testing phase.

The following steps can be used to integrate security into the architecture during design. Perform risk assessment of the application's proposed architecture and deployment environment. Application threat modeling discussed in section 12.3.3.1 can be used for this assessment in which you identify the threats, rank them, and determine the countermeasures to address them. Threats with high- and medium-level rankings should be addressed first, but you should pay attention to all threats. You should document all threats and countermeasures implemented as well as any threats that were not addressed. You should also document any context-specific vulnerabilities that are dependent on how and where the application is deployed that will need to be addressed during rollout in the production environment. Your design will be passed on to the coding team who should use the secure coding practices discussed in section 12.3.4 to implement the design. Developers can also make use of security design patterns (Gamma et al. 1994) to deal with security-related issues and solve known security problems.

12.3.4 Secure Coding Practices

The major objective of secure coding practices is to increase awareness about software security among the developer community. Programs that are developed following secure coding standards are much more secure than programs that do not follow any standards. Writing secure code is an essential part of secure software development. Unfortunately, neither a novice nor an experienced programmer necessarily knows how to write code securely. A developer's unintentional ignorance of known vulnerabilities and insecure coding practices can generate security flaws in a program. Such flaws could be responsible for crashing a program, enabling a denial-of-service attack or other kinds of attacks. We therefore need to adopt effective approaches to detect and eliminate such flaws. Hence, adopting secure coding practices helps avoid most of the software defects responsible for causing vulnerabilities and improves the quality of the software. However, we should always remember that in order to develop a secure application, practicing secure coding techniques alone is not sufficient. Security should be incorporated into every phase of the SDLC, as we have stated before.

Secure coding techniques include both general guidelines, which can be used to improve software security independent of the programming language that is used for software development, and techniques specific to a programming language used for software development. Our discussion will focus mainly on programming language independent recommendations. Guidelines for specific languages such as Java and C++ can be found on vendor websites. For example, see Oracle (2018) for securing coding guidelines in Java. For further discussions and recommendations on secure coding techniques, see also Shiralkar and Grove (2009) and OWASP (2010).

Efficient input validation is a must as most vulnerabilities emanate from inadequate input validation or none at all. Every input entered by a user should be sanitizing, but the best practice is to create a "white list" of expected known good input parameters and formats instead of relying on a "black list" of known bad inputs. SQL injection is of particular concern whenever user requests need data from a database via a web server, which is very common. In order for a SQL injection to be successful, the attacker will need to manipulate the standard SQL query and exploit non-validated input vulnerabilities (Incapsula 2018). To prevent SQL injections, we must practice input validation and utilize a web application firewall to filter out SQL injections (Incapsula 2018).

We should never trust the input to SQL statements but must use parameterized SQL statements. Parameterized SQL statements or queries (also known as prepared statements) are a technique of query execution that separates a query string from query parameter values. This is done to avoid SQL injections and to speed up query executions in certain scenarios. Because query parameters are passed separately, parameter values cannot modify (and break) the query string. We are effectively insisting that all input to SQL statements must be sanitized.

We should use libraries (e.g., anti-cross site scripting library) to protect against security bugs during web application development and ensure to test the code with a web application scanner to detect vulnerabilities. We must make sure to use modern compilers, which often include defenses against coding errors; for example, GCC protects code from buffer overflows. Use proper error/exception handling when coding and check the return values of every function, especially security-related functions. In addition, use a secure coding checklist. In order to track changes made to the code or document, use version/configuration control; this enables easy rollback to a previous version in case of errors. Remember that version control facilitates accountability and saves development time.

Your organization should implement peer reviews and sound security testing, as explained in section 12.3.5. Having an organization security policy that prohibits the use of banned functions in coding is also important. Don't store sensitive data in cookies and encrypt all confidential data using strong cryptographic techniques. Encoding of every user-supplied input is recommended because attackers use malicious input to conduct XSS types of attacks and doing so can prevent the client's web browser from interpreting these as executable code. Manage cryptographic keys carefully by having a key management plan. Use long keys and strong cryptographic algorithms to make it harder for adversaries. Use of FIPS-approved cryptographic algorithms is recommended and security protocols with inherent cryptographic weakness should not be used (e.g., SSL V2).

Education and training are also important and every organization should educate its developers on how to write secure code, for example, by offering training seminars or courses. The training should cover strategies and best practices to mitigate common threats as well as emphasis on security features of programming languages (especially those used by the organization), and how to implement those features to build secure applications. Finally, secure coding practices must include keeping informed about known and new vulnerabilities as well as software bugs by reading security forums and newsgroups, magazines, research papers, and newsletters. See examples of well-known vulnerability resources (Mitre 2018b, NIST 2018, VulnDB 2019). Remember new vulnerabilities and malware are discovered every day, and you need to stay informed so that you can take proactive action as necessary.

12.3.5 Security in Testing

Software testing is an important phase in SDLC before any software is deployed in the production environment. Various kinds of testing are performed, such as unit testing, system testing, integration testing, and user acceptance testing. The purpose of testing is to uncover software bugs and then remove them, hopefully creating a better software product. Similarly, software security testing is an indispensable phase of a secure SDLC, and it performs sanity checking before the software is deployed in the production environment. The idea behind it is to discover security flaws in code and then fix them. We now discuss some approaches for conducting software security testing.

12.3.5.1 Code Reviews

During code reviews, an objective group of experts reviews both your code and its documentation for misunderstandings, inconsistencies, and other faults. A team composed of the developer (you) and three or four other technical experts is created. The team studies the program in a consistent way and looks for faults. A recorder or scribe records any faults found, and these are thereafter rectified by the developer after the review. All the findings and errors resulting from code review must be documented and maintained. Code reviews can be manual or automated.

From a security perspective, code review is a process of software security testing in which the developer of the program, other technical experts, and a team of quality assurance testers review the code together. The developer will be required to fix any security flaws after the end of the review. Peer reviews can also be conducted in which software developers check each other's code to find security bugs and other kinds of bugs before handing the code to the QA team. In addition to peer code review, it is a best practice to engage a third-party with special expertise in security to perform source code review before your application is deployed.

During a security code review, we look for known common vulnerabilities and system calls that could be used for malicious purposes. Code reviews have been shown to be very successful in detecting faults and are often included in an organization's list of mandatory or best practices. Remember that the earlier in the development process a fault is spotted, the easier and less expensive it is to correct.

12.3.5.2 Vulnerability Testing and Penetration Testing

Application vulnerability scanning tools can be used to discover vulnerabilities in software so that these vulnerabilities can be fixed before they can be discovered by nefarious actors. There are various tools available to identify vulnerabilities in code, such as Nessus, OpenVAS, Tripwire IP360, Wireshark, AirCrack, Retina, Nikto, Nexpose, and Microsoft Baseline Security Analyzer (MBSA) (cWatch 2018, STC Admin 2018). Some of them scan for vulnerabilities and carry out a wide range of network checks. The vulnerabilities that can be found by these application vulnerability scanning tools include cross-site scripting (XSS), injection flaws, malicious file execution, and cross sire request forgery (CSRF).

Penetration testing, also called pen testing, is a procedure that is used to discover defects at the operating system or server level. Tools specifically designed for pen testing are used to scan one or more target systems in order to discover open ports that may indicate the presence of vulnerabilities. The scanning consists of transmitting TCP/IP packets to the target system in attempts to communicate with various common services, in order to discover which services are operating on the target system. Most of the tools mentioned above can also be used for pen testing.

Vulnerability scanners automate security auditing and play a vital role in security applications and the network by scanning your network and websites for different security risks. These scanners are also capable of generating a prioritized list of those applications that should be patched, describing the vulnerabilities, and providing steps on how to remediate them. These tools make pen testing a lot easier by automating the scanning and analysis of scan results, and is it even possible for some of them to even automate the patching process. Pen testers should have no knowledge of the source code or architecture of the application. It is recommended that a third-party pen tester conduct the penetration testing, and the results of this assessment process are documented and presented to the organization.

12.3.5.3 Fuzz Testing

Fuzz testing means testing software against malformed data by feeding it with unexpected, invalid, or random test inputs to see how it reacts. The system is then monitored for crashes and other undesirable behavior. Fuzz testing can be effective for finding security vulnerabilities. Every time software is changed or updated, fuzz testing should be conducted. Fuzzing techniques use black-box testing, and they allow the detection of most of the common vulnerabilities, for example, buffer overflow, cross-site scripting, and SQL injections.

12.4 Conclusion

Security vulnerabilities in software products are an increasing concern to organizations, developers, and, of course, consumers. Ultimately, when data breaches occur, the consumers are affected, and so are the organizations that use those software products to conduct business operations.

There is no doubt that the cost to an organization that releases software containing security vulnerabilities is very high. Avoiding the introduction of such vulnerabilities requires awareness and commitment on the part of an organization that develops software products to secure software development practices (secure SDLC), as discussed in this chapter. It is expected that a software product developed using secure SDLC will cost more, and it will impact the project schedule, but this is much better than paying the high costs of software breaches later. An organization's reputation is also affected by such events and that is also important to consider.

It is important that developers are knowledgeable in known security vulnerabilities so that they can avoid writing code that is exploitable. We also know that the sooner a vulnerability is discovered, the easier and cheaper it is to fix it, as discussed in section 12.3.3. Adopting a secure SDLC that considers security at every stage of development contributes to the early identification of potential vulnerabilities. This chapter discussed secure SDLC, including discussions on the activities at each phase of the development cycle. Developers need to have a sound understanding of threat modeling, secure designs, secure coding practices, and security in testing in order to minimize security bugs and efficiently develop secure applications.

Software developers must make the extra effort and time to improve the security of software products by insisting on security by design. I am aware that this will increase the costs and schedules of software products. One of the challenges software development organizations have been facing for decades is delivering working software that meets the customer requirements on time and on budget. Budgetary constraints and market forces, for example, sometimes force companies to reduce the amount of time dedicated to testing, which implies that the cost of maintenance will be higher because ultimately more bugs ship with the software and will have to be dealt with during the production stage of the software. Such bugs could be vulnerabilities that increase the risk of using the software. Senior management in software development organizations must embrace and champion the "secure by design" philosophy for their software products. It will cost a bit more and take longer; but more successful software development companies are those that develop quality and secure software that meets the users/stakeholders' business requirements.

Today's business environment is characterized by regular data breaches. These data breaches are very expensive to the enterprises that are attacked. After such attacks, customers lose trust and run with their money, resulting in a huge negative impact to the organizations that were attacked. This includes damage to the organization's reputation and possible violations of regulatory compliance. For these reasons, spending more effort, time, and money is well worth it in the long run. In future, robust and resilient cyber defenses are what will differentiate the more successful organizations from the less successful ones. However, we must always remember that deploying secure software applications alone is only part of the solution to defending cyberspace. The core pillars for robust and resilient cyber defenses include secure software (by design), user/customer training and education, and risk-driven management practices.

Further Readings

Readers interested in learning more about secure software development can refer to the sources listed in this section. The OWASP Secure SDLC Project website provides regular updates on any developments in this area (OWASP 2018b). Nikolaenya (2017) provides a useful guide for secure software development stage by stage. Othmane et al. (2014) have discussed proposals for extending the Agile development process with security engineering activities.

The Software Assurance Forum for Excellence in Code (SAFECode), a non-profit organization, regularly publishes on different aspects of security processes and practices in order to advance software assurance methods and positively impacts the security and reliability of the technology systems (SAFECode 2018b). It hosts many publications based on the experiences of its members, but one publication of interest is the paper titled "Fundamental Practices for Secure Software Development" that includes updates to the fundamental practices to reflect current best practices, new technical considerations, and broader practices now considered foundational to a successful Secure Development Life Cycle (SDL) program (SAFECode 2018a). The third edition of the paper was published in March 2018. This paper is an excellent source of best practices and new technical considerations. Scanlon (2018a, 2018b) has provided excellent discussions on ten types of application security testing tools, and decision-making factors for selecting application security testing tools.

References

Advisers, The Council of Economic. 2018. *The Cost of Malicious Cyber Activity*. February. Accessed June 4, 2018. www.whitehouse.gov/wp-content/uploads/2018/02/The-Cost-of-Malicious-Cyber-Activity-to-the-U.S.-Economy.pdf.

Alberts, C, A Dorofee, J Stevens, and C Woody. 2003. "Introduction to the OCTAVE approach." *SEI, CMU*. August. Accessed May 23, 2015. https://resources.sei.cmu.edu/asset_files/UsersGuide/2003_012_001_51556.pdf.

Alliance, A. 2001. *Manifesto for Agile Software Development*. Accessed July 25, 2018. www.agilemanifesto.org/.

Amber, S. 2012. "Examining the Agile Manifesto." Accessed August 12, 2018. www.ambysoft.com/essays/agileManifesto.html.

Boehm, B W. 1988. "A spiral model for software development and enhancement." *IEEE Computer, 21(5)* 61–72.

cWatch. 2018. *Top 10 Vulnerability Assessment Scanning Tools.* March 16. Accessed June 2, 2018. https://cwatch.comodo.com/blog/website-security/top-10-vulnerability-assessment-scanning-tools/.

Gamma, E, R Helm, R Johnson, and J Vlissides. 1994. *Design Patterns: Elements of Object-Oriented Software Architecture.* Reading: Addison-Wesley.

Goldenson, D, and D Gibson. 2003. *Demonstrating the Impact and Benefits of CMMI: An Update and Preliminary Results. (CMU/SEI-2003-SR-009, ADA418491).* Pittsburgh: Software Engineering Institute, Carnegie Mellon University.

Hoffer, J A, J George, and J Valacich. 2014. *Modern Systems Analysis and Design,* 7th Edition. Boston, MA: Pearson.

Hoo, K S, A W Sudbury, and A R Jaquith. 2001. "Tangible ROI through secure software engineering." *Secure Business Quarterly, 1(2)* (Fourth Quarter).

Howard, M, and D LeBlanc. 2007. *Best Practices Writing Secure Code for Windows Vista.* Redmond: Microsoft Press.

Howard, M, and S Lipner. 2006. *The Security Development Lifecycle.* Redmond: Microsoft Press.

Humphrey, W S. 2000a. *The Personal Software Process (PSP).* Report. Pittsburgh: Software Engineering Institute. CMU/SEI Report Number: CMU/SEI-2000-TR-022.

Humphrey, W S. 2000b. *The Team Software Process (TSP).* Report. Pittsburgh: Software Engineering Institute. CMU/SEI Report Number: CMU/SEI-2000-TR-023.

Incapsula, I. 2018. *SQL Injection.* Accessed October 20, 2018. www.incapsula.com/web-application-security/sql-injection.html.

InfoBase, FFIEC IT Examination Handbook. 2013. *Development and Acquisition, Rapid Application Development.* Accessed August 12, 2018. https://ithandbook.ffiec.gov/it-booklets/development-and-acquisition/development-procedures/software-development-techniques/rapid-application-development.aspx.

Kissel, R, K Stine, M Scholl, H Rossman, J Fahlsing, and J Gulick. 2008. "Security considerations in the system development life cycle." *nvlpubs.nist.gov.* October. Accessed July 14, 2018. http://nvlpubs.nist.gov/nistpubs/Legacy/SP/nistspecialpublication800-64r2.pdf.

Marasco, J. 2007. *What Is the Cost of a Requirement Error?* June 26. Accessed August 12, 2018. www.stickyminds.com/article/what-cost-requirement-error.

McGraw, G. 2006. *Software Security: Building Security in.* Boston, MA: Addison-Wesley.

McManus, J. 2018. "Security by design: teaching secure software design and development techniques." *Journal of Computing Sciences in Colleges, 33(3)* 75–82.

Mead, M. 2006. *SQUARE Process.* January 30. Accessed May 4, 2016. www.us-cert.gov/bsi/articles/best-practices/requirements-engineering/square-process.

Mead, N. 2013. *SQUARE Process. United States Computer Emergency Readiness Team.* July 5. Accessed September 23, 2018. www.us-cert.gov/bsi/articles/best-practices/requirements-engineering/square-process.

Mead, N R, E D Hough, and T R Stehney. 2005. *Security Quality Requirements Engineering.* May 1. Accessed May 23, 2018. http://resources.sei.cmu.edu/asset_files/technicalreport/ 2005_005_001_14594.pdf.

Meier, J D, A Mackman, M Dunner, S Vasireddy, R Escamilla, and A Murukan. 2003. *Improving Web Application Security: threats and Countermeasures, Chapter 3 – Threat Modeling.* June. Accessed March 2, 2018. http://msdn.microsoft.com/en-us/library/ ff648644.aspx.

Microsoft. 2018. *Microsoft Threat Modeling Tool 2016.* Accessed March 20, 2018. www. microsoft.com/en-us/download/details.aspx?id=49168.

Mitre. 2018a. *CAPEC: Common Attack Patterns Enumeration and Classification.* July 2. Accessed September 12, 2018. http://capec.mitre.org/.

Mitre. 2018b. *Common Vulnerabilities and Exposures.* Accessed November 1, 2018. https:// cve.mitre.org.

Mukesh, S. 2009. "Defect prevention: reducing costs and enhancing." *i Six Sigma.* December 14. Accessed June 23, 2017. http://arsavir.free.fr/Cost%20of%20Defect/Defect% 20Prevention_%20Reducing%20Costs%20and%20Enhancing%20Quality.pdf.

Nikolaenya, D. 2017. *ScienceSoft.* October 11. Accessed January 24, 2019. www.scnsoft. com/blog/secure-software-development-guide.

NIST. 2018. *National Vulnerability Database.* November 6. Accessed December 1, 2018. https://nvd.nist.gov/.

Noopur, D. 2013. *Secure Software Development Life Cycle Processes.* July 31. Accessed August 24, 2017. www.us-cert.gov/bsi/articles/knowledge/sdlc-process/secure-software-development-life-cycle-processes.

O'Connor, B. 2018. *Cybercrime: The $1.5 Trillion Problem.* May 9. Accessed July 30, 2018. www.experian.com/blogs/ask-experian/cybercrime-the-1-5-trillion-problem.

Oracle. 2018. *Secure Coding Guidelines for Java SE.* September 27. Accessed October 12, 2018. www.oracle.com/technetwork/java/seccodeguide-139067.html.

Othmane, L B, P Angin, H Weffers, and B Bhargava. 2014. "Extending the agile development process to develop acceptably secure software." *IEEE Transactions on Dependable and Secure Computing, 11(6)* 497–509.

OWASP. 2010. "OWASP." November. Accessed August 12, 2018. www.owasp.org/ images/0/08/OWASP_SCP_Quick_Reference_Guide_v2.pdf.

———. 2018a. *Application Threat Modeling.* October 14. Accessed October 28, 2018. www.owasp.org/index.php/Application_Threat_Modeling.

———. 2018b. *OWASP Secure Software Development Lifecycle Project.* December 28. Accessed January 31, 2019. www.owasp.org/index.php/OWASP_Secure_ Software_Development_Lifecycle_Project.

———. 2018c. *Secure SDLC Cheat Sheet.* August 25. Accessed October 22, 2018. www. owasp.org/index.php/Secure_SDLC_Cheat_Sheet.

Phleeger, S L, and J M Atlee. 2010. *Software Engineering: Theory and Practice,* 4th Edition. Upper Saddle River, NJ: Pearson.

Royce, W W. 1970. *Managing the Development of Large Software Systems.* Accessed July 20, 2018. http://leadinganswers.typepad.com/leading_answers/files/original_waterfall_ paper_winston_royce.pdf.

SAFECode. 2018a. "SAFECode." *SAFECode Organization Web Site.* March 1. Accessed January 8, 2019. https://safecode.org/wp-content/uploads/2018/03/SAFECode_Fundamental_Practices_for_Secure_Software_Development_March_2018.pdf.

———. 2018b. *SAFECode: Driving Security and Integrity.* December. Accessed January 6, 2019. https://safecode.org/.

Scanlon, T. 2018a. August 20. Accessed January 30, 2019. https://insights.sei.cmu.edu/sei_blog/2018/08/decision-making-factors-for-selecting-application-security-testing-tools.html.

———. 2018b. July 9. Accessed January 20, 2019. https://insights.sei.cmu.edu/sei_blog/2018/07/10-types-of-application-security-testing-tools-when-and-how-to-use-them.html.

Schwaber, K, and M Beedle. 2002. *Agile Software Development with Scrum.* Upper Saddle River: Prentice Hall.

Schwaber, K, and J Sutherland. 2017. *The Scrum Guide™.* November. Accessed October 7, 2018. www.scrumguides.org/scrum-guide.html.

Shiralkar, T, and B Grove. 2009. *Guidelines for Secure Coding.* Accessed June 12, 2016. www.atsec.com.

STC Admin. 2018. *Top 12 Vulnerability Assessment Scanning Tools.* December 4. Accessed January 16, 2019. www.softwaretestingclass.com/top-12-vulnerability-assessment-scanning-tools/.

VulnDB. 2019. *Know Your Vulnerabilities Before They are Used Against You.* January 30. Accessed January 12, 2019. https://vulndb.cyberriskanalytics.com/.

Wells, D. 2009. *Extreme Programming: A Gentle Introduction.* Accessed June 4, 2017. www.extremeprogramming.org/.

Yuan, X, E B Nuakoh, J S Beal, and H Yu. 2014. "Retrieving relevant CAPEC attack patterns for secure software development." *9th Cyber and Information Security Research Conference.* Oak Ridge: ACM.

Zain. 2017. *Understanding the Agile Software Development Lifecycle and Process Work-flow.* January 12. Accessed July 12, 2018. www.smartsheet.com/understanding-agile-software-development-lifecycle-and-process-workflow.

Chapter 13

Data Security and Privacy

Biodun Awojobi and Junhua Ding

Department of Information Science, University of North Texas

Introduction

Data security is about the technology and policy for protecting confidentiality, integrity, and availability of data during its entire life cycle. Confidentiality means data are only accessible by authorized users and are protected from unauthorized disclosure; integrity means data are correct and prevented from improper and unauthorized modification; and availability means data are always accessible whenever needed and are prevented from denying of service (Bertino 2016, 400–407). The emerging of big data and Internet of Things (IoT) brings the great attention of data privacy. While data confidentiality is the basic requirement to data privacy, it has additional requirements for complying legal privacy regulations and managing individual privacy preferences (Bertino 2016, 400–407).

The technology for ensuring data security and privacy can be summarized as four controls: access controls, flow controls, inference controls, and cryptographic controls (Denning and Denning 1979, 227–249). The four controls for protecting data security were introduced by D. Denning and P. Denning in their data security paper published four decades ago (Denning and Denning 1979, 227–249). The four controls are still applied to modern data security and privacy, although many new control models have been invented. For example, role-based access control (RBAC) (Ferraiolo, Kuhn, and Chandramouli 2003; Ding and Mo 2012, 92–101) has been widely implemented in modern computing systems, and differential privacy (Dwork et al. 2006, 265–284) has been implemented for inference controls by a few large data consumers including Google and Apple.

Computers are increasingly becoming more portable and always connected to the internet. The rapid technological advancement in the smartphone market and sensor-based devices that are connected to the internet exposes opportunities for new privacy and security threats. Generally, IoT establishes the concept that virtually every physical thing or sensor-based device in the world that can be connected to a power source can become a computer that is connected to the internet (Fleisch 2010, 125–157). It is expected that in year 2020, there will be at least 25 billion connected devices (O'Neill 2016, 48–49). These always-on and internet-connected devices that are in our homes and offices will collect data about us and store the data in a database on remote servers. A privacy or security breach of the connected devices can be costly.

Mobile applications have revolutionized the user experience with smartphones. Most smartphones are equipped with sensor and connectivity capabilities like a GPS, WiFi, bluetooth, memory, data storage, camera, and lots of sensors like an accelerometer, light sensors, and so on. In most of the developed countries, smartphones are also always connected to the Internet (Gramlich 2019). IoT devices are similar to smartphone devices because, like smartphone devices used in most developed countries are always connected to the Internet, IoT devices are generally always connected to the Internet at all times. Not all mobile applications require the Internet to run on a smartphone. Mobile applications often make a user to create an account on their server for recognition and tagging, they are also capable of controlling the sensor and connectivity capabilities of the smartphone by default. This means that if the mobile application is owned by a third-party developer, the app developer can collect sensitive user data without the knowledge of the end user. Smartphone makers such as Apple and Google have features built in their software to limit and in most cases prevent privacy and breaches; however, the enforcement is based on a shared responsibility model. The device owner needs to implement the control to protect against a security breach. Enforcing password protection policies such as implementing biometric authentication, use of strong password complexity, age, and reuse requirement is important.

The use of social media platforms such as Facebook, Instagram, YouTube continues to grow because of the wide adoption of mobile applications on smartphones (Gramlich 2019). Social media platforms introduce many unknown privacy risks to the users of the platform, especially risks around user profiling (Isaak and Hanna 2018, 56–59). The privacy issue gets heightened, especially when the social media platform owner is unaware of the sensitive user data being collected through its platform. Users should have the right to delete mobile application from their smartphones and also completely erase their presence on a social media platform. Security and privacy need to be protected during the four phases of the data life cycle. The five phases of the data life cycle are data generation, data sharing, data usage, data storage, and data destruction. The social media platform owner should create explicit security and privacy policies and guidelines for the platform users and developers.

Technology for Data Security and Privacy

Access controls are the selective restriction of data access for regulating who can access the data and what can be accessed. Flow controls regulate the data propagation process to protect data from being accessed by unauthorized users and prevent a data flow if it is unauthorized (Bacon et al. 2014, 76–89). Inference controls prevent data from leaking specific information that is reconstructed through deduction of confidential information by inference (Denning and Denning 1979, 227–249). Cryptographic controls are to protect data from unauthorized accessing through encryption of data, which are scrambled into unintelligible ciphertext (Denning and Denning 1979, 227–249). In the following section, we describe some representative models for each control.

Access Controls

The main access control models include discretionary access control (DAC), mandatory access control (MAC), RBAC, and attribute-based access control (ABAC). Each organization may use different access control models according to its security and privacy requirements.

DAC asks the system administrator or the owner of data to set the access control policy for each data object. The access control policy defines access rights for each user to a data object such as a file. An example of DAC is the access control of Unix files, which defines the read, write, and execute permissions for the owner, users in the same group, and other users of a file. DAC allows a user to transfer its ownership to another user, which could cause security problems.

MAC is a hierarchical approach to control the access to data objects. Only the system administrator can set the access control policy for each user to access every data object, but an end user of a data object cannot change the access control to the object. Each data object is assigned with a security label, which contains a classification of security information such as top secret or confidential, and a category of management information such as department or college. Each user in the MAC system is also assigned with the classification and category properties from the same set of properties applied to the data objects. When a user accesses a data object, the MAC system compares the user's classification and category information to the one that is assigned to the data object to grant or decline the access request. An example of MAC is the access controls of Windows operating system for admins, ordinary users, and guests.

RBAC creates a group of roles such as Administrator or Developer based on their business responsibilities in an organization that the RBAC system belongs to. Each role is assigned with a set of access rights such as read, write, and execute, and then each user is assigned with one or multiple roles. The user has all access rights for its roles. Each data object contains a set of access control attributes. Attributes are sets of access rights that are used to describe the access control. A user can only access the data object when its role has the access rights that meet the control requirements set by the access control attributes.

ABAC uses attributes as building blocks in a structured language that defines access polices. The attributes include user attributes, resource attributes, object attributes, environment attributes, and other attributes. When a user accesses a data object, ABAC system grants or declines the access based on the access policy, the user's attributes, the attributes of the object, and the environmental attributes. Although ABAC provides dynamic, context-aware and risk-intelligent access controls, ensuring the completeness and soundness of the access policy could be a grand challenging task.

Flow Controls

Flow controls can be implemented with the access controls (i.e., MAC, DAC, RBAC, and ABAC) discussed above. The access control rights include one for controlling whether a user can transfer a data object and an object can be transferred or not. MAC, RBAC, and ABAC require a central authority to specify the flow controls or policies, but DAC can allow each user to specify the flow controls for each data object. System support is still needed for continuously checking the data flows at runtime (Bacon et al. 2014, 76–89). A system may use a dynamic technique to enforce security policies during runtime by terminating data transferring when a possible leak is about to happen.

Inference Controls

A statistical database is a database for statistical purposes through offering queries for finding statistical information from the database. The statistical information is calculated from individual information collected in the database, but the individual information should be kept confidential. However, an adversary could learn the confidential contents of a statistical database by creating a series of targeted queries and remembering the results. An adversary can also learn the private information though cross-checking queried data from different data sources (Narayanan and Shmatikov 2006). Differential privacy (Dwork et al. 2006, 265–284) is an effective way of inference control to protect data from inference attacks. Differential privacy is an approach for preventing leaking of privacy information of individuals that are in a statistics database that publishes aggregate information. For example, many companies and government agencies publish statistical aggregates while protecting confidentiality of individual survey responses through inference controls with differential privacy. One of the advantages from differential privacy is that it provides acceptable accuracy of the statistical aggregates while prevents privacy information of individuals from being visible to anyone. The basic idea of differential privacy is to add randomized "noise" into an aggregate query result so that an adversary learns "virtually nothing" more about an individual than they would learn when the individual is absent from the dataset. The "virtually nothing" means an adversary can only get an insignificantly small change in belief about an individual in the database. The E-differential privacy proposed by Dwork et al. (2006, 265–284) is used to control the extent of the change. The lower is the value of E, the less is the change of the benefit of the individual. A randomized algorithm A is E-differential private when the probability of result x returned by algorithm A that is applied to dataset D_1 to the probability of the same result x returned by algorithm A that is applied to dataset D_2 is significantly small, which is defined by E, for all x and for all pairs of datasets D_1 and D_2, where D_1 and D_2 are different with only one record (Dwork et al. 2006, 265–284). Laplace mechanism (Dwork et al. 2006, 265–284) can be implemented to process results of aggregate queries with "noise" to make them differential private.

Cryptographic Controls

Cryptographic controls are critical for protecting data confidentiality and integrity from accidental leaking or modification. There are two classes of encryption: symmetric encryption and asymmetric encryption. Symmetric encryption uses only one key or called "secret key" to encrypt and decrypt the targeted data. Key exchange between sender and receiver of encrypted data is necessary and it is important to keep its confidentiality. Examples of symmetric encryption algorithms include Advanced Encryption Standard (AES) and Data Encryption Standard (DES). DES was developed by IBM and was the first standardized cipher for securing electronic communications (Denning and Denning 1979, 227–249). It is used in variations such as 2-key DES or 3-key DES. AES is the most widely used symmetric encryption algorithm. National Institute of Standards and Technology (NIST) set AES as the standard for the encryption of electronic data in 2001. AES cipher has a block size of 128 bits with three different key lengths as AES-128, AES-192, and AES-256.

Asymmetric encryption uses matched paired keys called "public key" and "secret key." The public key is published so that it is publicly accessible, but the secret key is kept as secret by the owner who created the paired public key and secret key. Both public key and secret key can be used for encrypting and decrypting data. When data is encrypted with a public key, then only the one who owns the paired secret key can decrypt the data. If data is encrypted with a secret key, then anyone can decrypt the data with the paired public key. The latter encryption scenario is to ensure the encrypted data are only from an expected sender. Since public keys are published for public, it is important to ensure a public key is published by the declared owner. Public key certification authority is the entity to certify the ownership of a public key by the named subject of the certificate.

Encryption is also used for ensuring data integrity. For example, Message Authentication Code (MAC) is used for protecting data integrity by allowing a user to check any changes to the data content. The sender of data produces a MAC according to the data content using an MAC algorithm, then encrypts the MAC and attaches the encrypted MAC to the data. The receiver of the data produces a MAC from the received data using the same MAC algorithm, and then decrypts the received MAC. If the decrypted MAC and the one produced by the receiver are same, then received data is not modified.

Cryptographic controls are also implemented in many data communication protocols such as https to ensure the security and privacy of the communication, which won't be discussed in this section.

IoT Implication for Cybersecurity

Smartphones might be currently the most widely connected devices to the internet. However, this is changing with the emerging IoT applications. IoT devices, which are sensor-based devices that help with home automation, are always connected to the Internet. Like smartphones, our interactions with IoT devices generate data points, which can be converted into intelligent information. Gartner believes that by 2020 we will have 25 billion connected devices, while Cisco believes that we will have 50 billion connected devices by 2020 (O'Neill 2016, 48–49). Other analysts, like Morgan Stanley, believe that the number of connected devices will be as high as 75 billion. IoT will not only dramatically change the way we operate in the world today, but it will also open new cybersecurity challenges. For example, we can wake-up in the morning and realize that our car has driven itself to our most recently visited location because of a hack or a malfunction. IoT is the future; it is part of humanity's journey to digital transformation. In fact, some argue that the benefits outweigh the security and privacy risks that IoT poses currently.

The three key principles of cybersecurity are explained with the CIA triad, which are confidentiality, integrity, and availability (Aldossary and Allen 2016). IoT privacy is a confidentiality issue, while integrity and availability are IoT security issues. When an unauthorized user gains access to the data on an IoT device, it means that there has been a breach of privacy. A breach of security often occurs prior to a breach of privacy. For example, the password to the IoT device may have been compromised and used to maliciously collect data from the device. In this scenario, a security breach preceded a privacy breach. An example of a breach of integrity with IoT devices is manipulating device settings such that the user data is sent to another server. Preventing the device from being accessible to the owner either through physically or programmatically tampering with the device is an Availability issue. IoT has a significant technological intensity. Tech intensity refers to the adoption rate of new digital solutions. Imagine if you can receive a package with your IoT-powered doorbell system, while you are in another country. This could have been a package that would have had to be returned if it weren't for IoT. IoT was listed on the top-three list of disruptive technologies, second only to Mobile Internet and Automation of knowledge work, which are both technologies heavily influenced by IoT (Manyika et al. 2013).

The issue of IoT security is one that we cannot underplay or ignore. IoT devices are typically low-powered devices. This means they lack robust hardware resources like memory and computing power to load a full network stack – features that are often included in high-powered devices like a standard computer. Most IoT devices only have the module required to perform its core functionality. This implies that the devices are naturally prone and vulnerable to hacks (Jaswal et al. 2017, 1277). For most IoT device manufacturers, creating an easy to setup experience trumps security. Most consumer IoT devices do not follow common security best practices such as using strong passwords or enforcing high-security wireless encryption standards. If an IoT device is compromised, the data collected by the bad actors from the device or group of devices is often personal and sensitive which can pose a huge privacy threat. It is difficult to have an end-to-end security design for IoT devices such that the entire information collection, transmission, processing, and retrieval process are captured. Some legacy devices have vulnerabilities that cannot be patched through software and can only be mitigated with hardware. Smartphones are commonly used to set up and control IoT devices, typically with a mobile application. If the mobile application developer is collecting information about the usage of the application, then there are some security vulnerabilities that need to be assessed and addressed.

Mobile Applications Security Issues

Smartphones and tablets are low-powered devices compared to desktop and laptop computers. Device manufacturers of smartphones and tablets install an operating system that is capable of maximizing the device features and capabilities of the hardware like a battery, sensor, touchscreen, and so on. Smartphones and tablet by default have a mobile version of a web browser installed. As the operating system of a mobile device is not designed to run a full-featured browser as you have on a Microsoft Windows machine or a Macintosh device, running one can potentially drain the devices' resources like the battery. The mobile web browser is an example of a mobile application. We, therefore, define a mobile app as a software application that can be downloaded wirelessly from an app store or through any other software vendor. It is designed to run specifically on smartphones, tablets, and other mobile devices. Mobile apps are designed to take advantage of mobile device software and hardware capabilities and resources. iPhones, for example, have inbuilt components such as an accelerometer, touchscreen, front-facing cameras and so on.

Mobile device manufacturers preinstall mobile apps on their mobile devices. Examples of such apps from the Apple iOS ecosystem are Safari browser, mail app, photos app, and so on. Mobile apps preinstalled and developed by mobile device manufacturers are called native apps. Users are also able to download other apps from other developers like games, social apps, and so on from the device manufacturers' apps store. Apps developed by other developers are referred to as third-party apps. Some of these mobile apps are offered as free apps while others need to be paid for.

Unlike a web browser that relies on the Internet to download the required website data, most mobile apps do not require the Internet as they already have the data required to run the application downloaded. A significant design advantage for mobile app developers is that the user experience and behavior of the app user are controlled by the app developer. In terms of security, app developers continually provide security and feature updates for the apps. Unfortunately, mobile apps are not immune from zero-day attacks. Sometimes device exploits may have been carried out successfully on several devices before the app developer discovers it and subsequently releases a security fix.

Ensuring that a mobile device is secure and that our sensitive data does not fall in the wrong hands is not the sole responsibility of the software developers and hardware manufacturers; we have a part to play. The software in mobile devices allows for its users to enforce a certain level of security. A mobile device operating system like the Apple iOS or the Alphabets Android has been preloaded with the standard security framework and toolsets. It is the responsibility of the user to implement them. For example, passwords on mobile devices are not mandatory but are recommended. Some mobile device users prefer not to have a password on their device because of the hassle of unlocking the device. It is important to note that passwords are not highly respected in the security community as some argue that policies need to be enforced to ensure that passwords are secure. Policies like:

- Password age requirements, where a password can only be used for a certain period, after which the user needs to change the password;
- Password complexity;
- Password reuse requirements also help bolster the security posture of our mobile devices.

Knowing that your username and password is all that is required to gain access to any financial or health information typically stored on your mobile device further highlights the importance of these policies. More so when websites like BuyAcc (https://buyaccs.com/) that specializes in selling compromised usernames and passwords obtained from the dark web exist.

That being said, some security experts argue that the more policies there are, the harder it becomes for a user to use the mobile device; hence that argument for the eradication of passwords completely. There has been support for other means of authentication that are considered more secure like biometrics (fingerprint, facial recognition, voice recognition, and so on) and pattern recognition. The good news is that modern day mobile devices provide an alternative to passwords as a primary means of authentication. Users just need to enable the features and start using them. Research has shown that while authenticating using biometrics provides a unique authentication mechanism and better security, they aren't totally foolproof. Combining more than one authentication method based on the value of the asset is a common practice especially with organizations that want to protect their assets. For example, some organizations implement multifactor authentication on some of their assets such that whenever an employee wants to access a corporate asset, the employee will be asked for an additional layer of authentication like the code from a hardware token, a phone call, or a fingerprint. We shall cover the topic of security in further detail later in this chapter.

Data Security and Social Media

The popularity of social media platforms has grown over the years and continues to grow. Almost all the social media services like Facebook, YouTube, Snapchat, Instagram, WhatsApp, and so on have a mobile app available for download for free on the popular mobile device platforms like Android and iOS. According to a 2018 Pew Research survey of U.S. adults who use social media sites through the website or on their cellphone, two-thirds of American adults use Facebook, only YouTube has a higher adoption rate, which was 73%. Two other companies owned by Facebook, Instagram, and WhatsApp are used by 35% and 22% of U.S. adults, respectively. 74% of American women and 62% of men use Facebook, also 74% of the survey responders indicate that they visit Facebook once a day and 45% of the responders get news from Facebook (Gramlich 2019). From the survey results, we can infer that Facebook's influence as a social media platform is very strong, beyond that, the user data that Facebook has access to is huge. When you include some of Facebook's notable acquisitions in the social media space like WhatsApp and Instagram, you can only begin to imagine how close Facebook is from identifying you. WhatsApp is a social platform that uses your primary phone number as a means of identification and communicating with other WhatsApp users. This implies that Facebook technically has a record of your valid phone number plus other tracking data. Instagram, on the other hand, has a record of your pictures, videos, and the pictures that you like. This is in addition to the geolocation features Instagram has.

By creating an open platform such as an Application Programming Interface (API) that provides consumers and third-party developers with the ability to get information from the social media platform, it can be said that a platform like Facebook's motive was to democratize the data. While it is good at democratizing a platform, there should be controls as well, especially when the data of the authorizing users and their friends are being shared. Facebook's motives can also be questioned as a bigger play for more advertisement revenue, partly because the results of the psychoanalysis campaign by the third-party apps are used to create targeted ads to Facebook users.

In 2013, Cambridge Analytica (a third-party company) created an application "this is your digital life." The app connected with Open Graph asked Facebook's users to complete a psychological quiz mostly through the user's mobile phone browser, Facebook app, or through the web browser. About 300,000 users were reported to have been paid to complete the quiz (Isaak and Hanna 2018, 56–59). While there has been no public announcement regarding what the selection criteria are for the Facebook users that were paid to complete the psychological test, the data reported on the resulting number of user personal data harvested was 87 million (Isaak and Hanna 2018). Smartphone apps make giving access to apps like Facebook and other third-party developers very easy.

Our use of mobile applications comes with some undisclosed risks. In the case of the Facebook scenario described above, by signing in to an application or authorizing an application to access your Facebook profile, you disclose your personal information and that of your Facebook friends. As a best practice, you should only accept a friend request from the people you know. Ghost accounts are popular on social media platforms like Twitter and Facebook. These ghost accounts might clone the persona of a celebrity or a brand, so accepting the friend request doesn't raise an eyebrow. The ghost account will then send messages to the gained connections; hence the targeted harvesting of data will commence. Mobile app developers devise various methods to capture user data. A technology called Silverpush, which is a software used by mobile app developers as part of their code to eavesdrop on television through the use of "audio beacons" emitted by TVs. With Silverpush, mobile apps can capture the viewing habits of the users through their mobile phone even when the app is not being used, hence determining the targeted ad that the user should be served. In 2016, the U.S. Federal Trade Commission (FTC) issued warning letters to 12 app developers in the Google Play Store who used the Silverpush code with their applications (Federal Trade Commission 2016). It can be argued that the apps seek consent from the user before gaining access to the user's microphone.

However, not all users take the time to read the pop-ups that show up on the phone display. Silverpush publicly stated that the app is not being used in the U.S. – this, therefore, raises concerns about privacy, information exposure, and limitations outside the United States for apps like this. The ability to track sound around a mobile device user without the user knowing is a severe privacy concern, especially because there is no disclosure as to whom the data is being shared with.

Security and Privacy in Data Life Cycle

Data life cycle refers to the entire process of data life from data generation to end of life. There are a few of different ways for defining the phases of data life cycle. We will define the phases of data life cycle including data generation, data sharing, data usage, data storage, and data destruction. Each phase may have different security and privacy requirements.

Data Generation

Data can be generated in many different ways such as collecting data through crowdsourcing, data acquisition using sensors, or producing data from running software. The ownership of the generated data has to be clearly defined in this phase since it decides the regularities and policies for using, sharing, storing, and erasing the data, and the technology for protecting the data security and privacy. It is not easy to define the owner of the data. For example, who is the owner of the data that were taken through online survey regarding the drinking preference of adults? Therefore, legal policies are needed for defining the data ownership. In addition, it is important to have ways for appropriately managing the privacy preference as well as the change of preference for each data owner. It is necessary to have mechanisms for preventing data acquisition devices from illegal data acquisition, and ensuring the owner to control the data usage and sharing (Bertino 2016, 400–407).

Data Sharing

Data sharing may be implemented in three ways: transferring data from one site to another through network, transferring data using physical equipment, and sharing data in a server. The owner of the data should control or be aware of the data sharing, but it is not practical in many cases. Therefore, legal guidelines and policies are needed for regulating the sharing of the data that are owned by others (Bertino 2016, 400–407). (We define the owner of data as the subject of the data refers to. For example, an owner of the health information of a patient in a hospital is the patient.) Access controls and flow controls are important for ensuring the security and privacy of data sharing.

Data Usage

Data usage has to be monitored and controlled since it could easily cause problems in data confidentiality, integrity, and availability. The owner of the data should be able to control how the data is to be used and is aware of the potential security and privacy risk as well as should know how the data could be used. Data usage should protect data population and individual privacy according to owners' preference and data usage policies. Encryption of data is an effective way to protect the data privacy. However, using encrypted data in many cases such as analyzing data using machine learning algorithms or indexing data is challenging although it is possible (Agrawal and Srikant 2000, 439–450; Mendes and Vilela 2017, 10562–10582). Access controls and cryptographic controls could be used for ensuring the security and privacy of data usage.

Data Storage

Encryption is an important way to ensure data confidentiality and integrity for data storage. For large amount of data, it is necessary to consider the computational cost for encrypting the data and secure way for key management (Chen and Zhao 2012, 647–651). When the data is needed again, the environment for using the data should be available. Of course, protecting the security of physical media of the data storage is also critical, but it is beyond the scope of the discussion in this section.

Data Destruction

Data could be destroyed in a logical way or a physical way. When a datum item is destroyed in a logical way, it is only marked as deleted in a storage system such as a database or a file system but the data is still available, which is vulnerable for security attacks. Data still could be recovered entirely or partially even if they were physical destroyed. Therefore, it is important to follow a process for ensuring data is completely destroyed for data destruction.

Conclusion

In this chapter, we discussed the general concept of data security and privacy, the four controls: access controls, flow controls, inference controls, and cryptographic controls for protecting data security and privacy. We briefly discussed the regulations for protecting data security and privacy in the European Union and the United States. This chapter provided an overview of IoT, the security and privacy implications of the technology. We explained the core concept of cybersecurity using the CIA triad. The chapter discussed the core notions of security and privacy with mobile applications such as passwords complexity and reuse requirements, and data exfiltration using mobile applications. It also discussed the data security issues with social media and provided guidance on how to be safe on social media platforms. Finally, we discussed the security and privacy issues on each phase in data life cycle. The chapter only gives an overview of the technology and regulation for protecting data security and privacy in general.

Bibliography

Agrawal, Rakesh, and Ramakrishnan Srikant. "Privacy-Preserving Data Mining." In Proceedings of the 2000 ACM SIGMOD international conference on Management of data, pp. 439–450. 2000.

Aldossary, Sultan and William Allen. 2016. "Data Security, Privacy, Availability and Integrity in Cloud Computing: Issues and Current Solutions." *International Journal of Advanced Computer Science and Applications* 7 (4). doi:10.14569/IJACSA.2016.070464.

Bacon, Jean, David Eyers, Thomas FJ-M Pasquier, Jatinder Singh, Ioannis Papagiannis, and Peter Pietzuch. 2014. "Information Flow Control for Secure Cloud Computing." *IEEE Transactions on Network and Service Management* 11 (1): 76–89.

Bertino, Elisa. "Data Security and Privacy: Concepts, Approaches, and Research Directions," 1:400–407. 1. IEEE, 2016.

Chen, Deyan, and Hong Zhao. 2012. "Data Security and Privacy Protection Issues in Cloud Computing," 1:647–651. 1. IEEE.

Denning, Dorothy E. and Peter J. Denning. 1979. "Data Security." *ACM Computing Surveys (CSUR)* 11 (3): 227–249.

Ding, Junhua, and Lian Mo. 2012. "Enforcement of Role Based Access Control in Social Network Environments." 92–101. IEEE.

Dwork, Cynthia, Frank McSherry, Kobbi Nissim, and Adam Smith. 2006. "Calibrating Noise to Sensitivity in Private Data Analysis." 265–284. Springer.

Federal Trade Commission. 2016. "FTC Issues Warning Letters to App Developers Using 'Silverpush' Code." *FTC Issues Warning Letters to App Developers Using'silverpush'-code*. www.ftc.gov/news-events/press-releases/2016/03/ftc-issues-warning-letters-app-developers-using-silverpush-code.

Ferraiolo, David F., D. Richard Kuhn, and Ramaswamy Chandramouli. *Role-Based Access Control*. Boston, MA: Artech House, 2007. http://bvbr.bib-bvb.de:8991/F?func=service&doc_library=BVB01&local_base=BVB01&doc_number=015003926&sequence=000005&line_number=0001&func_code=DB_RECORDS&service_type=MEDIA.

Fleisch, Elgar. 2010. "What Is the Internet of Things? An Economic Perspective." *Economics, Management, and Financial Markets* 5 (2): 125–157. www.ceeol.com/search/article-detail?id=267154.

Gramlich, John. 2019. "10 Facts about Americans and Facebook." Last modified February 12, www.pewresearch.org/fact-tank/2019/02/01/facts-about-americans-and-facebook/.

Isaak, Jim and Mina J. Hanna. 2018. "User Data Privacy: Facebook, Cambridge Analytica, and Privacy Protection." *Computer* 51 (8): 56–59. doi:10.1109/MC.2018.3191268. https://ieeexplore.ieee.org/document/8436400.

Jaswal, Komal, Tanupriya Choudhury, Roshan Lal Chhokar, and Sooraj Randhir Singh. 2017. "Securing the Internet of Things: A Proposed Framework." 1277. The Institute of Electrical and Electronics Engineers, Inc. (IEEE) Conference Proceedings. Piscataway: The Institute of Electrical and Electronics Engineers, Inc. (IEEE).

Manyika, James, Michael Chui, Jacques Bughin, Richard Dobbs, Peter Bisson, and Alex Marrs. 2013. *Disruptive Technologies: Advances that Will Transform Life, Business, and the Global Economy.* Vol. 180. McKinsey Global Institute, San Francisco, CA.

Mendes, Ricardo and João P. Vilela. 2017. "Privacy-Preserving Data Mining: Methods, Metrics, and Applications." *IEEE Access* 5 10562–10582.

Narayanan, Arvind and Vitaly Shmatikov. 2006. "How to Break Anonymity of the Netflix Prize Dataset." *arXiv Preprint Cs/0610105.*

O'Neill, Maire. 2016. "Insecurity by Design: Today's IoT Device Security Problem." *Engineering* 2 (1): 48–49.

PrivacyTrust. "Privacy by Design GDPR." https://privacytrust.com/gdpr/privacy-by-design-gdpr.html.

Index

C

California Consumer Privacy Act, 166
California School Library Association (CSLA), 91
Career and Technical Education (CTE) program, 58–59
Categorize requirements, 279
Center for Medicare and Medicaid Services (CMS), 188
Certified Information System Security Professional (CISSP), 60
Children's Online Privacy Protection Act (COPPA), 98
China, 117
Cloud Access Security Brokers (CASB), 187
Cloud computing service, 120
Coaching, 103
COBIT 5, 123
Code reviews, 284–285
Coding/programming errors, 191
Collect, analyze, and investigate big data, 28–29
Collect and Operate (CO), 62
Collection security, 132
Common Attack Pattern Enumeration and Classification (CAPEC), 275
Common vulnerabilities and exposures (CVE), 205
Communications protocols, 11
Compliance of information policy, 28
Comprehensive test plan, 269
Computer-based behaviors, 231, 233
Computer emergency response team (CERT), 229
Confidentiality, 119, 291
Confidentiality, integrity, and availability (CIA), 168
Conflation, 97
Connecticut Four, 116
Connection curation, 167
Continual system and security monitoring, 195
Coordinated attacks, 27–28
Covered Entity, 175, 177
Cross-site scripting (XSS) attack, 24–25
Cryptocurrency, 8
Cryptographic controls, 295–296
Cryptography, 105
Cryptojacking, 8–9
Cultural issue, 12
Current status, 167
Cyberattacks, 13, 118

coordinated attacks, 27–28
insider threat, 27
internet-based attacks on websites, 24–25
malware attacks, 25–26
network-based attacks, 22–24
social engineering, 26–27
Cyber breaches, 145–146
Cyberbullying, 162
Cybercriminals, 146–147, 149
Cyber-extremism, 161–162
Cyber hygiene practices, 28
Cyber information professionals, 21–22
Cyber physical attacks, 9
Cyberpunk genre, 93–94
Cybersecurity
    Act, 13
    threat, 79
    training, 79–81
Cybersecurity education, 58–60
    North America iSchools, 62, 64–68
Cybersecurity Learning Continuum, 81, 82
Cybersecurity Workforce Study, 57–58
CyberSeek, 10, 82
Cyberspace, 21–22
Cyber theft, 13

D

Darknet, 94
Data
    backup, 194
    breaches, 190–191
    confidentiality, 140–141
    destruction, 302
    encryption, 28
    fusion, 253–258
    generation, 301
    governance, 124
    protection, 221
    quality, 124
    sharing, 301
    storage, 302
    usage, 302
Data, information, and knowledge privacy, 141–142
Data Encryption Standard (DES), 295
Data life cycle, security and privacy, 301–302
Data Loss Prevention/Cloud Access Security Broker (DLP/CASB), 194–195
Data Loss Prevention (DLP), 187